TAKING SIDES

Clashing Views on

Environmental Issues

FOURTEENTH EDITION

TAKING SIDES

Clashing Views on

Environmental Issues

FOURTEENTH EDITION

Selected, Edited, and with Introductions by

Thomas Easton
Thomas College

The **McGraw·Hill** Companies

TAKING SIDES: CLASHING VIEWS ON ENVIRONMENTAL ISSUES, FOURTEENTH EDITION

Published by McGraw-Hill, a business unit of The McGraw-Hill Companies, Inc., 1221 Avenue of the Americas, New York, NY 10020. Copyright © 2011 by The McGraw-Hill Companies, Inc. All rights reserved. Previous edition(s) 2009, 2007, and 2005. No part of this publication may be reproduced or distributed in any form or by any means, or stored in a database or retrieval system, without the prior written consent of The McGraw-Hill Companies, Inc., including, but not limited to, in any network or other electronic storage or transmission, or broadcast for distance learning.

Some ancillaries, including electronic and print components, may not be available to customers outside the United States.

Taking Sides® is a registered trademark of the McGraw-Hill Companies, Inc.
Taking Sides is published by the **Contemporary Learning Series** group within the McGraw-Hill Higher Education division.

1 2 3 4 5 6 7 8 9 0 DOC/DOC 1 0 9 8 7 6 5 4 3 2 1 0

MHID: 0-07-351446-2
ISBN: 978-0-07-351446-8
ISSN: 1091-8825

Managing Editor: *Larry Loeppke*
Director, Specialized Production: *Faye Schilling*
Senior Developmental Editor: *Jill Meloy*
Editorial Coordinator: *Mary Foust*
Production Service Assistant: *Rita Hingtgen*
Permissions Coordinator: *Lenny J. Behnke*
Editorial Assistant: *Cindy Hedley*
Senior Marketing Manager: *Julie Keck*
Senior Marketing Communications Specialist: *Mary Klein*
Marketing Coordinator: *Alice Link*
Project Manager: *Erin Melloy*
Design Specialist: *Brenda Rolwes*
Cover Graphics: *Rick D. Noel*

Compositor: MPS Limited, A Macmillan Company
Cover Image: Photograph courtesy National Park Service

Library of Congress Cataloging-in-Publication Data

Main entry under title:
    Taking sides: clashing views on environmental issues/selected, edited, and with introductions by Thomas A. Easton.—14th ed.

    Includes bibliographical references.
    1. Environmental policy. 2. Environmental protection. I. Easton, Thomas A. *comp.*
                                                                                    363.7

www.mhhe.com

Editors/Academic Advisory Board

Members of the Academic Advisory Board are instrumental in the final selection of articles for each edition of TAKING SIDES. Their review of articles for content, level, and appropriateness provides critical direction to the editors and staff. We think that you will find their careful consideration well reflected in this volume.

TAKING SIDES: Clashing Views on ENVIRONMENTAL ISSUES
Fourteenth Edition

EDITOR

Thomas Easton
Thomas College

Preface

Most fields of academic study evolve over time. Some evolve in turmoil, for they deal in issues of political, social, and economic concern. That is, they involve controversy.

It is the mission of the Taking Sides series to capture current, ongoing controversies and make the opposing sides available to students. This book focuses on environmental issues, from the philosophical to the practical. It does not pretend to cover all such issues, for not all provoke controversy or provoke it in suitable fashion. But there is never any shortage of issues that can be expressed as pairs of opposing essays that make their positions clearly and understandably.

The basic technique—presenting an issue as a pair of opposing essays—has risks. Students often display a tendency to remember best those essays that agree with the attitudes they bring to the discussion. They also want to know what the "right" answers are, and it can be difficult for teachers to refrain from taking a side, or from revealing their own attitudes. Should teachers so refrain? Some do not, but of course they must still cover the spectrum of opinion if they wish to do justice to the scientific method and the complexity of an issue. Some do, though rarely so successfully that students cannot see through the attempt.

For any Taking Sides volume, the issues are always phrased as yes/no questions. Which answer—yes or no—is the correct answer? Perhaps neither. Perhaps both. Perhaps we will not be able to tell for another hundred years. Students should read, think about, and discuss the readings and then come to their own conclusions without letting my or their instructor's opinions dictate theirs. The additional readings mentioned in the introductions and postscripts should prove helpful.

This edition of *Taking Sides: Clashing Views on Environmental Issues* contains 38 readings arranged in pro and con pairs to form 19 issues. For each issue, an *introduction* provides historical background and a brief description of the debate. The *postscript* after each pair of readings offers recent contributions to the debate, additional references, and sometimes a hint of future directions. Each part is preceded by an *Internet References* page that lists several links that are appropriate for further pursuing the issues in that part.

Changes to this edition   Over half of this book consists of new material. Two issues, *Should North America's Landscape Be Restored to Its Pre-Human State?* (Issue 4) and *Is Carbon Capture Technology Ready to Limit Carbon Emissions?* (Issue 8), were added for the 2010 partial revision of the thirteenth edition. *Should the Military Be Exempt from Environmental Regulations* (Issue 5) has been brought back from the eleventh edition because of a recent Supreme Court decision. The global warming debate is now represented by *Will Restricting*

Carbon Emissions Damage the U.S. Economy? (Issue 6). Because of the new interest in energy issues, we have added *Should We Drill for Offshore Oil?* (Issue 7) and *Is It Time to Put Geothermal Energy Development on the Fast Track?* (Issue 9). We renamed the biofuels issue *Are Biofuels Responsible for Rising Food Prices?* (Issue 11), the population issue *Are Improved Aid Policies the Best Way to Improve Global Food Supply and Protect World Population?* (Issue 13), and the organic agriculture issue *Can Organic Farming Feed the World?* (Issue 15).

In addition, for three of the issues retained from the previous edition, one or both of the readings have been replaced. In all, 21 of the readings in this edition were not in the thirteenth edition.

A word to the instructor An *Instructor's Resource Guide with Test Questions* (multiple-choice and essay) is available through the publisher for the instructor using Taking Sides in the classroom. Also available is a general guidebook, *Using Taking Sides in the Classroom,* which offers suggestions for adapting the pro-con approach in any classroom setting. An online version of *Using Taking Sides in the Classroom* and a correspondence service for Taking Sides adopters can be found at www.mhhe.com/cls.

Taking Sides: Clashing Views on Environmental Issues is only one title in the Taking Sides series. If you are interested in seeing the table of contents for any of the other titles, please visit the Taking Sides Web site at www.mhhe.com/cls.

Thomas A. Easton
Thomas College

Contents In Brief

Contents

John E. Losey and Mace Vaughan argue that even conservative estimates
of the value of the services provided by wild insects are enough to justify
increased conservation efforts. They say that "everyone would benefit
from the facilitation of the vital services these insects provide." Professors
of applied ecology Marino Gatto and Giulio A. De Leo contend that the
pricing approach to valuing nature's services is misleading because it
falsely implies that only economic values matter.

UNIT 2   PRINCIPLES VERSUS POLITICS   63

C. Josh Donlan proposes that because the arrival of humans in the
Americas some 13,000 years ago led to the extinction of numerous large
animals (including camels, lions, and mammoths) with major effects on
local ecosystems, restoring these animals (or their near-relatives from
elsewhere in the world) holds the potential to restore health to these
ecosystems. There would also be economic and cultural benefits. Dustin
R. Rubenstein, Daniel I. Rubenstein, Paul W. Sherman, and Thomas A.
Gavin argue that bringing African and Asian megafauna to North America
is unlikely to restore pre-human ecosystem function and may threaten
present species and ecosystems. It would be better to focus resources on
restoring species where they were only recently extinguished.

Benedict S. Cohen argues that environmental regulations interfere with
military training and other "readiness" activities, and that though the U.S.
Department of Defense will continue "to provide exemplary stewardship
of the lands and natural resources in our trust" those regulations must be
revised to permit the military to do its job without interference. Jamie
Clark argues that reducing the Department of Defense's environmental
obligations is dangerous because both people and wildlife would be
threatened with serious, irreversible, and unnecessary harm.

David G. Hawkins, director of the Climate Center of the Natural Resources Defense Council, argues that we know enough to implement large-scale carbon capture and sequestration for new coal plants. The technology is ready to do so safely and effectively. Charles W. Schmidt argues that the technology is not yet technically and financially feasible, research is stuck in low gear, and the political commitment to reducing carbon emissions is lacking.

Susan Petty, president of AltaRock Energy, Inc., argues that the technology already exists to greatly increase the production and use of geothermal energy. Supplying 20 percent of U.S. electricity from geothermal energy by 2030 is a very realistic goal. Alexander Karsner, Assistant Secretary for Energy Efficiency and Renewable Energy at the U.S. Department of Energy, argues that it is not feasible to supply 20 percent of U.S. electricity from geothermal energy by 2030.

David Friedman, Research Director at the Union of Concerned Scientists, argues that the technology exists to improve the fuel efficiency standards for new cars and trucks and requiring improved efficiency can cut oil imports, save money, create jobs, and help with global warming. Charli E. Coon, Senior Policy Analyst with the Heritage Foundation, argues that the 1975 Corporate Average Fuel Economy (CAFE) program failed to meet its goals of reducing oil imports and gasoline consumption and has endangered human lives. It needs to be abolished and replaced with market-based solutions.

Donald Mitchell argues that although many factors contributed to the increase in internationally traded food prices from January 2002 to June 2008, the most important single factor was the large increase in biofuels production from grains and oilseeds in the U.S. and EU. Keith Kline, Virginia H. Dale, Russell Lee, and Paul Leiby argue that the impact of biofuels production on food prices is much less than alarmists claim. There would be greater impact if biofuels development focused on converting biowastes and fast-growing trees and grasses into fuels.

Iain Murray argues that the world's experience with nuclear power has shown it to be both safe and reliable. Costs can be contained, and if one is concerned about global warming, the case for nuclear power is unassailable. Professor Kristin Shrader-Frechette argues that nuclear power is one of the most impractical and risky of energy sources. Renewable energy sources such as wind and solar are a sounder choice.

UNIT 4   FOOD AND POPULATION   241

Professor Robert Paarlberg argues that global hunger, which afflicts nearly a billion people worldwide, many of them in Africa, calls for increased aid directed toward agricultural education, science, and research, and infrastructure development. Lester R. Brown argues that the problem is due more to water shortages, soil losses, rising population, and rising temperatures from global warming than to failures of aid policies. What is needed is immediate attention to the world's environmental problems, lacking which the result will be increased hunger, political conflict, and perhaps even the collapse of civilization.

Gerald D. Coleman argues that genetically engineered crops are useful, healthful, and nonharmful, and although caution may be justified, such crops can help satisfy the moral obligation to feed the hungry. Sean McDonagh argues that those who wish to feed the hungry would do better to address land reform, social inequality, lack of credit, and other social issues.

Catherine Badgley, et al., argue that organic methods could produce enough food to sustain a global human population that is even larger than today's, and without requiring additional farmland. Organic agriculture would also decrease the undesirable environmental effects of conventional farming. John J. Miller argues that organic farming is not productive enough to feed today's population, much less larger future populations; it is prone to dangerous biological contamination; and it is not sustainable.

UNIT 5 TOXIC CHEMICALS 295

Anne Platt McGinn, a senior researcher at the Worldwatch Institute, argues that although DDT is still used to fight malaria, there are other, more effective and less environmentally harmful methods. She maintains that DDT should be banned or reserved for emergency use. Donald R. Roberts argues that the scientific evidence regarding the environmental hazards of DDT has been seriously misrepresented by anti-pesticide activists. The hazards of malaria are much greater and, properly used, DDT can prevent them and save lives.

Professor of biological sciences Michele L. Trankina argues that a great many synthetic chemicals behave like estrogen, alter the reproductive functioning of wildlife, and may have serious health effects—including cancer—on humans. Michael Gough, a biologist and expert on risk assessment and environmental policy, argues that only "junk science" supports the hazards of environmental estrogens.

Correlation Guide

The *Taking Sides* series presents current issues in a debate-style format designed to stimulate student interest and develop critical thinking skills. Each issue is thoughtfully framed with an issue summary, an issue introduction, and a postscript. The pro and con essays—selected for their liveliness and substance—represent the arguments of leading scholars and commentators in their fields.

Taking Sides: Clashing Views on Environmental Issues, 14/e is an easy-to-use reader that presents issues on important topics such as *offshore drilling*, *nuclear power*, and *global food*. For more information on *Taking Sides* and other *McGraw-Hill Contemporary Learning Series* titles, visit www.mhhe.com/cls.

This convenient guide matches the issues in *Taking Sides: Environmental Issues*, 14/e with the corresponding chapters in two of our best-selling McGraw-Hill Environmental Science textbooks by Enger/Smith and Cunningham/Cunningham.

Taking Sides: Environmental Issues, 14/e	Environmental Science: A Study of Interrelationships, 12/e by Enger/Smith	Environmental Science: A Global Concern, 11/e by Cunningham/Cunningham
Issue 1: Is the Precautionary Principle a Sound Approach to Risk Analysis?	**Chapter 3:** Environmental Risk: Economics, Assessment, and Management	**Chapter 8:** Environmental Health and Toxicology **Chapter 24:** Environmental Policy, Law, and Planning
Issue 2: Is Sustainable Development Compatible With Human Welfare?	**Chapter 5:** Interactions: Environments and Organisms	**Chapter 1:** Understanding Our Environment **Chapter 23:** Ecological Economics **Chapter 25:** What Then Shall We Do?
Issue 3: Should a Price Be Put on the Goods and Services Provided by the World's Ecosystems?	**Chapter 6:** Kinds of Ecosystems and Communities	**Chapter 23:** Ecological Economics
Issue 4: Should North America's Landscape Be Restored to Its Pre-Human State?	**Chapter 12:** Land-Use Planning	**Chapter 12:** Biodiversity: Preserving Landscapes **Chapter 13:** Restoration Ecology
Issue 5: Should the Military Be Exempt from Environmental Regulations?	**Chapter 18:** Environmental Regulations: Hazardous Substances and Wastes **Chapter 19:** Environmental Policy and Decision Making	**Chapter 24:** Environmental Policy, Law, and Planning
Issue 6: Will Restricting Carbon Emissions Damage the U.S. Economy?	**Chapter 16:** Air Quality Issues	**Chapter 1:** Understanding Our Environment **Chapter 15:** Air, Weather, and Climate **Chapter 19:** Conventional Energy **Chapter 20:** Sustainable Energy

(Continued)

Taking Sides: Environmental Issues, 14/e	Environmental Science: A Study of Interrelationships, 12/e by Enger/Smith	Environmental Science: A Global Concern, 11/e by Cunningham/Cunningham
Issue 7: Should We Drill for Offshore Oil?	**Chapter 3:** Environmental Risk: Economics, Assessment, and Management **Chapter 8:** Energy and Civilization **Chapter 9:** Energy Sources	**Chapter 19:** Conventional Energy
Issue 8: Is Carbon Capture Technology Ready to Limit Carbon Emissions?	**Chapter 3:** Environmental Risk: Economics, Assessment, and Management **Chapter 8:** Energy and Civilization **Chapter 9:** Energy Sources	**Chapter 15:** Air, Weather, and Climate **Chapter 19:** Conventional Energy
Issue 9: Is It Time to Put Geothermal Energy Development on the Fast Track?		
Issue 10: Should Cars Be More Efficient?	**Chapter 8:** Energy and Civilization	**Chapter 20:** Sustainable Energy
Issue 11: Are Biofuels Responsible for Rising Food Prices?	**Chapter 11:** Biodiversity Issues	**Chapter 12:** Biodiversity: Preserving Landscapes **Chapter 20:** Sustainable Energy
Issue 12: Is It Time to Revive Nuclear Power?	**Chapter 10:** Nuclear Energy	**Chapter 19:** Conventional Energy
Issue 13: Are Improved Aid Policies the Best Way to Improve Global Food Supply and Protect World Population?	**Chapter 12:** Land-Use Planning **Chapter 13:** Soil and Its Uses **Chapter 14:** Agricultural Methods and Pest Management	**Chapter 1:** Understanding Our Environment **Chapter 9:** Food and Hunger
Issue 14: Is Genetic Engineering the Answer to Hunger?	**Chapter 14:** Agricultural Methods and Pest Management	**Chapter 9:** Food and Hunger
Issue 15: Can Organic Farming Feed the World?	**Chapter 14:** Agricultural Methods and Pest Management	**Chapter 9:** Food and Hunger **Chapter 10:** Farming: Conventional and Sustainable Practices
Issue 16: Should DDT Be Banned Worldwide?	**Chapter 14:** Agricultural Methods and Pest Management **Chapter 18:** Environmental Regulations: Hazardous Substances and Wastes	**Chapter 8:** Environmental Health and Toxicology **Chapter 10:** Farming: Conventional and Sustainable Practices **Chapter 18:** Water Pollution
Issue 17: Do Environmental Hormone Mimics Pose a Potentially Serious Health Threat?	**Chapter 18:** Environmental Regulations: Hazardous Substances and Wastes	
Issue 18: Is the Superfund Program Successfully Protecting Human Health from Hazardous Materials?	**Chapter 18:** Environmental Regulations: Hazardous Substances and Wastes **Chapter 19:** Environmental Policy and Decision Making	**Chapter 8:** Environmental Health and Toxicology **Chapter 21:** Solid, Toxic, and Hazardous Waste **Chapter 24:** Environmental Policy, Law, and Planning
Issue 19: Should the United States Reprocess Spent Nuclear Fuel?	**Chapter 10:** Nuclear Energy	**Chapter 19:** Conventional Energy

Introduction

Environmental Issues:
The Never-Ending Debate

One of the courses I teach is "Environmentalism: Philosophy, Ethics, and History." I begin by explaining the roots of the word "ecology," from the Greek *oikos* (house or household), and assigning the students to write a brief paper about their own household. How much, I ask them, do you need to know about the place where you live? And why?

The answers vary. Some of the resulting papers focus on people—roommates if the "household" is a dorm room, spouses and children if the students are older, parents and siblings if they live at home—and the needs to cooperate and get along, and perhaps the need not to overcrowd. Some pay attention to house-plants and pets, and occasionally even bugs and mice. Some focus on economics—possessions, services, and their costs, where the checkbook is kept, where the bills accumulate, the importance of paying those bills, and of course the importance of earning money to pay those bills. Some focus on maintenance—cleaning, cleaning supplies, repairs, whom to call if something major breaks. For some the emphasis is operation—garbage disposal, grocery shopping, how to work the lights, stove, fridge, and so on. A very few recognize the presence of toxic chemicals under the sink and in the medicine cabinet and the need for precautions in their handling. Sadly, a few seem to be oblivious to anything that does not have something to do with entertainment.

Not surprisingly, some students object initially that the exercise seems trivial. "What does this have to do with environmentalism?" they ask. Yet the course is rarely very old before most are saying, "Ah! I get it!" That nice, homey microcosm has a great many of the features of the macrocosmic environment, and the multiple ways people can look at the microcosm mirror the ways people look at the macrocosm. It's all there, as is the question of priorities: What is important? People or fellow creatures or economics or maintenance or operation or waste disposal or food supply or toxics control or entertainment? Or all of the above?

And how do you decide? I try to illuminate this question by describing a parent trying to teach a teenager not to sit on a woodstove. In July, the kid answers, "Why?" and continues to perch. In August, likewise. And still in September. But in October or November, the kid yells "Ouch!" and jumps off in a hurry.

That is, people seem to learn best when they get burned.

This is surely true in our homely *oikos,* where we may not realize our fellow creatures deserve attention until houseplants die of neglect or cockroaches invade the cupboards. Economics comes to the fore when the phone gets cut off, repairs when a pipe ruptures, air quality when the air conditioner breaks or strange fumes rise from the basement, waste disposal when the bags pile up and begin to stink or the toilet backs up. Toxics control suddenly matters when a child or pet gets into the rat poison.

In the larger *oikos* of environmentalism, such events are paralleled by the loss of a species, or an infestation by another, by floods and droughts, by lakes turned into cesspits by raw sewage, by air turned foul by industrial smokestacks, by groundwater contaminated by toxic chemicals, by the death of industries and the loss of jobs, by famine and plague and even war.

If nothing is going wrong, we are not very likely to realize there is something we should be paying attention to. And this too has its parallel in the larger world. Indeed, the history of environmentalism is in important part a history of people carrying on with business as usual until something goes obviously awry. Then, if they can agree on the nature of the problem (Did the floor cave in because the joists were rotten or because there were too many people at the party?), they may learn something about how to prevent recurrences.

The Question of Priorities

There is of course a crucial "if" in that last sentence: *If people can agree. . . .* It is a truism to say that agreement is difficult. In environmental matters, people argue endlessly over whether anything is actually wrong, what its eventual impact will be, what if anything can or should be done to repair the damage, and how to prevent recurrence. Not to mention who's to blame and who should take responsibility for fixing the problem! Part of the reason is simple: Different things matter most to different people. Individual citizens may want clean air or water or cheap food or a convenient commute. Politicians may favor sovereignty over international cooperation. Economists and industrialists may think a few coughs (or cases of lung cancer, or shortened lifespans) a cheap price to pay for wealth or jobs.

No one now seems to think that protecting the environment is not important. But different groups—even different environmentalists—have different ideas of what "environmental responsibility" means. To a paper company cutting trees for pulp, it may mean leaving a screen of trees (a "beauty strip") beside the road and minimizing erosion. To hikers following trails through or within view of the same tract of land, that is not enough; they want the trees left alone. The hikers may also object to seeing the users of trail bikes and all-terrain-vehicles on the trails. They may even object to hunters and anglers, whose activities they see as diminishing the wilderness experience. They may therefore push for protecting the land as limited-access wilderness. The hunters and anglers object to that, of course, for they want to be able to use their vehicles to bring their game home, or to bring their boats to their favorite rivers and lakes. They also argue, with some justification, that their license fees support a great deal of environmental protection work.

To a corporation, dumping industrial waste into a river may make perfect sense, for alternative ways of disposing of waste are likely to cost more and diminish profits. Of course, the waste renders the water less useful to wildlife or downstream humans, who may well object. Yet telling the corporation it cannot dump may be seen as depriving it of property. A similar problem arises when regulations prevent people and corporations from using land—and making money—as they had planned. Conservatives have claimed that environmental regulations thus violate the Fifth Amendment to the U.S. Constitution, which says "No person shall . . . be deprived of . . . property, without due process of law; nor shall private property be taken for public use, without just compensation."

One might think the dangers of such things as dumping industrial waste in rivers are obvious. But scientists can and do disagree, even given the same evidence. For instance, a chemical in waste may clearly cause cancer in laboratory animals. Is it therefore a danger to humans? A scientist working for the company dumping that chemical in a river may insist that no such danger has been proven. Yet a scientist working for an environmental group such as Greenpeace may insist that the danger is obvious since carcinogens do generally affect more than one species.

Scientists are human. They have not only employers but also values, rooted in political ideology and religion. They may feel that the individual matters more than corporations or society, or vice versa. They may favor short-term benefits over long-term benefits, or vice versa.

And scientists, citizens, corporations, and government all reflect prevailing social attitudes. When America was expanding westward, the focus was on building industries, farms, and towns. If problems arose, there was vacant land waiting to be moved to. But when the expansion was done, problems became more visible and less avoidable. People could see that there were "trade-offs" involved in human activity: more industry meant more jobs and more wealth, but there was a price in air and water pollution and human health (among other things).

Nowhere, perhaps, are these trade-offs more obvious than in Eastern Europe. The former Soviet Union was infamous for refusing to admit that industrial activity was anything but desirable. Anyone who spoke up about environmental problems risked jail. The result, which became visible to Western nations after the fall of the Iron Curtain in 1990, was industrial zones where rivers had no fish, children were sickly, and life expectancies were reduced. The fate of the Aral Sea, a vast inland body of water once home to a thriving fishery and a major regional transportation route, is emblematic: Because the Soviet Union wanted to increase its cotton production, it diverted for irrigation the rivers which delivered most of the Aral Sea's fresh water supply. The Sea then began to lose more water to evaporation than it gained, and it rapidly shrank, exposing sea-bottom so contaminated by industrial wastes and pesticides that wind-borne dust is now responsible for a great deal of human illness. The fisheries are dead, and freighters lie rusting on bare ground where once waves lapped.

The Environmental Movement

The twentieth century saw immense changes in the conditions of human life and in the environment that surrounds and supports human life. According to historian J. R. McNeill, in *Something New Under the Sun: An Environmental History of the Twentieth-Century World* (W. W. Norton, 2000), the environmental impacts that resulted from the interactions of burgeoning population, technological development, shifts in energy use, politics, and economics in that period are unprecedented in both degree and kind. Yet a worse impact may be that we have come to accept as "normal" a very temporary situation that "is an extreme deviation from any of the durable, more 'normal,' states of the world over the span of human history, indeed over the span of earth history." We are thus not prepared for the inevitable and perhaps drastic changes ahead.

Environmental factors cannot be denied their role in human affairs. Nor can human affairs be denied their place in any effort to understand environmental change. As McNeill says, "Both history and ecology are, as fields of knowledge go, supremely integrative. They merely need to integrate with each other."

The environmental movement, which grew during the twentieth century in response to increasing awareness of human impacts, is a step in that direction. Yet environmental awareness reaches back long before the modern environmental movement. When John James Audubon (1785–1851), famous for his bird paintings, was young, he was an enthusiastic slaughterer of birds (a few of which he used as models for the paintings). Later in life, he came to appreciate that birds were diminishing in numbers, as were the American bison, and he called for conservation measures. His was a minority voice, however. It was not till later in the century that George Perkins Marsh warned in *Man and Nature* (1864) that "We are, even now, breaking up the floor and wainscoting and doors and window frames of our dwelling, for fuel to warm our bodies and seethe our pottage, and the world cannot afford to wait till the slow and sure progress of exact science has taught it a better economy." The Earth, he said, was given to man for "usufruct" (to use the fruit of), not for consumption or waste. Resources should remain to benefit future generations. Stewardship was the point, and damage to soil and forest should be prevented and repaired. He was not concerned with wilderness as such; John Muir (1838–1914; founder of the Sierra Club) was the first to call for the preservation of natural wilderness, untouched by human activities. Marsh's ideas influenced others more strongly. In 1890, Gifford Pinchot (1865–1946) found "the nation . . . obsessed by a fury of development. The American Colossus was fiercely intent on appropriating and exploiting the riches of the richest of all continents." Under President Theodore Roosevelt, he became the first head of the U.S. Forest Service and a strong voice for conservation (not to be confused with preservation; Gifford's conservation meant using nature but in such a way that it was not destroyed; his aim was "the greatest good of the greatest number in the long run"). By the 1930s, Aldo Leopold (1887–1948), best known for his concept of the "land ethic" and his book, *A Sand County Almanac*, could argue that we had a responsibility not only to maintain the environment but also to repair damage done in the past.

The modern environmental movement was kick-started by Rachel Carson's *Silent Spring* (Houghton Mifflin,1962). In the 1950s, Carson realized that the use of pesticides was having unintended consequences—the death of non-pest insects, food-chain accumulation of poisons and the consequent loss of birds, and even human illness—and meticulously documented the case. When her book was published, she and it were immediately vilified by pesticide proponents in government, academia, and industry (most notably, the pesticides industry). There was no problem, the critics said; the negative effects if any were worth it, and she—a *woman* and a nonscientist—could not possibly know what she was talking about. But the facts won out. A decade later, DDT was banned and other pesticides were regulated in ways unheard of before Carson spoke out.

Other issues have followed or are following a similar course.

The situation before Rachel Carson and *Silent Spring* is nicely captured by Judge Richard Cudahy, who in "Coming of Age in the Environment," *Environmental Law,* Winter 2000, writes, "It doesn't seem possible that before 1960 there was no 'environment'—or at least no environmentalism. I can even remember the Thirties, when we all heedlessly threw our trash out of car windows, burned coal in the home furnace (if we could afford to buy any), and used a lot of lead for everything from fishing sinkers and paint to no-knock gasoline. Those were the days when belching black smoke meant a welcome end to the Depression and little else."

Historically, humans have felt that their own well-being mattered more than anything else. The environment existed to be used. Unused, it was only wilderness or wasteland, awaiting the human hand to "improve" it and make it valuable. This is not surprising at all, for the natural tendency of the human mind is to appraise all things in relation to the self, the family, and the tribe. An important aspect of human progress has lain in enlarging our sense of "tribe" to encompass nations and groups of nations. Some now take it as far as the human species. Some include other animals. Some embrace plants as well, and bacteria, and even landscapes.

The more limited standard of value remains common. Add to that a sense that wealth is not just desirable but a sign of virtue (the Puritans brought an explicit version of this with them when they colonized North America; see Lynn White, Jr., "The Historical Roots of Our Ecological Crisis," *Science,* 1967), and it is hardly surprising that humans have used and still use the environment intensely. People also tend to resist any suggestion that they should restrain their use out of regard for other living things. Human needs, many insist, must come first.

The unfortunate consequences include the loss of other species. Lions vanished from Europe about 2000 years ago. The dodo of Mauritius was extinguished in the 1600s (see the American Museum of Natural History's account at http://www.amnh.org/exhibitions/expeditions/treasure_fossil/Treasures/Dodo/dodo.html?acts). The last of North America's passenger pigeons died in a Cincinnati zoo in 1914 (see http://www.amnh.org/exhibitions/expeditions/treasure_fossil/Treasures/Passenger_Pigeons/pigeons.html?acts). Concern for such species was at first limited to those of obvious value to humans. In 1871, the U.S. Commission on Fish and Fisheries was created and charged with finding solutions to the decline in food fishes and promoting aquaculture. The first federal legislation designed to protect game animals was the Lacey Act of 1900. It was not until 1973 that the U.S. Endangered Species Act was adopted to shield all species from the worst human impacts.

Other unfortunate consequences of human activities include dramatic erosion, air and water pollution, oil spills, accumulations of hazardous (including nuclear) waste, famine, and disease. Among the many "hot stove" incidents that have caught public attention are the following:

- The Dust Bowl—in 1934 wind blew soil from drought-stricken farms in Oklahoma all the way to Washington, DC;
- Cleveland's Cuyahoga River caught fire in the 1960s;
- The Donora, Pennsylvania, smog crisis—in one week of October 1948, 20 died and over 7000 were sickened;
- The London smog crisis in December 1952—4000 dead:
- The Torrey Canyon and Exxon Valdez oil spills, which fouled shores and killed seabirds, seals, and fish;
- Love Canal, where industrial wastes seeped from their burial site into homes and contaminated ground water;
- Union Carbide's toxics release at Bhopal, India—3800 dead and up to 100,000 ill, according to Union Carbide; others claim a higher toll;
- The Three Mile Island and Chornobyl nuclear accidents;
- The decimation of elephants and rhinoceroses to satisfy a market for tusks and horns;
- The loss of forests—in 1997, fires set to clear Southeast Asian forest lands produced so much smoke that regional airports had to close;
- Ebola, a virus which kills nine tenths of those it infects, apparently first struck humans because growing populations reached into its native habitat;
- West Nile Fever, a mosquito-borne virus with a much less deadly record, was brought to North America by travelers or immigrants from Egypt;
- Acid rain, global climate change, and ozone depletion, all caused by substances released into the air by human activities.

The alarms have been raised by many people in addition to Rachel Carson. For instance, in 1968 (when world population was only a little over half of what it is today), Paul Ehrlich's *The Population Bomb* (Ballantine Books) described the ecological threats of a rapidly growing population and Garrett Hardin's influential essay, "The Tragedy of the Commons," *Science* (December 13, 1968) described the consequences of using self-interest alone to guide the exploitation of publicly-owned resources (such as air and water). (In 1974, Hardin introduced the unpleasant concept of "lifeboat ethics," which says that if there are not enough resources to go around, some people must do without). In 1972, a group of economists, scientists, and business leaders calling themselves "The Club of Rome" published *The Limits to Growth* (Universe Books), an analysis of population, resource use, and pollution trends that predicted difficult times within a century; the study was redone as *Beyond the Limits to Growth: Confronting Global Collapse, Envisioning a Sustainable Future* (Chelsea Green, 1992) and again as *Limits to Growth: The 30-Year Update* (Chelsea Green, 2004), using more powerful computer models, and came to very similar conclusions. Among the most recent books is Jared Diamond's *Collapse: How Societies Choose to Fail or Succeed* (Viking, 2005), which uses historical cases to illuminate the roles of human biases and choices in dealing with environmental problems. Among Diamond's important points is the idea that in order to cope successfully with such problems, a society may have to surrender cherished traditions.

The following list of selected U.S. and U.N. laws, treaties, conferences, and reports illustrates the national and international responses to the various cries of alarm:

1967   The U.S. Air Quality Act set standards for air pollution.

1968   The U.N. Biosphere Conference discussed global environmental problems.

1969   The U.S. Congress passed the National Environmental Policy Act, which (among other things) required federal agencies to prepare environmental impact statements for their projects.

1970   The first Earth Day demonstrated so much public concern that the Environmental Protection Agency (EPA) was created; the Endangered Species Act, Clean Air Act, and Safe Drinking Water Act soon followed.

1971   The U.S. Environmental Pesticide Control Act gave the EPA authority to regulate pesticides.

1972   The U.N. Conference on the Human Environment, held in Stockholm, Sweden, recommended government action and led to the U.N. Environment Programme.

1973   The Convention on International Trade in Endangered Species of Wild Fauna and Flora (CITES) restricted trade in threatened species; because enforcement was weak, however, a black market flourished.

1976   The U.S. Resource Conservation and Recovery Act and the Toxic Substances Control Act established control over hazardous wastes and other toxic substances.

1979   The Convention on Long-Range Transboundary Air Pollution addressed problems such as acid rain (recognized as crossing national borders in 1972).

1982   The Law of the Sea addressed marine pollution and conservation.

1982   The second U.N. Conference on the Human Environment (the Stockholm +10 Conference) renewed concerns and set up a commission to prepare a "global agenda for change," leading to the 1987 Brundtland Report (*Our Common Future*).

1983   The U.S. Environmental Protection Agency and the U.S. National Academy of Science issued reports calling attention to the prospect of global warming as a consequence of the release of greenhouse gases such as carbon dioxide.

1987   The Montreal Protocol (strengthened in 1992) required nations to phase out use of chlorofluorocarbons (CFCs), the chemicals responsible for stratospheric ozone depletion (the "ozone hole").

1987   The Basel Convention controlled cross-border movement of hazardous wastes.

1988   The U.N. assembled the Intergovernmental Panel on Climate Change, which would report in 1995, 1998, and 2001 that the dangers of global warming were real, large, and increasingly ominous.

1992   The U.N. Convention on Biological Diversity required nations to act to protect species diversity.

1992   The U.N. Conference on Environment and Development (also known as the Earth Summit), held in Rio de Janeiro, Brazil, issued a broad call for environmental protections.

1992   The U.N. Convention on Climate Change urged restrictions on carbon dioxide release to avoid climate change

1994   The U.N. Conference on Population and Development, held in Cairo, Egypt, called for stabilization and reduction of global population growth, largely by improving women's access to education and health care.

1997   The Kyoto Protocol attempted to strengthen the 1992 Convention on Climate Change by requiring reductions in carbon dioxide emissions, but U.S. resistance limited success.

2001   The U.N. Stockholm Convention on Persistent Organic Pollutants required nations to phase out use of many pesticides and other chemicals. It took effect May 17, 2004, after ratification by over fifty nations (not including the United States and the European Union).

2002   The U.N. World Summit on Sustainable Development, held in Johannesburg, South Africa, brought together representatives of governments, nongovernmental organizations, businesses, and other groups to examine "difficult challenges, including improving people's lives and conserving our natural resources in a world that is growing in population, with ever-increasing demands for food, water, shelter, sanitation, energy, health services and economic security."

2003   The World Climate Change Conference held in Moscow, Russia, concluded that global climate is changing, very possibly because of human activities, and the overall issue must be viewed as one of intergenerational justice. "Mitigating global climate change will be possible only with the coordinated actions of all sectors of society."

2005   The U.N. Millennium Project Task Force on Environmental Sustainability released its report, *Environment and Human Well-Being: A Practical Strategy.*

2005   The U.N. Millennium Ecosystem Assessment released its report, *Ecosystems and Human Well-Being: Synthesis* (http://www.millenniumassessment.org/en/index.aspx) (Island Press).

2005   The U.N. Climate Change Conference held in Montreal, Canada, marked the taking effect of the Kyoto Protocol, ratified in 2004 by 141 nations (not including the U.S. and Australia, which finally ratified the Protocol in 2007).

2007   The Intergovernmental Panel on Climate Change (IPCC) released its Fourth Assessment Report, asserting that global warming is definitely due to human releases of carbon dioxide, the effects on both nature and humanity will be profound, and mitigation, though possible, will be expensive.

2009   In December, the U.N. Climate Change Conference sought increased commitments to reducing carbon dioxide emissions. The pro-science attitude of the Obama Administration in Washington, D.C., gave many hope that the U.S. would finally ratify the Kyoto Protocol.

Rachel Carson would surely have been pleased by these responses, for they suggest both concern over the problems identified and determination to solve those problems. But she would just as surely have been frustrated, for a simple listing of laws, treaties, and reports does nothing to reveal the endless wrangling and the way political and business forces try to block progress whenever it is seen as interfering with their interests. Agreement on banning chlorofluorocarbons was relatively easy to achieve because CFCs were not seen as essential to civilization and there were substitutes available. Restraining greenhouse gas emissions is harder because we see fossil fuels as essential and though substitutes may exist, they are so far more expensive.

The Globalization of the Environment

Years ago, it was possible to see environmental problems as local. A smokestack belched smoke and made the air foul. A city sulked beneath a layer of smog. Bison or passenger pigeons declined in numbers and even vanished. Rats flourished in a dump where burning garbage produced clouds of smoke and runoff contaminated streams and groundwater and made wells unusable. Sewage, chemical wastes, and oil killed the fish in streams, lakes, rivers, and harbors. Toxic chemicals such as lead and mercury entered the food chain and affected the health of both wildlife and people.

By the 1960s, it was becoming clear that environmental problems did not respect borders. Smoke blows with the wind, carrying one locality's contamination to others. Water flows to the sea, carrying sewage and other wastes with it. Birds migrate, carrying with them whatever toxins they have absorbed with their food. In 1972, researchers were able to report that most of the acid rain falling on Sweden came from other countries. Other researchers have shown that the rise and fall of the Roman Empire can be tracked in Greenland, where glaciers preserve lead-containing dust deposited over the millennia—the amount rises as Rome flourished, falls with the Dark Ages, and rises again with the Renaissance and Industrial Revolution. Today we know that pesticides and other chemicals can show up in places (such as the Arctic) where they have never been used, even years after their use has been discontinued. The 1979 Convention on Long-Range Transboundary Air Pollution has been strengthened several times with amendments to address persistent organic pollutants, heavy metals, and other pollutants.

We are also aware of new environmental problems that exist only in a global sense. Ozone depletion, first identified in the stratosphere over Antarctica, threatens to increase the amount of ultraviolet light reaching the ground, and thereby increase the incidence of skin cancer and cataracts, among other things. The cause is the use by the industrialized world of chlorofluorocarbons (CFCs) in refrigeration, air conditioning, aerosol cans, and electronics (for cleaning grease off circuit boards). The effect is global. Worse yet, the cause is rooted in northern lands such as the United States and Europe, but the worst effects may be felt where the sun shines brightest—in the tropics, which are dominated by developing nations. According to the Global Humanitarian Forum's "Human Impact Report: Climate Change—The Anatomy of a Silent Crisis" (May 29, 2009;

see http://www.ghf-ge.org/programmes/human_impact_report/index.cfm) global warming is already affecting over 300 million people and is responsible for 300,000 deaths per year. A serious issue of justice or equity is therefore involved.

A similar problem arises with global warming, which is also rooted in the industrialized world and its use of fossil fuels. The expected climate effects will hurt worst the poorer nations of the tropics, and perhaps worst of all those on low-lying South Pacific islands, which are expecting to be wholly inundated by rising seas. People who depend on the summertime melting of winter snows and mountain glaciers (including the citizens of California) will also suffer, for already the snows are less and the glaciers are vanishing.

Both the developed and the developing world are aware of the difficulties posed by environmental issues. In Europe, "green" political parties play a major and growing part in government. In Japan, some environmental regulations are more demanding than those of the United States. Developing nations understandably place dealing with their growing populations high on their list of priorities, but they play an important role in U.N. conferences on environmental issues, often demanding more responsible behavior from developed nations such as the United States (which often resists these demands; it has refused to ratify international agreements such as the Kyoto Protocol, for example).

Western scholars have been known to suggest that developing nations should forgo industrial development because if their huge populations ever attain the same per-capita environmental impact as the populations of wealthier lands, the world will be laid waste. It is not hard to understand why the developing nations object to such suggestions; they too want a better standard of living. Nor do they think it fair that they suffer for the environmental sins of others.

Are global environmental problems so threatening that nations must surrender their sovereignty to international bodies? Should the U.S. or Europe have to change energy supplies to protect South Pacific nations from inundation by rising seas? Should developing nations be obliged to reduce birth rates or forgo development because their population growth and industrialization are seen as exacerbating pollution or threatening biodiversity?

Questions such as these play an important part in global debates today. They are not easy to answer, but their very existence says something important about the general field of environmental studies. This field is based in the science of ecology. Ecology focuses on living things and their interactions with each other and their surroundings. It deals with resources and limits and coexistence. It can see problems, their causes, and even potential solutions. And it can turn its attention to human beings as easily as it can to deer mice.

Yet human beings are not mice. We have economies and political systems, vested interests, and conflicting priorities and values. Ecology is only one part of environmental studies. Other sciences—chemistry, physics, climatology, epidemiology, geology, and more—are involved. So are economics, history, law, and politics. Even religion can play a part.

Unfortunately, no one field sees enough of the whole to predict problems (the chemists who developed CFCs could hardly have been expected to realize what would happen if these chemicals reached the stratosphere). Environmental studies is a field for teams. That is, it is a holistic, multidisciplinary field.

This gives us an important basic principle to use when evaluating arguments on either side of any environmental issue: Arguments that fail to recognize the complexity of the issue are necessarily suspect. On the other hand, arguments that endeavor to convey the full complexity of an issue may be impossible to understand; a middle ground is essential for clarity, but any reader or student must realize that something important may be being left out.

Current Environmental Issues

In 2001, the National Research Council's Committee on Grand Challenges in Environmental Sciences published *Grand Challenges in Environmental Sciences* (National Academy Press, 2001) in an effort to reach "a judgment regarding the most important environmental research challenges of the next generation—the areas most likely to yield results of major scientific and practical importance if pursued vigorously now." These areas include the following:

- Biogeochemical cycles (the cycling of plant nutrients, the ways human activities affect them, and the consequences for ecosystem functioning, atmospheric chemistry, and human activities)
- Biological diversity
- Climate variability
- Hydrologic forecasting (groundwater, droughts, floods, etc.)
- Infectious diseases
- Resource use
- Land use
- Reinventing the use of materials (e.g., recycling)

Similar themes appeared when *Issues in Science and Technology* celebrated its twentieth anniversary with its Summer 2003 issue. The editors noted that over the life of the magazine to date, some problems have hardly changed, nor has our sense of what must be done to solve them. Others have been affected, sometimes drastically, by changes in scientific knowledge, technological capability, and political trends. In the environmental area, the magazine paid special attention to:

- Biodiversity
- Overfishing
- Climate change
- The Superfund program
- The potential revival of nuclear power
- Sustainability

Many of the same basic themes were reiterated when *Science* magazine (published weekly by the American Association for the Advancement of Science) published in November and December 2003 a four-week series on the "State of the Planet," followed by a special issue on "The Tragedy of the Commons." In the introduction to the series, H. Jesse Smith began with these words: "Once in a while, in our headlong rush toward greater prosperity, it is wise to ask ourselves whether or not we can get there from here. As global population increases, and the demands we make on our natural resources grow even faster, it becomes ever more clear that the well-being we seek is imperiled by what we do."

Among the topics covered in the series were:

- Human population
- Biodiversity
- Tropical soils and food security
- The future of fisheries
- Freshwater resources
- Energy resources
- Air quality and pollution
- Climate change
- Sustainability
- The burden of chronic disease

Many of the topics on these lists are covered in this book. There are of course a great many other environmental issues—many more than can be covered in any one book such as this one. I have not tried to deal here with invasive species, the removal of dams to restore populations of anadromous fishes such as salmon, the depletion of aquifers, floodplain development, urban planning, or many others. My sample of the variety available begins with the more philosophical issues. For instance, there is considerable debate over the "precautionary principle," which says in essence that even if we are not sure that our actions will have unfortunate consequences, we should take precautions just in case (see Issue 1). This principle plays an important part in many environmental debates, from those over the value of restoring damaged ecosystems (Issue 4) or the wisdom of offshore oil drilling (Issue 7) to the folly (or wisdom) of reprocessing nuclear waste (Issue 19).

I said above that many people believed (and still believe) that nature has value only when turned to human benefit. One consequence of this belief is that it may be easier to convince people that nature is worth protecting if one can somehow calculate a cash value for nature "in the raw." Some environmentalists object to even trying to do this, on the grounds that economic value is not the only value, or even the value that should matter. (See Issue 3.) Related to this question of value is that of whether certain human activities should take precedence over environmental protection. Should the military be able to ignore environmental regulations (Issue 5)? Should profits and jobs come before combating global warming (Issue 6)?

Should we be concerned about the environmental impacts of specific human actions or products? Here too we can consider offshore oil drilling (Issue 7), as well as the conflict between the value of DDT for preventing malaria and its impact on ecosystems (Issue 16), the hormone-like effects of some pesticides and other chemicals on both wildlife and humans (Issue 17). World hunger is a major problem, with people arguing that it may be eased by improved foreign aid policies (Issue 13), genetic engineering (Issue 14), or organic farming (Issue 15).

Waste disposal is a problem area all its own. It encompasses carbon dioxide from fossil fuel combustion (Issue 8), hazardous waste (Issue 18), and nuclear waste (Issue 19). A new angle on hazardous waste comes from the popularity of the personal computer—or more specifically, from the huge numbers of PCs that are discarded each year.

What solutions are available? Some are specific to particular issues, as geothermal power (Issue 9), automobile efficiency (Issue 10), biofuels (Issue 11), or a revival of nuclear power (Issue 12) may be to the problems associated with fossil fuels. Some are more general, as we might expect as soon as we hear someone speak of population growth as a primary cause of environmental problems (there is some truth to this, for if the human population were small enough, its environmental impact—no matter how sloppy people were—would also be small).

Some analysts argue that whatever solutions we need, government need not impose them all. Private industry may be able to do the job if government can find a way to motivate industry, as with the idea of tradable pollution rights (see Issue 6).

The overall aim, of course, is to avoid disaster and enable human life and civilization to continue prosperously into the future. The term for this is "sustainable development" (see Issue 2), and it was the chief concern of the U.N. World Summit on Sustainable Development, held in Johannesburg, South Africa, in August 2002. Exactly how to avoid disaster and continue prosperously into the future are the themes of the U.N. Millennium Ecosystem Assessment report, *Ecosystems and Human Well-Being: Synthesis* (http://www. millenniumassessment.org/en/index.aspx) (Island Press, 2005). The main findings of this report are that over the past half century meeting human needs for food, fresh water, fuel, and other resources has had major negative effects on the world's ecosystems; those effects are likely to grow worse over the next half century and will pose serious obstacles to reducing global hunger, poverty, and disease; and although "significant changes in policies, institutions, and practices can mitigate many of the negative consequences of growing pressures on ecosystems, . . . the changes required are large and not currently under way." Also essential will be improvements in knowledge about the environment, the ways humans affect it, and the ways humans depend upon it, as well as improvements in technology, both for assessing environmental damage and for repairing and preventing damage, as emphasized by Bruce Sterling in "Can Technology Save the Planet?" *Sierra* (July/August 2005). Sterling concludes, perhaps optimistically, that "When we see our historical predicament in its full, majestic scope, we will stir ourselves to great and direly necessary actions. It's not beyond us to think and act in a better way. Yesterday's short-sighted habits are leaving us, the way gloom lifts with the dawn." George Musser, introducing the special September 2005 "Crossroads for Planet Earth" issue of *Scientific American* in "The Climax of Humanity," notes that the next few decades will determine our future. Sterling's optimism may be fulfilled, but if we do not make the right choices, the future may be very bleak indeed.

Thomas A. Easton
Thomas College

Internet References . . .

The Earth Day Network

The Earth Day Network (EDN) promotes environmental citizenship and helps activists in their efforts to change local, national, and global environmental policies. It also makes the Ecological Footprint quiz available.

http://www.earthday.net/

The Natural Resources Defense Council

The Natural Resources Defense Council is one of the most active environmental research and advocacy organizations. Its home page lists its concerns as clean air and water, energy, global warming, toxic chemicals, ocean health, and much more.

http://www.nrdc.org/

The United Nations Environment Programme

The United Nations Environment Programme "works to encourage sustainable development through sound environmental practices everywhere. Its activities cover a wide range of issues, from atmosphere and terrestrial ecosystems, the promotion of environmental science and information, to an early warning and emergency response capacity to deal with environmental disasters and emergencies."

http://www.unep.org/

The International Institute for Sustainable Development

The International Institute for Sustainable Development advances sustainable development policy and research by providing information and engaging in partnerships worldwide. It says it "promotes the transition toward a sustainable future. We seek to demonstrate how human ingenuity can be applied to improve the well-being of the environment, economy and society."

http://www.iisd.org/

The Ecosystems Services Project

The Ecosystems Services Project studies the services people obtain from their environments, the economic and social values inherent in these services, and the benefits of considering these services more fully in shaping land management policies.

http://www.ecosystemservicesproject.org/

Environmental Philosophy

*E*nvironmental debates are rooted in questions of values—what is right? what is just?—and inevitably political in nature. It is worth stressing that people who consider themselves to be environmentalists can be found on both sides of most of the issues in this book. They differ in what they see as their own self-interest and even in what they see as humanity's long-term interest.

Understanding the general issues raised in this section is useful preparation for examining the more specific controversies that follow in later sections.

- Is the Precautionary Principle a Sound Approach to Risk Analysis?
- Is Sustainable Development Compatible with Human Welfare?
- Should a Price Be Put on the Goods and Services Provided by the World's Ecosystems?

ISSUE 1

Is the Precautionary Principle a Sound Approach to Risk Analysis?

YES: Nancy Myers, from "The Rise of the Precautionary Principle: A Social Movement Gathers Strength," *Multinational Monitor* (September 2004)

NO: Bernard D. Goldstein, from "The Precautionary Principle: Is It a Threat to Toxicological Science?" *International Journal of Toxicology* (January/February 2006)

ISSUE SUMMARY

YES: Nancy Myers, communications director for the Science and Environmental Health Network, argues that because the Precautionary Principle "makes sense of uncertainty," it has gained broad international recognition as being crucial to environmental policy.

NO: Bernard D. Goldstein, professor of Environmental and Occupational Health at the University of Pittsburgh, argues that although the Precautionary Principle is potentially valuable, it poses a risk that scientific (particularly toxicological) risk assessment will be displaced to the detriment of public health, social justice, and the field of toxicology itself.

The traditional approach to environmental problems has been reactive. That is, first the problem becomes apparent—wildlife or people sicken and die, or drinking water or air tastes foul. Then researchers seek the cause of the problem, and regulators seek to eliminate or reduce that cause. The burden is on society to demonstrate that harm is being done and that a particular cause is to blame.

An alternative approach is to presume that *all* human activities—construction projects, new chemicals, new technologies, etc.—have the potential to cause environmental harm. Therefore, those responsible for these activities should prove in advance that they will not do harm and should take suitable steps to prevent any harm from happening. A middle ground is occupied by the "Precautionary Principle," which has played an increasingly important part in environmental law ever since it first appeared in Germany in the mid-1960s. On the international scene, it has been applied to climate change, hazardous waste management, ozone depletion, biodiversity, and fisheries management. In 1992 the Rio Declaration on Environment and Development, listing it as Principle 15, codified it thus:

> In order to protect the environment, the precautionary approach shall be widely applied by States according to their capabilities. When there are threats of serious or irreversible damage, lack of full scientific certainty shall not be used as a reason for postponing cost-effective measures to prevent environmental degradation.

Other versions of the principle also exist, but all agree that when there is reason to think—not absolute proof—that some human activity is or might be harming the environment, precautions should be taken. Furthermore, the burden of proof should be on those responsible for the activity, not on those who may be harmed. This has come to be broadly accepted as a basic tenet of ecologically or environmentally sustainable development. See Marco Martuzzi and Roberto Bertollini, "The Precautionary Principle, Science and Human Health Protection," *International Journal of Occupational Medicine and Environmental Health* (January 2004).

The Precautionary Principle also contributes to thinking in the areas of risk assessment and risk management in general. Human activities can damage health and the environment. Some people insist that action need not be taken against any particular activity until and unless there is solid, scientific proof that it is doing harm, and even then risks must be weighed against each other. Others insist that mere suspicion should be grounds enough for action.

Since solid, scientific proof can be very difficult to obtain, the question of just how much proof is needed to justify action is vital. Not surprisingly, if action threatens an industry, that industry's advocates will argue against taking precautions, generally saying that more proof is needed. Those who feel threatened by an industry or a new technology are more likely to favor the Precautionary Principle; see John S. Dryzek, Robert E. Goodin, et. al., "Promethean Elites Encounter Precautionary Publics: The Case of GM Foods," *Science, Technology & Human Values* (May 2009). The "Promethean Elites" are those who—like the Prometheus of myth—favor progress over the status quo and may argue that the Precautionary Principle holds back progress; see Helene Guldberg, "Challenging the Precautionary Principle," *Spiked-Online* (July 1, 2003) (http://www.spiked-online. com/Articles/00000006DE2F.htm). Yet, says Charles Weiss in "Defining Precaution," a review of *The Precautionary Principle*, UNESCO's World Commission on the Ethics of Scientific Knowledge and Technology Report, *Environment* (October 2007), the principle "is an important corrective to the pressure from enthusiasts and vested interests to push technology in unnecessarily risky directions."

In the following selections, Nancy Myers argues that the precautionary principle has gained broad international recognition as crucial to environmental policy for good reason. "The essence of the Precautionary Principle is that when lives and the future of the planet are at stake, people must act on . . . clues and prevent as much harm as possible, despite imperfect knowledge and even ignorance." Bernard Goldstein argues that although the Precautionary Principle is potentially valuable, it poses a risk that scientific (particularly toxicological) risk assessment will be displaced to the detriment of public health, social justice, and the field of toxicology itself.

YES

Nancy Myers

The Rise of the Precautionary Principle: A Social Movement Gathers Strength

Ed Soph is a jazz musician and professor at the University of North Texas in Denton, a growing town of about 100,000 just outside Dallas, Texas. In 1997, Ed and his wife Carol founded Citizens for Healthy Growth, a Denton group concerned about the environment and future of their town. The Sophs and their colleagues—the group now numbers about 400—are among the innovative pioneers who are implementing the Precautionary Principle in the United States.

The Sophs first came across the Precautionary Principle in 1998, in the early days of the group's campaign to prevent a local copper wire manufacturer, United Copper Industries, from obtaining an air permit that would have allowed lead emissions. Ed remembers the discovery of the Wingspread Statement on the Precautionary Principle—a 1998 environmental health declaration holding that "When an activity raises threats of harm to human health or the environment, precautionary measures should be taken even if some cause and effect relationships are not fully established scientifically"—as "truly a life-changing experience." Using the Precautionary Principle as a guide, the citizens refused to be drawn into debates on what levels of lead, a known toxicant, might constitute a danger to people's health. Instead, they pointed out that a safer process was available and insisted that the wise course was not to issue the permit. The citizens prevailed.

The principle helped again in 2001, when a citizen learned that the pesticides 2,4-D, simazine, Dicamba and MCPP were being sprayed in the city parks. "The question was, given the 'suspected' dangers of these chemicals, should the city regard those suspicions as a reassurance of the chemicals' safety or as a warning of their potential dangers?" Ed recalls. "Should the city act out of ignorance or out of common sense and precaution?"

Soph learned that the Greater Los Angeles School District had written the Precautionary Principle into its policy on pesticide use and had turned to Integrated Pest Management (IPM), a system aimed at controlling pests without the use of toxic chemicals. The Denton group decided to advocate for a similar policy. They persuaded the city's park district to form a focus group of park users and organic gardening experts. The city stopped spraying the four problem chemicals and initiated a pilot IPM program.

The campaign brought an unexpected economic bonus to the city. In the course of their research, parks department staff discovered that corn gluten was a good turf builder and natural broadleaf herbicide. But the nearest supplier of corn gluten was in the Midwest, and that meant high shipping costs for the city. Meanwhile, a corn processing facility in Denton was throwing away the corn gluten it produced as a byproduct. The parks department made the link, and everyone was pleased. The local corn company was happy to add a new product line; the city was happy about the expanded local business and the lower price for a local product; and the environmental group chalked up another success.

The citizens of Denton, Texas, did not stop there. They began an effort to improve the community's air pollution standards. They got arsenic-treated wood products removed from school playgrounds and parks and replaced with nontoxic facilities. "The Precautionary Principle helped us define the problems and find the solutions," Ed says.

But, as he wrote in an editorial for the local paper, "The piecemeal approach is slow, costly and often more concerned with mitigation than prevention." Taking a cue from Precautionary Principle pioneers in San Francisco, they also began lobbying for a comprehensive new environmental code for the community, based on the Precautionary Principle.

In June 2003, San Francisco's board of supervisors had become the first government in the United States to embrace the Precautionary Principle. A new environmental code drafted by the city's environment commission put the Precautionary Principle at the top, as Article One. Step one in implementing the code was a new set of guidelines for city purchasing, pointing the way toward "environmentally preferable" purchases by careful analysis and choice of the best alternatives. The White Paper accompanying the ordinance pointed out that most of the city's progressive environmental policies were already in line with the Precautionary Principle, and that the new code provided unity and focus to the policies rather than a radically new direction.

That focus is important; too often, environmental matters seem like a long, miscellaneous and confusing list of problems and solutions.

Likewise in Denton, the Precautionary Principle has not been a magic wand for transforming policy, but it has put backbone into efforts to enact truly protective and far-sighted environmental policies. Ed Soph points out that, in his community as in others, growth had often been dictated by special interests in the name of economic development, and the environment got short shrift.

"Environmental protection and pollution prevention in our city have been a matter, not of proactive policy, but of reaction to federal and state mandates, to the threat of citizens' lawsuits, and to civic embarrassment. Little thought is given to future environmental impacts," he told the city council when he argued for a new environmental code.

He added, "The toxic chemical pollution emitted by area industries has been ignored or accepted for all the ill-informed or selfish reasons that we are too familiar with. The Precautionary Principle dispels that ignorance and empowers concerned citizens with the means to ensure a healthier future."

The Precautionary Principle has leavened the discussion of environmental and human health policy on many fronts—in international treaty negotiations and global trade forums, in city resolutions and national policies, among conservationists and toxicologists, and even in corporate decision making.

Two treaties negotiated in 2000 incorporated the principle for the first time as an enforceable measure. The Cartagena Protocol on Biosafety allows countries to invoke the Precautionary Principle in decisions on admitting imports of genetically modified organisms. It became operative in June 2003. The Stockholm Convention on Persistent Organic Pollutants prescribes the Precautionary Principle as a standard for adding chemicals to the original list of 12 that are banned by the treaty. This treaty went into force in February 2004.

Making Sense of Uncertainty

Understanding the need for the Precautionary Principle requires some scientific sophistication. Ecologists say that changes in ecological systems may be incremental and gradual, or surprisingly large and sudden. When change is large enough to cause a system to cross a threshold, it creates a new dynamic equilibrium that has its own stability and does not change back easily. These new interactions become the norm and create new realities.

Something of this new reality is evident in recently observed changes in patterns of human disease:

- Chronic diseases and conditions affect more than 100 million men, women, and children in the United States—more than a third of the population. Cancer, asthma, Alzheimer's disease, autism, birth defects, developmental disabilities, diabetes, endometriosis, infertility, multiple sclerosis and Parkinson's disease are becoming increasingly common.
- Nearly 12 million children in the United States (17 percent) suffer from one or more developmental disabilities. Learning disabilities alone affect at least 5 to 10 percent of children in public schools, and these numbers are increasing. Attention deficit hyperactivity disorder conservatively affects 3 to 6 percent of all school children. The incidence of autism appears to be increasing.
- Asthma prevalence has doubled in the last 20 years.
- Incidence of certain types of cancer has increased. The age-adjusted incidence of melanoma, non-Hodgkins lymphoma, and cancers of the prostate, liver, testis, thyroid, kidney, breast, brain, esophagus and bladder has risen over the past 25 years. Breast cancer, for example, now strikes more women worldwide than any other type of cancer, with rates increasing 50 percent during the past half century. In the 1940s, the lifetime risk of breast cancer was one in 22. Today's risk is one in eight and rising.
- In the United States, the incidence of some birth defects, including male genital disorders, some forms of congenital heart disease and obstructive disorders of the urinary tract, is increasing. Sperm density is declining in some parts of the United States and elsewhere in the world.

These changes in human health are well documented. But proving direct links with environmental causative factors is more complicated.

Here is how the scientific reasoning might go: Smoking and diet explain few of the health trends listed above. Genetic factors explain up to half the population variance for several of these conditions—but far less for the majority of them—and in any case do not explain the changes in disease incidence rates. This suggests that other environmental factors play a role. Emerging science suggests this as well. In laboratory animals, wildlife and humans, considerable evidence documents a link between environmental contamination and malignancies, birth defects, reproductive disorders, impaired behavior and immune system dysfunction. Scientists' growing understanding of how biological systems develop and function leads to similar conclusions.

But serious, evident effects such as these can seldom be linked decisively to a single cause. Scientific standards of certainty (or "proof") about cause and effect are high. These standards may never be satisfied when many different factors are working together, producing many different results. Sometimes the period of time between particular causes and particular results is so long, with so many intervening factors, that it is impossible to make a definitive link. Sometimes the timing of exposure is crucial—a trace of the wrong chemical at the wrong time in pregnancy, for example, may trigger problems in the child's brain or endocrine system, but the child's mother might never know she was exposed.

In the real world, there is no way of knowing for sure how much healthier people might be if they did not live in the modern chemical stew, because the chemicals are everywhere—in babies' first bowel movement, in the blood of U.S. teenagers and in the breastmilk of Inuit mothers. No unexposed "control" population exists. But clearly, significant numbers of birth defects, cancers and learning disabilities are preventable.

Scientific uncertainty is a fact of life even when it comes to the most obvious environmental problems, such as the disappearance of species, and the most potentially devastating trends, such as climate change. Scientists seldom know for sure what will happen until it happens, and seldom have all the answers about causes until well after the fact, if ever. Nevertheless, scientific knowledge, as incomplete as it may be, provides important clues to all of these conditions and what to do about them.

The essence of the Precautionary Principle is that when lives and the future of the planet are at stake, people must act on these clues and prevent as much harm as possible, despite imperfect knowledge and even ignorance.

Environmental Failures

A premise of Precautionary Principle advocates is that environmental policies to date have largely not met this challenge. Part of the explanation for why they have not is that the dimensions of the emerging problems are only now becoming apparent. The limits of the earth's assimilative capacity are much clearer now than they were when the first modern environmental legislation was enacted 30 years ago.

Another part of the explanation is that, although some environmental policies are preventive, most have focused on cleaning up messes after the fact—what environmentalists call "end of pipe" solutions. Scrubbers on power plant stacks, catalytic converters on tailpipes, recycling and super-sized funds dedicated to detoxifying the worst dumps have not been enough. The Precautionary Principle holds that earlier, more comprehensive and preventive approaches are necessary. Nor is it enough to address problems only after they have become so obvious that they cannot be ignored—often, literally waiting for the dead bodies to appear or for coastlines to disappear under rising tides.

The third factor in the failure of environmental policies is political, say Precautionary Principle proponents. After responding to the initial burst of concern for the environment, the U.S. regulatory system and others like it were subverted by commercial interests, with the encouragement of political leaders and, increasingly, the complicity of the court system. Environmental laws have been subjected to an onslaught of challenges since the 1980s; many have been modified or gutted, and all are enforced by regulators who have been chastened by increasing challenges to their authority by industry and the courts.

The courts, and now increasingly international trade organizations and agreements like the World Trade Organization (WTO) and the North American Free Trade Agreement (NAFTA), have institutionalized an anti-precautionary approach to environmental controls. They have demanded the kinds of proof and certainty of harms and efficacy of regulation that science often cannot provide.

False Certainties

Ironically, one tool that has proved highly effective in the battle against environmental regulations was one that was meant to strengthen the enforcement of such laws: quantitative risk assessment. Risk assessment was developed in the 1970s and 1980s as a systematic way to evaluate the degree and likelihood of harmful side effects from products and technologies. With precise, quantitative risk assessments in hand, regulators could more convincingly demonstrate the need for action. Risk assessments would stand up in court. Risk assessments could "prove" that a product was dangerous, would cause a certain number of deaths per million, and should be taken off the market.

Or not. Quantitative risk assessment, which became standard practice in the United States in the mid-1980s and was institutionalized in the global trade agreements of the 1990s, turned out to be most useful in "proving" that a product or technology was not inordinately dangerous. More precisely, risk assessments presented sets of numbers that purported to state definitively how much harm might occur. The next question for policymakers then became: How much harm is acceptable? Quantitative risk assessment not only provided the answers; it dictated the questions.

As quantitative risk assessment became the norm, commercial and industrial interests were increasingly able to insist that harm must be proven "scientifically"—in the form of a quantitative risk assessment demonstrating harm in excess of acceptable limits—before action was taken to stop a process or product. These exercises were often linked with cost-benefit assessments that heavily weighted the immediate monetary costs of regulations and gave little, if any, weight to costs to the environment or future generations.

Although risk assessments tried to account for uncertainties, those projections were necessarily subject to assumptions and simplifications. Quantitative risk assessments usually addressed a limited number of potential harms, often missing social, cultural or broader environmental factors. These risk assessments have consumed enormous resources in strapped regulatory agencies and have slowed the regulatory process. They have diverted attention from questions that could be answered: Do better alternatives exist? Can harm be prevented?

The slow pace of regulation, the insistence on "scientific certainty," and the weighting toward immediate monetary costs often give the benefit of doubt to products and technologies, even when harmful side effects are suspected. One result is that neither international environmental agreements nor national regulatory systems have kept up with the increasing pace and cumulative effects of environmental damage.

A report by the European Environment Agency in 2001 tallied the great costs to society of some of the most egregious failures to heed early warnings of harm. Radiation, ozone depletion, asbestos, Mad Cow disease and other case studies show a familiar pattern: "Misplaced 'certainty' about the absence of harm played a key role in delaying preventive actions," the authors conclude.

They add, "The costs of preventive actions are usually tangible, clearly allocated and often short term, whereas the costs of failing to act are less tangible, less clearly distributed and usually longer term, posing particular problems of governance. Weighing up the overall pros and cons of action, or inaction, is therefore very difficult, involving ethical as well as economic considerations."

The Precautionary Approach

As environmentalists looked at looming problems such as global warming, they were appalled at the inadequacy of policies based on quantitative risk assessment. Although evidence was piling up rapidly that human activities were having an unprecedented effect on global climate, for example, it was difficult to say when the threshold of scientific certainty would be crossed. Good science demanded caution about drawing hard and fast conclusions. Yet, the longer humanity waited to take action, the harder it would be to reverse any effect. Perhaps it was already too late. Moreover, action would have to take the form of widespread changes not only in human behavior but also in technological development. The massive shift away from fossil fuels that might yet mitigate the effects of global warming would require rethinking the way humans produce and use energy. Nothing in the risk-assessment-based approach to policy prepared society to do that.

The global meetings called to address the coming calamity were not helping much. Politicians fiddled with blame and with protecting national economic interests while the globe heated up. Hard-won and heavily compromised agreements such as the 1997 Kyoto agreement on climate change were quickly mired in national politics, especially in the United States, the heaviest fossil-fuel user of all.

In the United States and around the globe, a different kind of struggle had been going on for decades: the fight for attention to industrial pollution in communities. From childhood lead poisoning in the 1930s to Love Canal in the 1970s, communities had always faced an uphill battle in proving that pollution and toxic products were making them sick. Risk assessments often made the case that particular hazardous waste dumps were safe, or that a single polluting industry could not possibly have caused the rash of illnesses a community claimed. But these risk assessments missed the obvious fact that many communities suffered multiple environmental assaults, compounded by other effects of poverty. A landmark 1987 report by the United Church of Christ coined the term "environmental racism" and confirmed that the worst environmental abuses were visited on communities of color. This growing awareness generated the international environmental justice movement.

In early 1998, a small conference at Wingspread, the Johnson Foundation's conference center in Racine, Wisconsin, addressed these dilemmas head-on. Participants groped for a better approach to protecting the environment and human health. At that time, the Precautionary Principle, which had been named in Germany in the 1970s, was an emerging precept of international law. It had begun to appear in international environmental agreements, gaining reference in a series of protocols, starting in 1984, to reduce pollution in the North Sea; the 1987 Ozone Layer Protocol; and the Second World Climate Conference in 1990.

At the Rio Earth Summit in 1992, precaution was enshrined as Principle 15 in the Rio Declaration on Environment and Development: "In order to protect the environment, the precautionary approach shall be widely applied by states according to their capabilities. Where there are threats of serious or irreversible damage, lack of full scientific certainty shall not be used as a reason for postponing cost-effective measures to prevent environmental degradation."

In the decade after Rio, the Precautionary Principle began to appear in national constitutions and environmental policies worldwide and was occasionally invoked in legal battles. For example:

- The Maastricht Treaty of 1994, establishing the European Union, named the Precautionary Principle as a guide to EU environment and health policy.
- The Precautionary Principle was the basis for arguments in a 1995 International Court of Justice case on French nuclear testing. Judges cited the "consensus flowing from Rio" and the fact that the Precautionary Principle was "gaining increasing support as part of the international law of the environment."
- At the World Trade Organization in the mid-1990s, the European Union invoked the Precautionary Principle in a case involving a ban on imports of hormone-fed beef.

THE U.S. CHAMBER OF COMMERCE:
POLICY BRIEF AGAINST THE PRECAUTIONARY PRINCIPLE

Objective

To ensure that regulatory decisions are based on scientifically sound and technically rigorous risk assessments, and to oppose the adoption of the Precautionary Principle as the basis for regulation.

Summary of the Issue

The U.S. Chamber supports a science-based approach to risk management; where risk is assessed based on scientifically sound and technically rigorous analysis. Under this approach, regulatory actions are justified where there are legitimate, scientifically ascertainable risks to human health, safety, or the environment. That is, the greater the risk, the greater the degree of regulatory scrutiny. This standard has served the nation well, and has led to astounding breakthroughs in the fields of science, health care, medicine, biotechnology, agriculture, and many other fields. There is, however, a relatively new theory, known as the Precautionary Principle that is gaining popularity among environmentalists and other groups. The Precautionary Principle says that when the risks of a particular activity are unclear or unknown, assume the worst and avoid the activity. It is essentially a policy of risk avoidance.

The regulatory implications of the Precautionary Principle are huge. For instance, the Precautionary Principle holds that since the existence and extent of global warming and climate change are not known one should assume the worst and immediately restrict the use of carbon-based fuels. However the nature and extent of key environmental, health, and safety concerns requires careful scientific and technical analysis. That is why the U.S. Chamber has long supported the use of sound science, cost-benefit analysis, risk assessment and a full understanding of uncertainty when assessing a particular regulatory issue.

The Precautionary Principle has been explicitly incorporated into various laws and regulations in the European Union and various international bodies. In the United States, radical environmentalists are pushing for its adoption as a basis for regulating biotechnology, food and drug safety, environmental protection, and pesticide use.

U.S. Chamber Strategy

- Support a science-based approach to risk management, where risk is assessed based on scientifically sound and technically rigorous standards.
- Oppose the domestic and international adoption of the Precautionary Principle as a basis for regulatory decision making.
- Educate consumers, businesses, and federal policymakers about the implications of the Precautionary Principle.

The Wingspread participants believed the Precautionary Principle was not just another weak and limited fix for environmental problems. They believed it could bring far-reaching changes to the way those policies were formed and implemented. But action to prevent harm in the face of scientific uncertainty alone did not translate into sound policies protective of the environment and human health. Other norms would have to be honored simultaneously and as an integral part of a precautionary decision-making process. Several other principles had often been linked with the Precautionary Principle in various statements of the principle or in connection with precautionary policies operating in Northern European countries. The statement released at the end of the meeting, the Wingspread Statement on the Precautionary Principle, was the first to put four of these primary elements on the same page—acting upon early evidence of harm, shifting the burden of proof, exercising democracy and transparency, and assessing alternatives. These standards form the basis of what has come to be known as the overarching or comprehensive Precautionary Principle or approach:

> When an activity raises threats of harm to human health or the environment, precautionary measures should be taken even if some cause and effect relationships are not fully established scientifically.
>
> In this context the proponent of an activity, rather than the public, should bear the burden of proof.
>
> The process of applying the Precautionary Principle must be open, informed and democratic and must include potentially affected parties. It must also involve an examination of the full range of alternatives, including no action.

The conference generated widespread enthusiasm for the principle among U.S. environmentalists and academics as well as among some policymakers. That was complemented by continuing and growing support for the principle among Europeans as well as ready adoption of the concept in much of the developing world. And in the years following Wingspread, the Precautionary Principle has gained new international status.

Bernard D. Goldstein  **NO**

The Precautionary Principle: Is It a Threat to Toxicological Science?

The central figure in developing the *1970 US Clean Air Act,* Senator Edmund Muskie, rather plaintively asked that the scientific information underlying this act be obtained from a one-handed scientist. He was tired of hearing scientists saying "on the one hand . . . , yet on the other hand. . . ." The Senator was expressing the desire of lawmakers for indisputable scientific facts on which to base regulatory actions. Unfortunately, this is not the way science works, particularly in situations where there is uncertainty concerning the questions being asked, or the terms being used.

Lack of clarity in definition or consistency in use is a particular problem when considering the Precautionary Principle. My discussion of the impact of the Precautionary Principle will inevitably fall into the "on the one hand . . . , yet on the other hand . . ." category that made Senator Muskie so unhappy. I will argue that the Precautionary Principle is both good public health and that it is causing public health problems; that it will both increase and decrease the importance of toxicological science in decision making; and that the European Community (EC) is both using the Precautionary Principle in some innovative ways to protect public health and the environment, and that the EC is intentionally abusing the Precautionary Principle by using it as a means to erect artificial trade barriers not warranted by toxicological science and risk analysis.

Definitions of the Precautionary Principle

The Precautionary Principle was developed primarily in Europe, beginning in the 1980s in Germany. One of the first times it received major prominence was in 1992 at the United Nations Conference on Environment and Development in Rio de Janeiro. The primary goal of this meeting was to establish an environmental agenda that would guide nations toward sustainable development. At this meeting the Precautionary Principle was adopted in the following form:

From *International Journal of Toxicology,* vol. 25, no. 1, January/February 2006, pp. 3–7. Copyright © 2006 by American College of Toxicology. Reprinted by permission of Sage Publications via Rightslink.

"Nations shall use the precautionary approach to protect the environment. Where there are threats of serious or irreversible damage, scientific uncertainty shall not be used to postpone cost-effective measures to prevent environmental degradation."

It is informative to compare this definition with the more recent Wingspread Statement:

"When an activity raises threat [of harm] to human health or the environment, precautionary measures should be taken even if some cause and effect relationships are not fully established scientifically."

Wingspread is notable for a number of differences from the earlier Rio declaration. The Rio declaration includes a statement that the threat of harm should be serious and irreversible, but no threat of harm is specified in Wingspread; the Rio declaration states that the action should be cost-effective, but cost-effectiveness is not part of the Wingspread statement; the whole thrust of the Wingspread statement is positive in that measures should be taken, whereas in Rio it is negative in that the lack of uncertainty should not postpone taking measures; and finally the Rio statement focuses solely on the environment whereas the Wingspread statement goes beyond the environment to the broad field of public health.

The problem of appropriately defining the Precautionary Principle is not simply academic. Like sustainable development, as a concept the Precautionary Principle is supportable by everyone. But the Precautionary Principle also has been adopted in various national and international treaties and by governmental entities in a way that provides legal teeth to what is otherwise a nebulous policy concept. The Wingspread approach is beginning to take hold in the United States where the Precautionary Principle has been approved by the San Francisco and Berkeley City Councils and is under consideration by other local authorities. Where laws and actions are at stake, definitions become very important. The European approach, which often infuriates its trading partners including the United States, is to state that the Precautionary Principle is enshrined within European policy—but to never define it. Rather, a policy or action is judged as to whether it will result in a precautionary approach. This has led to what Carruth and I call the Cynical American Definition:

"The Precautionary Principle is a nebulous doctrine invented by Europeans as a means to erect a trade barrier against any item produced more efficiently elsewhere."

There are many different actions that can be taken under the heading of the Precautionary Principle, or the precautionary approach as it is often described. It is useful to consider such actions under the public health and prevention nomenclature of primary and secondary prevention. Simply put, primary prevention is an approach that focuses on there never being a problem in the first place; e.g., an individual does not start smoking or, even better, there are no cigarettes being produced. Secondary prevention depends upon the early detection of a problem, such as detecting high blood pressure before someone has a stroke or is otherwise symptomatic. Much of what is done as a result of environmental risk assessment and risk management is secondary prevention, particularly when it is related to the standard approaches to dealing with known toxic chemicals. Many risk assessors would argue that the Precautionary Principle is already part of risk assessment. Among the potentially precautionary standard risk assessment practices are 10-fold "safety" factors; the 95% upper confidence limit; and the use of exposure models containing prudent default assumptions. The primary preventive approaches of the Precautionary Principle go well beyond more stringent risk assessment. Its components include taking preventive action in the face of uncertainty; shifting the burden of proof to the proponents of an activity such as industry of government; exploring a wide range of alternatives; and increasing public involvement in decision making.

The Precautionary Principle and World Trade

The implications of how to define the Precautionary Principle have nowhere been more evident than for some of the trade issues to which it is being applied. In order to keep a level playing field, various world trade agreements depend upon harmonizing risk assessments so that a nation cannot just arbitrarily decide that another nation's product is unhealthy. Unfortunately, there are a variety of examples in which it appears that the European community has used the Precautionary Principle as a means to exclude items from elsewhere in the world. Before describing these as inappropriate uses of toxicological science, let me first emphasize that we in the United States have much to learn from the EC as it wrestles with core issues of democracy that impact on the control of environmental hazards. The current debates about the governance relationships among citizens, states, and the multistate European confederation are unparalleled in the US since the Federalist Papers; and are of profound global importance.

As I do not want current broader EU-US issues to cloud the discussion, I will start with an issue that does not involve the United States. An example of action based upon the Precautionary Principle that fits under the heading of secondary prevention is the adoption by the EC of a very stringent standard for aflatoxin. This was done in essence by adding an additional safety factor for the genotoxicity of aflatoxins. For all nondairy products designated for human consumption, the EC standard is one-fifth that of the United States and for milk it is one-tenth of that of the United States. This has resulted in the exclusion of an estimated $700 million yearly of trade from Sub-Saharan Africa to the advantage of European growers. These Sub-Saharan nations are among the poorest in the world, with relatively fixed trade routes following old colonial patterns. The difference in risk is less than one cancer per year in Europe. The Joint FAO/WHO Expert Committee on Food Additives (JECFA), the expert body of the Food and Agricultural Organization and the World Health Organization, at the request of the Codex Alimentarius Commission, reviewed the difference between the US and European standard and found

that it was of no health significance. There is also concern that the EC standard is halting the usual harvesting of Brazil nuts in the Amazon area with the
result that the trees are being cut down rather than providing resources for
indigenous groups.

An example of a trade barrier related to primary prevention aspects of
the Precautionary Principle is that of the EU's ban on beef from cattle that
have previously been treated with estrogenic growth agents. Canada and the
US brought this issue to the World Trade Organization (WTO) who decided in
their favor. The EU argued that despite the lack of recognition of risk by JECFA,
or even its own scientific bodies, the Precautionary Principle permitted them
to ban beef from hormone-treated animals as a risk could not be ruled out, and
that a formal risk assessment was not needed. Among the arguments advanced
by Canada and the US were that the EC was inconsistent in not having limits
on natural estrogens, and was inconsistent in permitting the use in swine of
antibiotics whose residues in pork were carcinogenic in laboratory animals.
The WTO ruling basically said that the Precautionary Principle was not yet
generally accepted as a legal doctrine capable of supplanting risk assessment.
Much of the subsequent EU activity in justifying the Precautionary Principle
can be seen as a means of reversing the WTO opinion. The use of the Precautionary Principle as a trade barrier will likely be central to the forthcoming
WTO hearings on the EU's ban of genetically modified foods.

The use of a primary preventive precautionary approach also frequently
occurs in US law. An interesting example demonstrating a switch to primary
precaution is the hazardous air pollutant (HAP) provisions of the *1990 Clean
Air Act* amendments. Prior to 1990, to regulate a compound as a HAP, the
Environmental Protection Agency (EPA) had to go through an involved and
time-consuming process. It began by listing a compound as one reasonably
anticipated to cause adverse health effects (and is not regulated by setting an
ambient standard). After suitable hearings, the EPA would then use risk-based
and other considerations to decide which sources of the listed compound to
regulate. The burden of proof was on the government to show the potential for
adverse consequences. Relatively few compounds had been regulated under
this approach, primarily those believed to be human carcinogens. In 1990,
faced with impatience at this slow approach, and the recognition that common air contaminants responsible for many tons of emissions were uncontrolled, Congress dramatically changed the HAP regulatory process. Congress
listed over 180 compounds to be regulated for which the only rule-making
procedure was the onerous one of attempting to remove them from the list by
demonstrating safety, i.e., the burden of proof was switched to the industry.
Further, Congress specified that emission sources should be required to use
the maximum available control technology (MACT), reasoning that all sources
should do as well as the best (defined operationally as the upper 12th per-
centile). Risk considerations are taken into account only secondarily. If, after
MACT, there is still a one in a million risk to the maximally exposed individual
(MEI), additional steps are to be taken. These steps are currently under consideration and likely will go through court challenges in the next few years. The
use of the MEI as the target in a public health bill is problematic—it does not

matter whether the source is in the middle of a populous major metropolitan area or upwind of the Mojave Desert.

If a similar switch in the burden of proof, and away from risk based approaches, was under consideration today, the debate would be phrased in terms of the Precautionary Principle. As it is now 15 years from the passage of the *1990 Clean Air Act* amendments, the change in the HAP provisions allows an opportunity to consider the effectiveness of the Precautionary Principle in action. Carruth and I have pointed out potential public health shortfalls to the precautionary approach taken in the HAP amendments. From the vantage point of a toxicologist interested in understanding the mechanism of chemical effects, the most important is that once a compound is listed, why would there be any interest in studying it further? In fact, there does appear to be a decline in EPA's research interest in HAPs in recent years. Another issue is that the compounds on the HAP list have varying degrees of toxicity, e.g., benzene and toluene. But if they are both on the list and subject to the same controls, what is the incentive to use the least toxic agent? Also, nonlisted compounds are likely to be in the large universe of compounds for which there is less toxicological information and even more likelihood for unwanted surprise. Yet the HAP amendments may well push toward use of nonlisted compounds as it allows the industry source to avoid regulation. There is also a concern that the MACT standards, which have taken years and much debate to establish, will inhibit investment in even better technology—for example, air pollution control technology that might result from an advance in materials science. All of these concerns can be counteracted by provisions enacted into the 1990 HAP amendments—but whether they will be is uncertain. Accordingly, despite the *1990 Clean Air Act* HAP amendments likely leading to a decrease in tonnage of air pollutants emitted, it is still questionable whether there will be any significant lessening of toxicity directly caused by HAPs.

Registration Evaluation and Authorization of Chemicals (Reach)

Currently under consideration by the EU is a new far-reaching act that is firmly based upon the Precautionary Principle (Registration, Evaluation, and Authorisation of Chemicals 2005). It is a work in progress, with amendments being acted upon that may change the details. REACH does have provisions for prioritization of chemicals based upon exposure, or based upon hazard— but not based upon risk. In many ways, REACH is the antithesis of the *US Toxic Substances Control Act* through its requirement of a substantial amount of safety data for all chemicals, new or existing in commerce. Through this requirement it presumably will counteract the major problems we have had in the US with existing chemicals, such as the counterproductive move toward using MTBE and other oxygenated fuels without adequate toxicological testing. Also, through its insistence on the testing of all chemicals, REACH presumably will avoid the potential problem posed by unlisted chemicals in the HAP provisions. Like the *1990 US Clean Air Act* HAP amendments, REACH is justified by its proponents in part on the basis of frustration with the slowness of existing regulatory approaches. This expectation of a speedier regulatory response has not been borne out for the US HAP amendments—15 years after the passage of the act much remains unfinished, including the inevitable court cases.

Alan Boobis has pointed out that there are far fewer toxicologists in Europe than in the US, and has questioned whether there are sufficient toxicologists to accomplish the task. Accordingly, REACH may greatly increase the demand for toxicologists; although in my view, much of these funds would be better spent on developing the scientific basis for newer equivalents of the Ames test that would be effective in primary prevention. A similar criticism can be made of the current approach to testing high production volume chemicals.

Toxicology as Primary Prevention

Toxicology should be given more recognition as a primary precautionary approach. An excellent example is the Ames test, which was developed based upon scientific advances in understanding basic bacterial genetics; the role of mutagenesis in carcinogenesis; and the importance of liver metabolism in activating indirect carcinogens. To develop new consumer and industrial products, the chemical industry routinely uses the Ames test, and similar tests, to weed out potentially harmful chemicals from their development processes. The focusing of new chemical development and marketing on less harmful products is an excellent example of a primary preventive precautionary approach achieved through advances in toxicological sciences.

Unfortunately, the contributions of toxicology to primary precautionary approaches through safety assessment are not well understood by many who advocate the Precautionary Principle. Some are distrustful of scientists and are opposed to what they see as a technocracy that is controlling their lives. This is not surprising, in view of the frequency with which industry or government has used scientific uncertainty as a basis to stall needed actions. The justifiable concern is that without a precautionary approach, industry will operate under the CATNIP principle—Cheapest Available Technology Not Involving Prosecution.

Communication with Those Who Advocate the Precautionary Principle

Potentially problematic is the lack of communication between those active in the field of toxicology or environmental risk assessment and those most strongly advocating the Precautionary Principle. Of note is that of the 34 signers of the Wingspread Statement on the Precautionary Principle described above, none appear to have been members of the Society of Toxicology or of the Society for Risk Analysis. This lack of communication among different disciplines concerned with environmental protection may well limit understanding of the role of toxicological science in preventing disease. As this could be of grave consequence to our field, our reaching out to those who are advocates of the Precautionary Principle is particularly important. Let me emphasize that many, and perhaps most, active supporters of the Precautionary Principle are well aware of the role of science in preventing disease, and that there is a growing literature on the subject of science and the Precautionary Principle.

Conclusion

Among the major external threats to the field of toxicology are animal rights activists who would rather put humans and animal pets at risk rather than subject any laboratory animal to study; so called ethicists who seem opposed to any controlled experimental exposures of human subjects to chemical agents, even at levels at which we are exposed in our homes and in gasoline stations; the whole field of toxic torts; and, I would add, the Precautionary Principle. Each of these threats has their potentially valuable side in promoting the use of toxicology or developing new approaches—such as in vitro techniques that can replace laboratory animals. Similarly, although the Precautionary Principle can be a threat to the field of toxicology, it also challenges us to think carefully about the goals of toxicology and of the importance of broadly communicating about what we do and why we do it.

POSTSCRIPT

Is the Precautionary Principle a Sound Approach to Risk Analysis?

Ronald Bailey, in "Precautionary Tale," *Reason* (April 1999), defines the Precautionary Principle as "precaution in the face of any actions that may affect people or the environment, no matter what science is able—or unable—to say about that action." "No matter what science says" is not quite the same thing as "lack of full scientific certainty." Indeed, Bailey turns the Precautionary Principle into a straw man and thereby endangers whatever points he makes that are worth considering. One of those points is that widespread use of the Precautionary Principle would hamstring the development of the Third World. Roger Scruton, in "The Cult of Precaution," *National Interest* (Summer 2004), calls the Precautionary Principle "a meaningless nostrum" that is used to avoid risk and says it "clearly presents an obstacle to innovation and experiment," which are essential. Bernard D. Goldstein and Russellyn S. Carruth remind us in "Implications of the Precautionary Principle: Is It a Threat to Science?" *International Journal of Occupational Medicine and Environmental Health* (January 2004), that there is no substitute for proper assessment of risk. Jonathan Adler, in "The Precautionary Principle's Challenge to Progress," in Ronald Bailey, ed., *Global Warming and Other Eco-Myths* (Prima, 2002), argues that because the Precautionary Principle does not adequately balance risks and benefits, "the world would be safer without it." Peter M. Wiedemann and Holger Schutz, "The Precautionary Principle and Risk Perception: Experimental Studies in the EMF Area," *Environmental Health Perspectives* (April 2005), report that "precautionary measures may trigger concerns, amplify . . . risk perceptions, and lower trust in public health protection." Cass R. Sunstein, *Laws of Fear: Beyond the Precautionary Principle* (Cambridge, 2005), criticizes the Precautionary Principle in part because, he says, people overreact to tiny risks. John D. Graham, dean of the Frederick S. Pardee RAND Graduate School, argues in "The Perils of the Precautionary Principle: Lessons from the American and European Experience," Heritage Lecture #818 (delivered January 15, 2004, at the Heritage Foundation, Washington, D.C.), that the Precautionary Principle is so subjective that it permits "precaution without principle" and threatens innovation and public and environmental health. It must therefore be used cautiously.

The 1992 Rio Declaration emphasized that the Precautionary Principle should be "applied by States according to their capabilities" and that it should be applied in a cost-effective way. These provisions would seem to preclude the draconian interpretations that most alarm the critics. Yet, say David Kriebel et al., in "The Precautionary Principle in Environmental Science," *Environmental Health Perspectives* (September 2001), "environmental scientists should be aware of the policy uses of their work and of their social responsibility to do science that protects human health and the environment." Businesses are also conflicted, writes Arnold Brown in "Suitable Precautions," *Across the Board* (January/February 2002), because the Precautionary Principle tends to slow decision making, but he maintains that "we will all have to learn and practice anticipation." Professor Goldstein mentions the European Union's REACH (Registration Evaluation and Authorization of Chemicals) program. It took effect on June 1, 2007. See http://ec.europa.eu/environment/chemicals/reach/reach_intro.htm for details.

Does the Precautionary Principle make us safer? The January 23, 2009, issue of *CQ Researcher* presents a debate, under that title, between Gary Marchant, who believes that the principle "fails to provide coherent or useful answers on how to deal with uncertain risks," and Wendy E. Wagner, who contends that the existing chemical regulatory system shows the consequences of not taking a precautionary approach.

ISSUE 2

Is Sustainable Development Compatible with Human Welfare?

YES: Jeremy Rifkin, from "The European Dream: Building Sustainable Development in a Globally Connected World," *E Magazine* (March/April 2005)

NO: Ronald Bailey, from "Wilting Greens," *Reason* (December 2002)

ISSUE SUMMARY

YES: Jeremy Rifkin, president of the Foundation on Economic Trends, argues that Europeans pride themselves on their quality of life, and their emphasis on sustainable development promises to maintain that quality of life into the future.

NO: Environmental journalist Ronald Bailey states that sustainable development results in economic stagnation and threatens both the environment and the world's poor.

Over the last 30 years, many people have expressed concerns that humanity cannot continue indefinitely to increase population, industrial development, and consumption. The trends and their impacts on the environment are amply described in numerous books, including historian J. R. McNeill's *Something New Under the Sun: An Environmental History of the Twentieth-Century World* (W. W. Norton, 2000).

"Can we keep it up?" is the basic question behind the issue of sustainability. In the 1960s and 1970s, this was expressed as the "Spaceship Earth" metaphor, which said that we have limited supplies of energy, resources, and room and that we must limit population growth and industrial activity, conserve, and recycle in order to avoid crucial shortages. "Sustainability" entered the global debate in the early 1980s, when the United Nations secretary general asked Gro Harlem Brundtland, a former prime minister and minister of environment in Norway, to organize and chair the World Commission on Environment and Development and produce a "global agenda for change." The resulting report, *Our Common Future* (Oxford University Press, 1987), defined *sustainable development* as "development that meets the needs of the present without compromising the ability of future generations to meet their own needs." It recognized that limits on population size and resource use cannot be known precisely; that problems may arise not suddenly but rather gradually, marked by rising costs; and that limits may be redefined by changes in technology. The report also recognized that limits exist and must be taken into account when governments, corporations, and individuals plan for the future.

The Brundtland report led to the UN Conference on Environment and Development held in Rio de Janeiro in 1992. The Rio conference set sustainability firmly on the global agenda and made it an essential part of efforts to deal with global environmental issues and promote equitable economic development. In brief, sustainability means such things as cutting forests no faster than they can grow back, using groundwater no faster than it is recharged by precipitation, stressing renewable energy sources rather than exhaustible fossil fuels, and farming in such a way that soil fertility does not decline. In addition, economics must be revamped to take into account environmental costs as well as capital, labor, raw materials, and energy costs. Many add that the distribution of the Earth's wealth must be made more equitable as well.

Given continuing growth in population and demand for resources, sustainable development is clearly a difficult proposition. Some think it can be done, but others think that for sustainability to work, either population or resource demand must be reduced. Not surprisingly, many people see sustainable development as in conflict with business and industrial activities, private property rights, and such human freedoms as the freedoms to have many children, to accumulate wealth, and to use the environment as one wishes. Economics professor Jacqueline R. Kasun, in "Doomsday Every Day: Sustainable Economics, Sustainable Tyranny," *The Independent Review* (Summer 1999), goes so far as to argue that sustainable development will require sacrificing human freedom, dignity, and material welfare on a road to tyranny. Lester R. Brown, "Picking up the Tab," *USA Today Magazine* (May 2007), suggests that because governments currently spend some $700 billion a year subsidizing environmentally destructive activities such as automobile driving, one essential step toward sustainability is to end the subsidies.

In the following selections, Jeremy Rifkin, president of the Foundation on Economic Trends, argues that Europeans pride themselves on a quality of life that in some ways exceeds that of Americans, and their emphasis on inclusivity, diversity, sustainable development, social rights, and individual human rights promises to maintain that quality of life into the future. Ronald Bailey argues that preserving the environment, eradicating poverty, and limiting economic growth are incompatible goals. Indeed, vigorous economic growth provides wealth for all and leads to environmental protection.

YES

Jeremy Rifkin

The European Dream: Building Sustainable Development in a Globally Connected World

A growing number of Americans are beginning to wonder why Europe has leaped ahead of the U.S. to become the most environmentally advanced political space in the world today. To understand why Europe has left America behind in the race to create a sustainable society, we need to look at the very different dreams that characterize the American and European frame of mind.

Ask Americans what they most admire about the U.S.A. and they will likely cite the individual opportunity to get ahead—at least until recently. The American Dream is based on a simple but compelling covenant: Anyone, regardless of the station to which they are born, can leverage a good public education, determination and hard work to become a success in life. We can go from "rags to riches."

Ask Europeans what they most admire about Europe and they will invariably say "the quality of life." Eight out of 10 Europeans say they are happy with their lives and when asked what they believe to be the most important legacy of the 20th century, 58 percent of Europeans picked their quality of life, putting it second only to freedom in a list of 11 legacies.

While the American Dream emphasizes individual success, the European Dream emphasizes collective well-being. The reason for this lies in the divergent histories of the two continents. America's founders came over from Europe 200 years ago in the waning days of the Protestant Reformation and the early days of the European Enlightenment. They took these two streams of European thought, froze them in time, and kept them alive in their purest form until today. Americans are the most devoutly Christian and Protestant people in the industrial world and the fiercest champions of the capitalist marketplace and the nation-state.

Both the Protestant Reformation and the Enlightenment emphasized the central role of the individual in history. John Calvin exhorted the faithful that every person stands alone with his or her God. Adam Smith, in turn, argued that all individuals pursue their own self interest in the marketplace. This individualist strain fit the American context far better than it did the European setting. In a wide-open frontier, every new immigrant did indeed

From *E/The Environmental Magazine*, March/April 2005, pp. 34–39. Copyright © 2005 by Jeremy Rifkin. Reprinted by permission of the author.

stand alone and had to secure his or her survival with little or no social supports. Americans, even today, are taught by his or her parents that to be free they must learn to be self-sufficient and independent, and that they cannot depend on others.

Europeans, however, never fully bought the idea of the individual alone in the universe. Europe was already densely populated and without a frontier by the late 18th century. Walled cities and tightly packed human settlement demanded a more communal way of life. While Americans defined freedom in terms of individual autonomy and mobility, Europeans defined freedom by their communal relationships.

In America there was enough cheap and free land and resources so that newcomers could become rich. In Europe, well-defined class boundaries—a remnant of the feudal aristocracy—made it far more difficult for an individual born in a lesser station of life to rise to the top and become wealthy. So while Americans preferred to pursue happiness individually, Europeans pursued happiness collectively by emphasizing the quality of life of the community. Today, Americans devote less than 11 percent of their Gross Domestic Product (GDP) to social benefits, compared to 26 percent in Europe.

Doing It Better

So, what does Europe do better than America? It works hard to create a remarkably high quality of life for all of its people. The European Dream focuses on inclusivity, diversity, sustainable development, social rights and universal human rights. And it works. While Americans are 28 percent wealthier per capita than Europeans, in many ways, Europeans experience a higher quality of life, clear evidence that, in the long run, cooperation rather than competition is sometimes a surer path to happiness.

Europe and the U.S. have nearly opposite approaches to the question of environmental stewardship. At the heart of the difference is the way Americans and Europeans perceive risk. We Americans take pride in being a risk-taking people. We come from immigrant stock, people who risked their lives to journey to the new world and start over, often with only a few coins in their pockets and a dream of a better life. When Europeans and others are asked what they most admire about Americans, our risk-taking, "can-do" attitude generally tops the list. Where others see difficulties and obstacles, Americans see opportunities.

Our optimism is deeply entwined with our faith in science and technology. It has been said that Americans are a nation of tinkerers. When I was growing up, the engineer was held in as high esteem as the cowboy, admired for his efforts to improve the lot of society and contribute to the progress and welfare of civilization.

On the other side of the water, the sensibilities are different. It's not that Europeans aren't inventive. One could make the case that over the course of history Europe has produced most of the great scientific insights and not a few of the major inventions. But with their longer histories, Europeans are far more mindful of the dark side of science and technology.

Saying No to GE Foods

In recent years, the European Union (EU) has turned upside down the stand-ard operating procedure for introducing new technologies and products into the marketplace and society, much to the consternation of the United States. The turnaround started with the controversy over genetically engineered (GE) foods and the introduction of genetically modified organisms (GMOs). The U.S. government gave the green light to the widespread introduction of GE foods in the mid 1990s, and by the end of the decade more than half of America's agricultural land was given over to GE crops. No new laws were enacted to govern the potential harmful environmental and health impacts. Instead, existing statutes were invoked, and no special handling or labeling of the products was required.

In Europe, massive opposition to GMOs erupted across the continent. Farmers, environmentalists and consumer organizations staged protests and political parties and governments voiced concern. A defacto moratorium on the planting of GE crops and sale of GE food products was put into effect. Meanwhile, the major food processors, distributors and retailers pledged not to sell any products containing GE traits.

The EU embarked on a lengthy review process to assess the environmen-tal and health risks of introducing GE food products. In the end, it established tough new protections designed to mitigate the potential harm. The meas-ures included procedures to segregate and track GE grain and food products from the fields to the retail stores to ensure against contamination; labeling of GMOs at every stage of the food process to ensure transparency; and inde-pendent testing as well as more rigorous testing requirements by the compa-nies producing GE seeds and other GMOs.

The EU is forging ahead on a wide regulatory front, changing the very conditions and terms by which new scientific and technological pursuits and products are introduced into the marketplace and the environment. Its bold initiatives put the EU far ahead of the rest of the world. Behind all of its new-found regulatory zeal is the looming question of how best to model global risks and create a sustainable and transparent approach to economic development.

Ensuring Safety

In May of 2003, the EU proposed sweeping new regulatory controls on chem-icals to mitigate toxic impacts on the environment and human and animal health. The proposed new law would require new companies to register and test for the safety of more than 30,000 chemicals at an estimated cost to the producers of nearly eight billion Euros. Under existing rules, 99 percent of the total volume of chemicals sold in Europe have not passed through any environmental and health testing and review process. In the past, there was no way to even know what kind of chemicals were being used by industry, making it nearly impossible to track potential health risks. The new regula-tions will change all of that. The "REACH" system—which stands for Regis-tration, Evaluation and Authorization of Chemicals—requires the companies to conduct safety and environmental tests to prove that the products they are producing are safe. If they can't, the products will be banned from the market.

The new procedures represent an about face to the way the chemical industry is regulated in the U.S. In America, new chemicals are generally assessed to be safe and the burden is primarily put on the consumer, the public or the government to prove that they cause harm. The EU has reversed the burden of proof. Former EU Environmental Commissioner Margot Wallstrom makes the point: "No longer do public authorities need to prove they [the products] are dangerous. The onus is now on industry to prove that the products are safe."

Making companies prove that their chemical products are safe before they are sold is a revolutionary change. It's very difficult to conceive of the U.S. entertaining the kind of risk prevention regulatory regime that the EU has rolled out. In a country where corporate lobbyists spend millions of dollars influencing congressional legislation, the chances of ever having a similar regulatory regime to the one being implemented in Europe would be nigh on impossible.

BUILDING THE HYDROGEN ECONOMY

At the very top of the list of environmental priorities for the EU is the plan to become a fully integrated renewable-based hydrogen economy by mid-century. The EU has led the world in championing the Kyoto Protocol on Climate Change, and to ensure compliance it has made a commitment to produce 22 percent of its electricity and 12 percent of all of its energy using renewable sources by 2010. Although a number of member states are lagging behind on meeting their renewable energy targets, the very fact that the EU has set benchmarks puts it far ahead of the U.S. in making the shift from fossil fuels to renewable energy sources. The Bush administration has consistently fought Congressional attempts to establish similar benchmarks for ushering in a U.S.-based renewable energy regime.

In June of 2003, EU President Romano Prodi said, "It is our declared goal of achieving a step-by-step shift toward a fully integrated hydrogen economy, based on renewable energy sources, by the middle of the century." He added that creating this economy would be the next critical step in integrating Europe after the introduction of the Euro.

The European hydrogen game plan is being implemented with a sense of history in mind. Great Britain became the world's leading power in the 19th century because it was the first country to harness its vast coal reserves with steam power. The U.S., in turn, became the world's preeminent power in the 20th century because it was the first country to harness its vast oil reserves with the internal-combustion engine. The multiplier effects of both energy revolutions were extraordinary. The EU is determined to lead the world into the third great energy revolution of the modern era.—*J.R.*

GMOs and chemical products represent just part of the new "risk prevention" agenda taking shape in Brussels. In early 2003, the EU adopted a new rule prohibiting electronics manufacturers from selling products in the EU that contain mercury, lead and other heavy metals. Another new regulation requires the manufacturers of all consumer electronics and household appliances to cover the costs for recycling their products. American companies complain that compliance with the new regulations will cost them hundreds of millions of dollars a year.

All of these strict new rules governing risk prevention would come as a shock to Americans who believe that the U.S. has the most vigilant regulatory oversight regime in the world for governing risks to the environment and public health. Although that was the case 30 years ago, it no longer is today.

The EU is the first governing institution in history to emphasize human responsibilities to the global environment as a centerpiece of its political vision. Europe's new sensitivity to global risks has led it to champion the Kyoto Protocol on climate change, the Biodiversity Treaty, the Chemical Weapons Convention and many others. The U.S. government has refused, to date, to ratify any of the above agreements.

A New Era

In Europe, intellectuals are increasingly debating the question of the great shift from a risk-taking age to a risk-prevention era. That debate is virtually non-existent in the U.S., where risk-taking is seen as a virtue. The new European intellectuals argue that vulnerability is the underbelly of risks. A sense of vulnerability can motivate people to band together in common cause. The EU stands as a testimonial to collective political engagement arising from a sense of risk and shared vulnerability.

What's changed qualitatively in the last half century since the dropping of the atomic bombs on Hiroshima and Nagasaki is that risks are now global in scale, open ended in duration, incalculable in their consequences and not compensational. Their impact is universal, which means that no one can escape their potential effects. Risks have now become truly democratized, making everyone vulnerable. When everyone is vulnerable, then traditional notions of calculating and pooling risks become virtually meaningless. This is what European academics call a risk society.

Americans aren't there yet. While some academics speak to global risks and vulnerabilities and a significant minority of Americans express their concerns about global risks, from climate change to loss of biodiversity, the sense of utter vulnerability just isn't as strong on this side of the Atlantic. Europeans say we have blinders on. In reality, it's more nuanced than that. Call it delusional, but the sense of personal empowerment is so firmly embedded in the American mind, that even when pitted against growing evidence of potentially overwhelming global threats, most Americans shrug such notions off as overly pessimistic and defeatist. "Individuals can move mountains." Most Americans believe that. Fewer Europeans do.

The EU has already institutionalized a litmus test that cuts to the core of the differences between America and Europe. It's called "the precautionary principle" and it has become the centerpiece of EU regulatory policy governing science and technology in a globalizing world.

The Precautionary Principle

In November 2002, the EU adopted a new policy on the use of the precautionary principle to regulate science and new products derived from technology innovations. According to the EU, reviews occur in "cases where scientific evidence is insufficient, inconclusive or uncertain and preliminary scientific evaluation indicates that there are reasonable grounds for concern that the potentially dangerous effects on the environment, human, animal or plant

---

THE TRANSITION TO ORGANIC AGRICULTURE

Europe is taking the lead in the shift to sustainable farming practices and organic food production. While the organic food sector is soaring in the U.S.—it represents the fastest-growing segment of the food industry—the government has done little to encourage it. Although the U.S. Department of Agriculture fields a small organic food research program, it amounts to only $3 million, less than .004 percent of its $74 billion budget. While American consumers are increasing their purchases of organic food, less than 0.3 percent of total U.S. farmland is currently in organic production.

By contrast, many of the EU member states have made the transition to organic agriculture a critical component of their economic development plans and have even set benchmarks. Germany, which has often been the leader in setting new environmental goals for the continent, has announced its intention to bring 20 percent of its agricultural output into organic production by the year 2020. (Organic agricultural output is now 3.2 percent of all farm output in Germany.)

The Netherlands, Sweden, Great Britain, Finland, Norway, Germany, Switzerland, Denmark, France and Austria also have national programs to promote the transition to organic food production. Denmark and Sweden enjoy the highest consumption of organic vegetables in Europe and both countries project that their domestic markets for organic food will soon reach or exceed 10 percent of domestic consumption.

Sweden has set a goal of having 20 percent of its total cultivated farm area in organic production by 2005. Italy already has 7.2 percent of its farmland under organic production while Denmark is close behind with seven percent.

Great Britain doubled its organic food production in 2002 and now boasts the second-highest sales of organic food in Europe, after Germany. According to a recent survey, nearly 80 percent of British households buy organic food. By comparison, only 33 percent of American consumers buy any organic food.—J.R.

health may be inconsistent with the high level of protection chosen by the EU." The key term is "uncertain." When there is sufficient evidence to suggest a potential negative impact, but not enough to know for sure, the precautionary principle allows regulatory authorities to err on the side of safety. They can suspend the activity altogether, modify it, employ alternative scenarios, monitor the activity or create experimental protocols to better understand its effects.

The precautionary principle allows governments to respond with a lower threshold of scientific certainty than in the past. "Scientific certainty" has been tempered by the notion of "reasonable grounds for concern." The precautionary principle gives authorities the flexibility to respond to events in real time, either before or while they are unfolding.

Advocates of the precautionary principle cite the introduction of halocarbons and the tear in the ozone hole in the Earth's upper atmosphere, the outbreak of mad cow disease in cattle, growing antibiotic resistant strains of bacteria caused by the over-administering of antibiotics to farm animals and the widespread deaths caused by asbestos, benzene and polychlorinated biphenyls (PCBs).

The precautionary principle has been finding its way into international treaties and covenants. It was first recognized in 1982 when the United Nations General Assembly incorporated it into the World Charter for Nature. The precautionary principle was subsequently included in the Rio Declaration on Environment and Development in 1992, the Framework Convention on Climate Change in 1992, the Treaty on EU (Maastricht Treaty) in 1992, the Cartagena Protocol on Biosafety in 2000 and the Stockholm Convention on Persistent Organic Pollutants (POPs) in 2001.

Valuing Nature

Americans, by and large, view nature as a treasure trove of useful resources waiting to be harnessed for productive ends. While Europeans share America's utilitarian perspective, they also have a love for the intrinsic value of nature. One can see it in Europeans' regard for the countryside and their determination to maintain natural landscapes, even if it means providing government assistance in the way of special subsidies, or foregoing commercial development. Nature figures prominently in Europeans' dream of a quality of life. Europeans spend far more time visiting the countryside on weekends and during their vacations than Americans.

The balancing of urban and rural time is less of a priority for most Americans, many of whom are just as likely to spend their weekends at a shopping mall, while their European peers are hiking along country trails. Anyone who spends significant time among Europeans knows that they have a great affinity for rural getaways. Almost everyone I know in Europe—among the professional and business class—has some small second home in the country somewhere—a dacha usually belonging to the family for generations. While working people may not be as fortunate, on any given weekend they can be seen exiting the cities en masse, motoring their way into the nearest rural enclave or country village for a respite from urban pressures.

The strongly held values about rural life and nature is one reason why Europe has been able to support green parties across the continent, with substantial representation in national parliaments as well as in the European Parliament. By contrast, not a single legislator at the federal level in the U.S. is a member of the Green Party.

There is another dimension to the European psyche that makes Europeans supportive of the precautionary principle—their sense of "connectedness."

Because we Americans place such a high premium on autonomy, we are far less likely to see the deep connectedness of things. We tend to see the world in terms of containers, each isolated from the whole and capable of standing alone. We like everything around us to be neatly bundled, autonomous, and self contained. The new view of science that is emerging in the wake of globalization is quite different. Nature is viewed as a myriad of symbiotic relationships, all embedded in a larger whole, of which they are an integral part. In this new vision of nature, nothing is autonomous, everything is connected.

By championing a host of global environmental treaties and accords taking the precautionary approach to regulation, the EU has shown a willingness to act on its commitment to sustainable development and global environmental stewardship. The fact that its commitments in most areas remain weak and are often vacillating is duly noted. But, at least Europe has established a new agenda for conducting science and technology that, if followed, could begin to wean the world from the old ways and toward a second Enlightenment.

Wilting Greens

It's clear that we've suffered a number of major defeats," declared Andrew Hewett, executive director of Oxfam Community Aid, at the conclusion of the World Summit on Sustainable Development, held in Johannesburg, South Africa, in September. Greenpeace climate director Steve Sawyer complained, "What we've come up with is absolute zero, absolutely nothing." The head of an alliance of European green groups proclaimed, "We barely kept our heads above water."

It wasn't supposed to be this way. Environmental activists hoped the summit would set the international agenda for sweeping environmental reform over the next 15 years. Indeed, they hoped to do nothing less than revolutionize how the world's economy operates. Such fundamental change was necessary, said the summiteers, because a profligate humanity consumes too much, breeds too much, and pollutes too much, setting the stage for a global ecological catastrophe.

But the greens' disappointment was inevitable because their major goals—preserving the environment, eradicating poverty, and limiting economic growth—are incompatible. Economic growth is a prerequisite for lessening poverty, and it's also the best way to improve the environment. Poor people cannot afford to worry much about improving outdoor air quality, let alone afford to pay for it. Rather than face that reality, environmentalists increasingly invoke "sustainable development." The most common definition of the phrase comes from the 1987 United Nations report *Our Common Future*: development that "meets the needs of the present without compromising the ability of future generations to meet their own needs."

For radical greens, sustainable development means economic stagnation. The Earth Island Institute's Gar Smith told Cybercast News, "I have seen villages in Africa . . . that were disrupted and destroyed by the introduction of electricity." Apparently, the natives no longer sang community songs or sewed together in the evenings. "I don't think a lot of electricity is a good thing," Smith added. "It is the fuel that powers a lot of multinational imagery." He doesn't want poor Africans and Asians "corrupted" by ads for Toyota and McDonald's, or by Jackie Chan movies.

Indian environmentalist Sunita Narain decried the "pernicious introduction of the flush toilet" during a recent PBS/BBC television debate hosted by Bill Moyers. Luckily, most other summiteers disagreed with Narain's curious disdain for sanitation. One of the few firm goals set at the confab was that adequate sanitation should be supplied by 2015 to half of the 2.2 billion people now lacking it.

Sustainable development boils down to the old-fashioned "limits to growth" model popularized in the 1970s. Hence Daniel Mittler of Friends of the Earth International moaned that "the summit failed to set the necessary economic and ecological limits to globalization." The *Jo'burg Memo*, issued by the radical green Heinrich Böll Foundation before the summit, summed it up this way: "Poverty alleviation cannot be separated from wealth alleviation."

The greens are right about one thing: The extent of global poverty is stark. Some 1.1 billion people lack safe drinking water, 2.2 billion are without adequate sanitation, 2.5 billion have no access to modern energy services, 11 million children under the age of 5 die each year in developing countries from preventable diseases, and 800 million people are still malnourished, despite a global abundance of food. Poverty eradication is clearly crucial to preventing environmental degradation, too, since there is nothing more environmentally destructive than a hungry human.

Most summit participants from the developing world understood this. They may be egalitarian, but unlike their Western counterparts they do not aim to make everyone equally poor. Instead, they want the good things that people living in industrialized societies enjoy.

That explains why the largest demonstration during the summit, consisting of more than 10,000 poor and landless people, featured virtually no banners or chants about conventional environmentalist issues such as climate change, population control, renewable resources, or biodiversity. Instead, the issues were land reform, job creation, and privatization.

The anti-globalization stance of rich activists widens this rift. Environmentalists claim trade harms the environment and further impoverishes people in the developing world. They were outraged by the dominance of trade issues at the summit.

"The leaders of the world have proved that they work as employees for the transnational corporations," asserted Friends of the Earth Chairman Ricardo Navarro. Indian eco-feminist Vandana Shiva added, "This summit has become a trade summit, it has become a trade show." Yet the U.N.'s own data underscore how trade helps the developing world. As fact sheets issued by the U.N. put it, "During the 1990s the economies of developing countries that were integrated into the world economy grew more than twice as fast as the rich countries. The 'non-globalizers' grew only half as fast and continue to lag further behind."

By invoking a zero sum version of sustainable development, environmentalists not only put themselves at odds with the developing world; they ignore the way in which economic growth helps protect the environment. The real commons from which we all draw is the growing pool of scientific, technological, and institutional concepts, and the capital they create. Past generations have left us far more than they took, and the result has been an explosion in human well-being, longer life spans, less disease, more and cheaper food, and expanding political freedom.

Such progress is accompanied by environmental improvement. Wealthier is healthier for both people and the environment. As societies become richer and more technologically adept, their air and water become cleaner, they set aside more land for nature, their forests expand, they use less land for agriculture, and more people cherish wild species. All indications suggest that the 21st century will be the century of ecological restoration, as humanity uses physical resources ever more efficiently, disturbing the natural world less and less.

In their quest to impose a reactionary vision of sustainable development, the disappointed global greens will turn next to the World Trade Organization, the body that oversees international trade rules. During the summit, the WTO emerged as the greens' bête noire. As Friends of the Earth International's Daniel Mittler carped, "Instead of using the [summit] to respond to global concerns over deregulation and liberalization, governments are pushing the World Trade Organization's agenda." "See you in Cancun!" promised Greenpeace's Steve Sawyer, referring to the location of the next WTO ministerial meeting in September 2003. That confab will build on the WTO's Doha Trade Round, launched last year, which is aimed at reducing the barriers to trade for the world's least developed countries.

The WTO may achieve worthy goals that eluded the Johannesburg summit, such as eliminating economically and ecologically ruinous farm and energy subsidies and opening developed country markets to the products of developing nations. Free marketeers and greens might even form an alliance on those issues.

But environmentalists want to use the WTO to implement their sustainable development agenda: global renewable energy targets, regulation based on the precautionary principle, a "sustainable consumption and production project," a worldwide eco-labeling scheme. According to Greenpeace's Sawyer, nearly everyone at the Johannesburg summit agreed "there is something wrong with unbridled neoliberal capitalism."

Let's hope the greens fail at the WTO just as they did at the U.N. summit. Their sustainable development agenda, supposedly aimed at improving environmental health, instead will harm the natural world, along with the economic prospects of the world's poorest people. The conflicting goals on display at the summit show that at least some of the world's poor are wise to that fact.

POSTSCRIPT

Is Sustainable Development Compatible with Human Welfare?

The first of the Rio Declaration's 22 principles states, "Human beings are at the centre of concerns for sustainable development. They are entitled to a healthy and productive life in harmony with nature." Any solution to the sustainability problem therefore should not infringe human welfare. This makes any solution that involves limiting or reducing human population or blocking improvements in standard of living very difficult to sell. Yet solutions may be possible. David Malin Roodman suggests in *The Natural Wealth of Nations: Harnessing the Market for the Environment* (W. W. Norton, 1998) that taxing polluting activities instead of profit or income would stimulate corporations and individuals to reduce such activities or to discover nonpolluting alternatives. In "Building a Sustainable Society," *State of the World 1999* (W. W. Norton, 1999), he adds recommendations for citizen participation in decision making, education efforts, and global cooperation, without which we are heading for "a world order [that] almost no one wants." (He is referring to a future of environmental crises, not the "new world order" feared by many conservatives, in which national policies are dictated by international [UN] regulators.) Roodman's recommendations may actually be on the way to reality. Arun Agrawal and Maria Carmen Lemos, "A Greener Revolution in the Making? Environmental Governance in the 21st Century," *Environment* (June 2007), argue that budget cuts and globalization are eroding the power of the state in favor of international "hybrid" arrangements that stress public-private partnerships, markets, and community and local participation.

Julie Davidson, in "Sustainable Development: Business as Usual or a New Way of Living?" *Environmental Ethics* (Spring 2000), notes that efforts to achieve sustainability cannot by themselves save the world. But such efforts may give us time to achieve new and more suitable values. It is thus heartening to see that the UN World Summit on Sustainable Development was held in Johannesburg, South Africa, in August 2002. Its aim was to strengthen partnerships between governments, business, nongovernmental organizations, and other stakeholders and to seek to eradicate poverty and make more equal the distribution of the benefits of globalization. See Gary Gardner, "The Challenge for Johannesburg: Creating a More Secure World," *State of the World 2002* (W. W. Norton, 2002), and the United Nations Environmental Programme's Global Environmental Outlook 3 (Earthscan, 2002), prepared as a "global state of the environment report" in preparation for the Johannesburg Summit.

The World Council of Churches brought to the Johannesburg Summit an emphasis on social justice. Martin Robra, in "Justice—The Heart of

Sustainability," *Ecumenical Review* (July 2002), writes that the dominant stress on economic growth "has served, first and foremost, the interests of the powerful economic players. It has further marginalized the poor sectors of society, simultaneously undermining their basic security in terms of access to land, water, food, employment, and other basic services and a healthy environment."

Is social justice or equity worth this emphasis? Or is sustainability more a matter of population control, of shielding the natural environment from human impacts, or of economics? A. J. McMichael, C. D. Butler, and Carl Folke, in "New Visions for Addressing Sustainability," *Science* (December 12, 2003), argue that it is wrong to separate—as did the Johannesburg Summit—achieving sustainability from other goals such as reducing fertility and poverty and improving social equity, living conditions, and health. They observe that human population and lifestyle affect ecosystems, ecosystem health affects human health, human health affects population and lifestyle. "A more integrated . . . approach to sustainability is urgently needed," they say, calling for more collaboration among researchers and other fields. Yet there remains reason to focus on single threats. In February 2007, Sigma Xi and the United Nations Foundation released the Scientific Expert Group Report on Climate Change and Sustainable Development, *Confronting Climate Change: Avoiding the Unmanageable and Managing the Unavoidable* (http://www.sigmaxi.org/about/news/UNSEGReport.shtml) (Executive Summary, *American Scientist,* February 2007). Among its many points is that climate change from global warming is a huge threat to sustainability. Even in spite of feasible attempts at mitigation and adaptation, there is a serious risk of "intolerable impacts on human well-being."

Lester R. Brown, "Could Food Shortages Bring Down Civilization?" *Scientific American* (May 2009), warns that unsustainable agricultural practices and use of ground water, along with global warming, threaten to diminish food supplies to the point where society may actually break down. Underlining this concern, in October 2008, the European Environment and Sustainable Development Advisory Councils held a conference on "Sustaining Europe for a Long Way Ahead." The EEAC Web site (http://www.eeac-net.org/conferences/sixteen/sixteen_frame.htm) noted that "Addressing the very long term through the lens of sustainable development is now a matter of urgency. The prospect of highly damaging, and extremely costly, effects of global change in climate, in natural hazards caused by human intervention, the loss of biodiversity and disruption of food security, poses serious threats to personal and collective human health and wellbeing. . . . The long term is indeed here already." See also Timothy O'Riordan, "On the Politics of Sustainability a Long Way Ahead," *Environment* (March 2009).

ISSUE 3

Should a Price Be Put on the Goods and Services Provided by the World's Ecosystems?

YES: John E. Losey and Mace Vaughan, from "The Economic Value of Ecological Services Provided by Insects," *BioScience* (April 2006)

NO: Marino Gatto and Giulio A. De Leo, from "Pricing Biodiversity and Ecosystem Services: The Never-Ending Story," *BioScience* (April 2000)

ISSUE SUMMARY

YES: John E. Losey and Mace Vaughan argue that even conservative estimates of the value of the services provided by wild insects are enough to justify increased conservation efforts. They say that "everyone would benefit from the facilitation of the vital services these insects provide."

NO: Professors of applied ecology Marino Gatto and Giulio A. De Leo contend that the pricing approach to valuing nature's services is misleading because it falsely implies that only economic values matter.

Human activities frequently involve trading a swamp or forest or mountainside for a parking lot or housing development or farm. People generally agree that these developments are worthwhile projects, for they have obvious benefits. But are there costs as well? Construction costs, labor costs, and material costs can easily be calculated, but what about the swamp? The forest? The species living there?

How much is a species worth? One approach to answering this question is to ask people how much they would be willing to pay to keep a species alive. If the question is asked when there are a million species in existence, few people will likely be willing to pay much. But if the species is the last one remaining, they might be willing to pay a great deal. Most people would agree that both answers fail to get at the true value of a species, for nature is not expressible solely in terms of cash values. Yet some way must be found to weigh the effects of human activities on nature against the benefits gained from those activities. If it is not, we will continue to degrade the world's ecosystems and threaten our own continued well-being.

Traditional economics views nature as a "free good." That is, forests generate oxygen and wood, clouds bring rain, and the sun provides warmth, all without charge to the humans who benefit. At the same time, nature has provided ways for people to dispose of wastes—such as dumping raw sewage into rivers or emitting smoke into the air—without paying for the privilege. This "free" waste disposal has turned out to have hidden costs in the form of the health effects of pollution (among other things), but it has been up to individuals and governments to bear the costs associated with those effects. The costs are real, but in general, they have not been borne by the businesses and other organizations that produced them. They have thus come to be known as "external" costs.

Environmental economists have recognized the problem of external costs, and government regulators have devised a number of ways to make those who are responsible accept the bill, such as instituting requirements for pollution control and fining those who exceed permitted emissions. Yet some would say that this approach does not help enough.

The *ecosystem services* approach recognizes that undisturbed ecosystems do many things that benefit us. A forest, for instance, slows the movement of rain and snowmelt into streams and rivers; if the forest is removed, floods may follow (a connection that recently forced China to deemphasize forest exploitation). Swamps filter the water that seeps through them. Food chains cycle nutrients necessary for the production of wood and fish and other harvests. Bees pollinate crops and make food production possible. These services are valuable—even essential—to us, and anything that interferes with them must be seen as imposing costs just as significant as the illnesses associated with pollution.

How can those costs be assessed? In 1997 Robert Costanza and his colleagues published an influential paper entitled "The Value of the World's Ecosystem Services and Natural Capital" in the May 15 issue of the journal *Nature*. In it, the authors listed a variety of ecosystem services and attempted to estimate what it would cost to replace those services if they were somehow lost. The total bill for the entire biosphere came to $33 trillion (the middle of a $16–54 trillion range), compared to a global gross national product of $25 trillion. Costanza et al. stated that this was surely an underestimate. Janet N. Abramovitz, "Putting a Value on Natures 'Free' Services," *WorldWatch* (January/February 1998), argues that nature's services are responsible for the vast bulk of the value in the world's economy and that attaching economic value to those services may encourage their protection.

In the following selections, John E. Losey, associate professor of entomology at Cornell University, and Mace Vaughan, Conservation Director at the Xerces Society for Invertebrate Conservation, argue that even conservative estimates of the value of the services provided by wild insects are enough to justify increased conservation efforts. Ecologists Marino Gatto and Giulio A. De Leo argue that the pricing approach to valuing nature's services is misleading because it ignores equally important "nonmarket" values.

YES

John E. Losey and
Mace Vaughan

The Economic Value of Ecological Services Provided by Insects

Natural systems provide ecological services on which humans depend. Countless organisms are involved in these complex interactions that put food on our tables and remove our waste. Although human life could not persist without these services, it is difficult to assign them even an approximate economic value, which can lead to their conservation being assigned a lower priority for funding or action than other needs for which values (economic or otherwise) are more readily calculated. Estimating even a minimum value for a subset of the services that functioning ecosystems provide may help establish a higher priority for their conservation.

In this article we focus on the vital ecological services provided by insects. Several authors have reviewed the economic value of ecological services in general, but none of these reviews focused specifically on insects. Insects comprise the most diverse and successful group of multicellular organisms on the planet, and they contribute significantly to vital ecological functions such as pollination, pest control, decomposition, and maintenance of wildlife species. . . . Our twofold goal is to provide well-documented, conservative estimates for the value of these services and to establish a transparent, quantitative framework that will allow the recalculation of the estimates as new data become available. We also should clarify that by "value" we mean documented financial transactions—mostly the purchase of goods or services—that rely on these insect-mediated services.

We restrict our focus to services provided by "wild" and primarily by native insects; we do not include services from domesticated species (e.g., pollination from domesticated honey bees) or pest control from mass-reared insect biological-control agents (e.g., *Trichogramma* wasps). We also exclude the value of commercially produced insect-derived products, such as honey, wax, silk, or shellac, and any value derived from the capture and consumption of insects themselves. The main reasons for these exclusions are that domesticated insects that provide services or products have been covered in many other forums, and they generally do not require the active conservation that we believe is warranted by those undomesticated insects that provide services. Furthermore, in the case of products or food derived directly from wild insects, we simply do not have data to report and therefore wish to maintain a focus on ecological services.

From *BioScience*, by John E. Losey and Mace Vaughan, vol. 56, no. 4, April 2006, pp. 311–316, 318–322. Copyright © 2006 by American Institute of Biological Sciences. Reprinted by permission via the Copyright Clearance Center.

The four insect services for which we provide value estimates were chosen not because of their importance, but because of the availability of data and an algorithm for their calculation. Three of these services (dung burial, pest control, and pollination) support the production of a commodity that has a quantifiable, published value. To be consistent in our analysis for all three of these commodities, we calculated an estimate for the amount of each commodity that depends on each service or on the amount saved in related expenses (e.g., the cost of fertilizer in our analysis of dung burial). We did not perform an in-depth analysis of how service-dependent changes in the quantity or quality of each commodity may have affected its per-unit price.

One way of looking at the economic implications of the removal of a service was provided by Southwick and Southwick, whose study involved crop pollination by honey bees. Because per-unit cost theoretically increases as supplies decrease, thus mitigating monetary losses, the costs of the service removal in the Southwick and Southwick study were lower than those calculated using our approach. However, all reported values are still within an order of magnitude of each other and, although our approach may not reflect what a consumer would pay for a commodity when these ecological services are *not* being performed, our calculations do provide a measure of the value of these crops at current estimated levels of service.

In the case of insect support of wildlife nutrition, we use a different approach to estimate costs. Instead of basing calculations on the money paid to producers for raw commodities, we use census data to find out how US consumers spent their money. By looking at the consumer end of this system, we immediately see an order-of-magnitude increase in the value reported. We believe it is important for this difference to be understood up front, because it both significantly affects our reported results and provides at least a hint of what happens when raw commodities are converted into value-added products. For example, consumers will spend potentially an order of magnitude more on jellies, pasta sauce, or hamburgers than the price paid to producers for blueberries, tomatoes, or beef.

Using the methods we describe in detail in the following sections, we estimate the annual value of four ecological services provided by primarily native insects in the United States to be more than $57 billion ($0.38 billion for dung burial, $3.07 billion for pollination, $4.49 billion for pest control of native herbivores, and $49.96 billion for recreation). We consider this estimate very conservative. If data were available to support more accurate estimates of the true value of these services (e.g., inclusion of value-added products and wages paid to those who produce such products) or to allow estimation of the value of other services, the results of our calculations would be much higher. In addition to the role of insects in the systems we analyze here, other potentially important services that insects provide could not be quantified, including suppression of weeds and exotic herbivorous species, facilitation of dead plant and animal decomposition, and improvement of the soil. Calculating the value of any of these services could add billions of dollars to our overall estimate. Nevertheless, we hope that even this minimum estimate for a subset of services provided by insects will allow these animals to be more correctly factored into land management and legislative decisions. In the following sections, we present a detailed description of how we calculated these estimates and discuss the implications of our results.

Dung Burial

Confining large mammals in small areas creates challenging waste-management problems. Cattle production in the United States provides a particularly pertinent example, because nearly 100 million head of cattle are in production, and each animal can produce over 9000 kilograms (kg), or about 21 cubic meters, of solid waste per year. Fortunately, insects—especially beetles in the family Scarabaeidae—are very efficient at decomposing this waste. In doing so, they enhance forage palatability, recycle nitrogen, and reduce pest habitat, resulting in significant economic value for the cattle industry.

Dung beetles process a substantial amount of the cattle dung accumulated annually in the United States. . . .

The importance of this service is illustrated by the success of dung beetles introduced into Australia to deal with the dung of nonnative cattle brought to that continent in 1788. Before the introduction of dung beetle species that were adapted to feed on cattle dung, Australia had no insect fauna to process cattle feces. Consequently, rangeland across the country was fouled by slowly decomposing dung. In addition, this dung provided fodder for pest species. Recent research in western Australia has revealed that populations of the pestiferous bush fly (*Musca vetustissima*) have been reduced by 80% following dung beetle introductions. . . .

Using data from Anderson and colleagues, we calculate that the average persistence—or time until complete decomposition—of an untreated dung pat on rangeland in California is 22.74 ± 0.64 months, while the average persistence of a pat treated with insecticides is 28.14 ± 0.71 months. This indicates that dung beetle activity results in a 19% decrease in the amount of time the average pat of dung makes forage unpalatable, which translates into substantial monetary savings. Note that, for the sake of this analysis, we must assume that the 19% decrease applies broadly across the United States, even though the rate of dung burial by beetles probably varies greatly depending upon the location.

Forage fouling.   Fincher estimated a potential value for enhanced palatability based on the concept that cattle will not consume plant material that is fouled with dung. If dung beetles were totally absent, forage fouling by dung would cause estimated annual losses of 7.63 kg of beef per head of cattle (L_{nb}). This level of loss is in comparison with the theoretical zero loss of production if no forage were ever fouled by dung. Fortunately, the cattle industry is not saddled with the full force of this potential loss because range fouling is reduced by the current action of dung beetles.

If we assume that the 19% decrease in dung persistence translates into a 19% decrease in lost beef, then, for cattle whose dung is processed by dung beetles, the per-animal loss would be 6.18 kg (L_b) each year as a result of forage fouling. This assumption seems justified, since for each increment of time a given patch of forage remains fouled, it also remains unavailable for grazing. By applying these estimated losses to the 32 million head that are untreated and on pasture or rangeland, we estimate that in the absence of dung beetles, beef losses due to forage fouling would be 244 million kg of beef per year ($C_p \times L_{nb}$), whereas losses at current levels of dung beetle function would be 198 million kg ($C_p \times L_b$). With an average price over 34 years (1970–2003, corrected for inflation) of live beef cattle at \$2.65 per kg (V_c), losses would be \$647 million ($V_c \times [C_p \times L_{nb}]$) in the absence of dung beetles and \$525 million ($V_c \times [C_p \times L_b]$) in the presence of dung beetles. Subtracting the estimated value at current levels of dung beetle activity from the theoretical value if no dung beetles were active, we estimate the value of the reduced forage fouling (V_{rf}) to be approximately \$122 million.

Nitrogen volatilization. Another important service provided by dung beetles is promoting decomposition of dung into labile forms of nitrogen that can be assimilated by plants and thus function as fertilizer when the dung is buried. In the absence of dung beetles, cattle feces that remain on the pasture surface until they are dry lose a large proportion of their inorganic nitrogen to the atmosphere. Experiments in South Africa and the United States have shown that approximately 2% of cattle dung is composed of nitrogen, and that 80% of this nitrogen is lost if the pats dry in the sun before they are buried.

Using Gillard's estimate of 27 kg of nitrogen produced annually per animal and assuming that 80% of this nitrogen is lost in the absence of dung beetle activity, we estimate that 21.6 kg would be lost per animal each year if dung beetles were not functioning (L_{nb}). On the basis of our interpretation of decomposition rates, we assume that these losses will be reduced 19% by the current level of dung beetle activity, compared with the estimate for no beetle activity. Thus, we estimate a loss of 17.5 kg per year (L_b) at current activity levels. Multiplying these per-animal values by the total number of cattle whose dung can potentially be buried by dung beetles (C_p, or 32 million), 691 million kg of nitrogen would be lost annually in the United States in the absence of dung beetle activity, compared with the 560 million kg lost at current levels of activity. With nitrogen valued at \$0.44 per kg (V_n), we estimate the value of nitrogen lost in the absence of dung beetles to be \$304 million and the value of nitrogen lost at current levels of dung beetle activity to be \$246 million. Subtracting the estimated value at current levels of dung beetle activity from the theoretical value if no dung beetles were active, the value of the reduction in nitrogen loss is approximately \$58 million.

Parasites. Many cattle parasites and pest flies require a moist environment such as dung to complete their development. Burying dung and removing this habitat can reduce the density of these pests. From field observations that reflected current levels of removal, Fincher estimated the annual losses due to mortality, morbidity, and medication of beef cattle, dairy cattle, and other livestock with internal parasites. To estimate the value of dung burial for reducing these losses, we will use only the losses associated with beef cattle,

FORMULAS USED TO ESTIMATE INSECT SERVICES

Formula used to estimate the number of cattle in the United States whose dung can be processed by dung beetles:

$C_p = (C_t \times P_r) \times P_{nt}$,

where

C_p = head of cattle producing dung that can be processed by dung beetles,

C_t = total head of cattle produced annually in the United States,

P_r = the proportion of cattle that are raised on range or pasture, and

P_{nt} = the proportion of cattle not treated with avermectins.

Formula used to estimate the value of beef saved because of reduced range fouling resulting from dung burial by dung beetles:

$V_{rf} = [V_c \times (C_p \times L_{nb})] - [V_c \times (C_p \times L_b)]$,

where

V_{rf} = value of reduced forage fouling,

V_c = value of cattle (per kilogram),

C_p = head of cattle producing dung that can be processed by dung beetles,

L_{nb} = losses (per animal) with no dung beetle activity, and

L_b = losses (per animal) at current levels of dung beetle activity.

Formula used to estimate the value of native insects for suppressing populations of potentially pestiferous native herbivorous insects:

$V_{ni} = (NC_{ni} - CC_{ni}) \times P_i$,

where

V_{ni} = the value of suppression of native insect pests by other insects,

NC_{ni} = the cost of damage from native insect pests with no natural control,

CC_{ni} = the cost of damage from native insect pests at current levels of natural control, and

P_i = the proportion of herbivorous insects controlled primarily by other insects.

because we do not have a good estimate for the proportion of dairy cattle or other livestock that live on open pasture or rangeland. Fincher reported that beef cattle ranchers lost $428 million annually because of parasites and pests.

Corrected for inflation, this is equal to $912 million in 2003 dollars. Given that 85% of beef cattle are on range or pasture and 44% of these cattle are not treated with insecticides, we calculate that 37% of the beef cattle in the United States have fewer parasites because of the facilitation of dung decomposition by dung beetles.

We go on to assume that cattle whose dung is processed by dung beetles suffer 19% fewer losses because of parasites, on the basis of our previous calculation that dung beetles accelerate decomposition by 19%. We also assume that cattle on rangeland, pasture, and feedlots all face the same level of loss from parasites in the absence of dung beetles. Following this logic, we estimate that damage from parasites is only 93% (100% − [37% × 19%]) of what it would be if dung beetles were not providing this service. In the absence of dung beetle activity, estimated losses would be $981 million instead of the current $912 million, and thus this service saves the cattle industry an estimated $70 million per year.

Pest flies.   Using a similar algorithm, we can calculate a value for the reduction in losses due to pest flies. Fincher estimated that losses due to horn flies and face flies cost ranchers $365 million and $150 million, respectively, for a total of $515 million. Corrected for inflation, this is the equivalent of $1.7 billion in 2003. Using the calculation described above for parasites, we assume that, as a result of the processing of dung by insects, damage from parasites is only 93% of what it would have been if the service were not being provided. We estimate that losses in the absence of dung beetle activity would be $1.83 billion instead of the current $1.7 billion, and thus this service is saving the cattle industry an estimated $130 million per year.

Adding the individual values of increased forage, nitrogen recycling, and reduced parasite and fly densities due to dung processing by beetles, we arrive at a combined annual total of $380 million. This is certainly an underestimate, since these same services are being provided to an unknown proportion of pasture-raised dairy cows, horses, sheep, goats, and pigs. Furthermore, what is said for dung recycling can also be said for burying beetles and flies that decompose carcasses. While the density of carcasses is much lower than the density of dung pats, their removal is important in rangeland, natural areas, and other public areas for returning nutrients to the soil, reducing potential spread of diseases, and increasing site utility.

Pollination by Native Insects

Pollination, especially crop pollination, is perhaps the best-known ecosystem service performed by insects. McGregor estimates that 15% to 30% of the US diet is a result, either directly or indirectly, of animal-mediated pollination. These products include many fruits, nuts, vegetables, and oils, as well as meat and dairy products produced by animals raised on insect-pollinated forage. While this estimate is probably high, it presents one of the best published measures of pollinator-dependant food in the US diet.

Here we attempt to calculate an estimate of the value of crops produced as a result of pollination by wild (i.e., unmanaged) native insects. The US government keeps records of the production of crops and, because of their value, their insect pollinators have been given some attention, especially pollination by managed insects such as the European honey bee (*Apis mellifera L.*). From these studies and personal accounts of crop scientists and entomologists, several authors make generalizations about the proportion of pollination attributed to various insect groups, mostly honey bees. These generalizations are essentially educated guesses of the percentage of necessary pollination provided by insects, and as such, they are likely to be inaccurate. The proportions that could be attributed to native, as opposed to managed, pollinators will vary widely for each crop, depending on geographic location, availability of natural habitat, and use of pesticides. In addition, cultivars of the same species can have drastically different dependencies on insect pollinators, further complicating any calculation of the value of pollinator insects.

To conduct a truly accurate economic analysis of the role of native insects in crop pollination, we would need a much better accounting of current levels of pollination by different species of managed bees (e.g., honey bee [*A. mellifera*], alfalfa leaf-cutter bee [*Megachile rotundata*], blue orchard bee [*Osmia lignaria*], alkali bee [*Nomia melanderi*]), and wild bees (e.g., bumble bees [*Bombus* spp.], southeastern blueberry bee [*Habropoda laboriosa*], squash bee [*Peponapis pruinosa*]) in crop pollination. . . . Although we still lack much of this information, the estimate we provide here for the value of crops produced as a result of wild native bee–mediated pollination is informative.

Several scientists have estimated the value of insect-pollinated crops that are dependent on honey bees, or the financial loss to society that could be expected if managed honey bees were removed from cropping systems. These authors make a variety of assumptions and take different approaches to calculating a value for honey bees. For example, Southwick and Southwick take into account the reduced crop output stemming from a lack of managed honey bees, adjusting their figures for the changes in value of each commodity as demand increases because of reduced supply. They also present a range of possible values based on assumptions of the pollination redundancy of managed honey bees and other bee pollinators, including feral honey bees and other native and normative bees. Taking all of this into account, they give a range of $1.6 billion ($2.1 billion when adjusted for inflation to represent 2003 dollars) to $5.2 billion ($6.8 billion in 2003 dollars) for the value of honey-bee pollinators. The lower estimate included effective pollination by other bees, making the managed honey bees redundant in some localities and thereby reducing their absolute value. On the high end, Southwick and Southwick estimate that honey bees are worth $5.2 billion if few or no other bees visit insect-pollinated crops.

Robinson and colleagues and Morse and Calderone take a simpler approach, summing the value of each commodity that they estimate is dependant on honey-bee pollinators. From this they generate a portion of the overall value of each crop that they attribute to pollination by honey bees and report values of $8.3 billion and $14.6 billion ($12.3 billion and $16.4 billion, respectively, when adjusted for inflation to represent 2003 dollars). This approach is more consistent with our other calculations of the value of ecosystem services, and so we choose to use it here to calculate the value of crop production that relies on native insect pollinators.

When we sum the average value of pollinator-dependent commodities reported in Morse and Calderone, we find that native pollinators—almost exclusively bees—may be responsible for almost $3.07 billion of fruits and vegetables produced in the United States. Here we must incorrectly assume that the proportion of honey bees to native species is constant in all settings. In some systems, such as agriculturally diverse, organic farms with nearby pockets of natural or seminatural habitat, native bees may be able to provide all of the pollination needs for certain crops. For example, Morse and Calderone assume that 90% of the insect pollinators of watermelon are honey bees. While this is probably true in most farms, some organic growers can rely on native bees for 100% of their melon pollination.

Pest Control

The best estimate available suggests that insect pests and their control measures cost the US economy billions of dollars every year, but this is only a fraction of the costs that would accrue if beneficial insects such as predators and parasitoids, among other forces, did not keep most pests below economically damaging levels. We calculate the value (V) of these natural forces by first estimating the cost of damage caused by insect pests at current levels of control (CC) and then subtracting this value from the estimated higher cost that would be caused by the greater damage from these insect pests if no controls were functioning (NC). Finally, we calculate a value for the specific action of insect natural enemies by multiplying the value of these natural forces by an estimate of the proportion (P_i) of pests that are controlled by beneficial insects as opposed to other mechanisms (e.g., pathogens or climate).

Because of data limitations, we restrict our estimate to the value derived from the suppression of insect pests that attack crop plants. Beneficial insects certainly suppress populations of both weeds and insects that attack humans and livestock, but the data were not available to calculate the value of these services. As with the rest of our analysis, we also limit our calculations to pest and beneficial insects native to the United States.

Our first step was to calculate the cost of damage due to insect pests at current levels of control from natural enemies. Drawing on previously published estimates, Yudelman and colleagues presented monetary values for total production of eight major crops and for the losses to these crops attributable to insects. Using these values, we calculated a ratio of insect loss to actual yield that allowed estimation of losses due to insects for any period for which yield values have been published. Assuming $50.5 billion for total production and $7.5 billion for losses due to insects in North America from 1988 through 1990, we calculated a ratio of 0.1485. . . .

Calkins found that only 35% of the exotic pests in the United States are pests in their home range. Extending this finding, we assume that the same relationship holds true in the United States, and thus only 35% of potential insect pest species that are native to the United States reach damaging levels. In other words, we assume that 65% of the potential damage from native pest species is being suppressed, and that 65% of the potential financial cost of this damage is being saved. We make this assumption based on (a) the abundant evidence of a strong correlation between pest density and the magnitude of loss due to pest damage, and (b) the lack of evidence of a correlation between the destructiveness of a pest and the probability that it will be suppressed.

To clarify, the pool of potential pest species—from which we assume 35% actually reach pest levels—is significantly smaller than the 90,000 described insect species in the United States, because many of the described species are not herbivores, and many of those that are herbivores do not feed on cultivated plants. Only 6000 (7%) of the described species in the United States and Canada cause any damage. For our estimate, we assume that these 6000 species, although they make up only 7% of the total species, account for 35% of the species that would be pests if they were not controlled. Following this logic, we assume that the pool of potential pests would be about 17,000 species, 11,000 of which (65%) are being kept below damage levels by biological or climatic controls.

These native species are estimated to comprise 39% of all pest species in the United States. Since native pests vary greatly in the amount of damage they cause, and include some of the most damaging pests in the United States (e.g., corn rootworm, Colorado potato beetle, and potato leafhopper), we assume that they are responsible for 39% of the cost of damage from all pests in the United States. Hence, we estimate that the cost associated with native pest species at current levels of suppression by natural enemies is 39% of $18.77 billion, or $7.32 billion. We designate this value current control by native insects (CC_{ni}).

On the basis of these assumptions, we estimate that the $7.32 billion lost annually to native insect pests (CC_{ni}) is 35% of what would be lost if natural controls were not functioning. If no natural forces were functioning to control native insect pests, we estimate that they would cause $20.92 billion in damage in the United States each year (NC_{ni}). By subtraction, the value of pest control by our native ecosystems is approximately $13.60 billion.

However, not all of this value for natural control of insect pests is attributable to beneficial insects. Some pest suppression comes from other causes, such as pathogens, climatic conditions, and host-plant resistance. One review of the factors responsible for suppression of 68 herbivore species reported that insects (e.g., predators and parasitoids) were primarily responsible for natural control in 33% of cultivated systems. On the basis of these findings, we estimate that insects are responsible for control of 33% of pests that are suppressed by natural controls, while pathogens or bottom-up forces control the rest. Using this average, we estimate the value of natural control attributable to insects to be $4.5 billion annually (33% of $13.6 billion).

Recreation and Commercial Fisheries

US citizens spend over $60 billion a year on hunting, fishing, and observing wildlife. Insects are a critical food source for much of this wildlife, including many birds, fish, and small mammals. Using 1996 US census data on the spending habits of Americans, adjusted for inflation to 2003 dollars, we estimated the amount of money spent on recreational activities that is dependent on services provided by insects. In this case, the predominant service is concentrating and moving nutrients through the food web.

Small game hunting.   Since most large game are either obligate herbivores or omnivores that are not substantially dependent on insects as a source of nutrition, we restrict our estimate of the value of insects for hunting to small game species. In 1996, expenditures for small game hunting totaled $2.5 billion ($2.9 billion in 2003 dollars). To calculate the proportion of this expenditure that is dependent on insects, we use the proportion of days spent hunting for each insectivorous small game species and the dependence of these birds on insects for food.

On the basis of published reports that most galliform chicks rely on insects as a source of protein and that many cannot even digest plant material, we assume that quail, grouse, and pheasant could not survive without insects as a nutritional resource. Therefore, multiplying the proportion of hunting days spent on each of these small game birds (0.15, 0.13, and 0.23, respectively, for a total of 0.51) by the total value for small game ($2.9 billion), we estimate that insects are required for $1.48 billion in expenditures.

Migratory bird hunting.   Insectivory in migratory birds—primarily waterfowl such as ducks and geese in the order Anseriformes—is not as predominant as in the primarily terrestrial galliform birds discussed above. According to Ehrlich and colleagues, 19 (43%) of the 44 species in this order are primarily insectivorous. Multiplying the total money spent on migratory bird hunting ($1.3 billion) by the 43% of species that are primarily insectivorous, we estimate the value of insects as food for hunted migratory birds at $0.56 billion in hunter expenditures.

Sport and commercial fishing.   The census also provides values for sport or recreational fishing. Since most recreational fishing is in fresh water and a majority of freshwater sport fish are insectivorous, we assume that the entire value of recreational fishing ($27.9 billion) is dependent on insects. In contrast to recreational fishing, the target of most commercial fishing is saltwater fish. There are very few marine insect species, but many fish that are caught in marine systems spend part of their life cycle in fresh water, and insects are often critical sources of nutrition during these periods. Commercial fishing is not covered by the census, but data are available on the number and value of fish landed annually in the United States by commercial operations. Twenty-five of these fish species are primarily insectivorous during at least one life stage. Summing their individual values, we estimate the total value of insects for commercial fishing to be approximately $225 million. Insectivorous fish account for more than 15% of the overall value of commercial fish.

Wildlife observation (bird watching).   The 1996 census reports that Americans spent $33.8 billion on wildlife observation. The census also asked respondents to note which types of wildlife they were watching (e.g., birds, mammals, reptiles, amphibians, insects). Because respondents were allowed to choose more than one category of wildlife, it was impossible to separate out observed groups of organisms that were dependent on insects from those that were not. Bird watching is the most inclusive category, with 96% of respondents indicating that they included birds in their observations. Thus, we assume that 96% of the budget for wildlife observation stems directly from birds, many of which are at least partly dependent on insects as a source of nutrition. Thus, we assume that bird watching accounts for 96% of $33.8 billion spent, or $32.4 billion a year, providing a conservative starting point for calculating the dependency of wildlife observation expenditures on insects.

Our next step is to estimate what proportion of this figure for bird observation was dependent on and attributable to insects. Using data from Ehrlich and colleagues, we calculate that 61% of the bird species known to breed in the United States are primarily insectivorous, and another 28% are at least partially insectivorous. To be conservative, we consider only bird species that are primarily insectivorous. This probably underestimates the importance of insectivory for birds, since many passerine and galliform birds that are listed as partially insectivorous could not survive without the vital protein that insects provide young chicks. This estimate is conservative also because it is based on bird species numbers rather than population numbers, and the passerines, which are overwhelmingly insectivorous, have relatively high population densities. Taking these factors into account, we estimate that insects are responsible for $19.8 billion, which is 61% of the $32.4 billion spent on bird observation annually in the United States.

Discussion

We estimate the value of those insect services we address to be almost $60 billion a year in the United States, which is only a fraction of the value for all the services insects provide. The implication of this estimate is that an annual investment of tens of billions of dollars would be justified to maintain these service-providing insects, were they threatened. And indeed, these beneficial insects are under ever increasing threat from a combination of forces, including habitat destruction, invasion of foreign species, and overuse of toxic chemicals.

Fortunately, no evidence suggests a short-term drastic decline in the insects that provide these services. What the evidence does indicate, however, is a steady decline in these beneficial insects, associated with an overall decline in biodiversity, accompanied by localized, severe declines in environments heavily degraded by human impacts. New evidence indicates that in some situations, the most important species for providing ecosystem services are lost first. The overall, gradual decline in species, coupled with nonlinear changes in service levels, makes it difficult to pinpoint an optimal level of annual investment to conserve beneficial insects and maintain the services they provide. . . .

[E]ven though we provide an estimate of the total value of certain insect services, the complications of redundancy and nonlinearity make it impossible to quantitatively gauge the level of resources that are justified for efforts aimed at conserving the services that insects provide. However, our findings lead us to espouse three qualitative guidelines. First, cost-free or relatively inexpensive measures are almost certainly justified to maintain and increase current service levels. Examples include volunteer construction of nest boxes for wild pollinators and the inclusion of a diverse variety of native plant species in plantings for bank or soil stabilization and site restoration. Second, actions or investments that are estimated to have an economic return at or slightly below the break-even point, such as the use of less toxic pesticides, are probably justified because of their nontarget benefits. Third, actions that lead to substantial decreases in biodiversity should be avoided because of the high probability of a major disruption in essential services.

Finally, although we cannot provide a quantitative formula to determine the optimal level of investment in the conservation of beneficial insects that provide essential services, we do feel justified, on the basis of our estimates, in making some specific recommendations. First, we recommend that conservation funding allocated via Farm Bill programs—such as the Conservation Security Program, Conservation Reserve Program, Wetlands Reserve Program, and Environmental Quality Incentives Program—pay specific attention to insects and the role they play in ecosystems. In particular, funding to provide habitat for beneficial insects such as predators, parasitoids, and pollinators in natural, seminatural, unproductive, or fallow areas in agricultural landscapes not only provides direct benefits to growers but, by focusing on the ecological needs of insects, results in habitat that supports a great diversity of wildlife.

Second, we recommend that ecosystem services performed by insects be taken into account in land-management decisions. Specifically, maintaining ecosystem services should be a goal of land management. With this goal in mind, specific practices such as grazing, burning, and pesticide use should be tailored to protect insect biodiversity. . . .

We believe it is imperative that some federal and local funds be directed toward the study of these beneficial insects and the vital services they provide so that conservation efforts can be optimally allocated, either through the agricultural programs listed above or through other means.

These steps are just a beginning. With greater attention, research, and conservation, the valuable services that insects provide can not only be sustained but increase in capacity. As a result, growers will be able to practice a more sustainable form of agriculture while spending less on managing pest insects or acquiring managed pollinators; ranchers will get more productivity out of their land; and wildlife lovers will find that the birds and fish they hunt occur in greater abundance than in the past few decades. In less direct but no less important ways, everyone would benefit from the facilitation of the vital services that insects provide. Judging from our estimate of the value of these four services, increased investment in the conservation of these services is justified.

Marino Gatto and
Giulio A. De Leo

 **NO**

Pricing Biodiversity and Ecosystem Services: The Never-Ending Story

In 1844, the French engineer Jules Juvénal Dupuit introduced cost–benefit analysis to evaluate investment projects. . . . The application of cost–benefit analysis to ecological issues fell out of favor three decades ago, and it was gradually replaced by multicriteria analysis in the decision-making process for projects that have an impact on the environment. Although multicriteria analysis is currently used for environmental impact assessments [EIA] in many nations, [recently] the concept of cost–benefit analysis has again become fashionable, along with the various pricing techniques associated with it, such as contingent valuation methods, hedonic prices, and costs of replacement of ecological services. . . . Economists have generated a wealth of virtuosic variations on the theme of assessing the societal value of biodiversity, but most of these techniques are invariably based on price—that is, on a single scale of values, that of goods currently traded on world markets.

Perhaps the most famous recent study on the issue of pricing biodiversity and ecological services is that by Costanza et al., who argued that if the importance of nature's free benefits could be adequately quantified in economic terms, then policy decisions would better reflect the value of ecosystem services and natural capital. Drawing on earlier studies aimed at estimating the value of a wide variety of ecosystem goods and services, Costanza et al. estimated the current economic value of the entire biosphere at $16–54 trillion per year, with an average value of approximately $33 trillion per year. By contrast, the gross national product of the United States totals approximately $18 trillion per year. The paper, as its authors intended, stimulated much discussion, media attention, and debate. A special issue of *Ecological Economics* (April 1998) was devoted to commentaries on the paper, which, with few exceptions, were laudatory. Some economists have questioned the actual numbers, but many scientists have praised the attempt to value biodiversity and ecosystem functions.

Although Costanza et al. acknowledged that their estimates were crude and imperfect, they also pointed the way to improved assessments. In particular, they noted the need to develop comprehensive ecological economic models that could adequately incorporate the complex interdependencies between ecosystems and economic systems, as well as the complex individual dynamics

From *BioScience*, vol. 50, no. 4, April 2000, by Marino Gatto and Giulio A. De Leo, pp. 347–354. Copyright © 2000 by American Institute of Biological Sciences. Reprinted by permission via the Copyright Clearance Center.

of both types of systems. Despite the authors' caveats and the fact that many economists have been circumspect in applying their own tools to decisions regarding natural systems, the monetary approach is perceived by scientists, policymakers, and the general public as extremely appealing; a number of biologists are also of the opinion that attaching economic values to ecological services is of paramount importance for preserving the biosphere and for effective decision-making in all cases where the environment is concerned.

In this article, we espouse a contrary view, stressing that, for most of the values that humans attach to biodiversity and ecosystem services, the pricing approach is inadequate—if not misleading and obsolete—because it implies erroneously that complex decisions with important environmental impacts can be based on a single scale of values. We contend that the use of cost–benefit analysis as the exclusive tool for decision-making about environmental policy represents a setback relative to the existing legislation of the United States, Canada, the European Union, and Australia on environmental impact assessment, which explicitly incorporates multiple criteria (technical, economic, environmental, and social) in the process of evaluating different alternatives. We show that there are sound methodologies, mainly developed in business and administration schools by regional economists and by urban planners, that can assist decision-makers in evaluating projects and drafting policies while accounting for the nonmarket values of environmental services.

The Limitations of Cost–Benefit Analysis and Contingent Valuation Methods

Historically, the first important implementation of cost–benefit analysis at the political level came in 1936, with passage of the US Flood Control Act. This legislation stated that a public project can be given a green light if the benefits, to whomsoever they accrue, are in excess of estimated costs. This concept implies that all benefits and costs are to be considered, not just actual cash flows from and to government coffers. However, public agencies (e.g., the US Army Corps of Engineers) quickly ran into a problem: They were not able to give a monetary value to many environmental effects, even those that were predictable in quantitative terms. For instance, engineers could calculate the reduction of downstream water flow resulting from construction of a dam, and biologists could predict the river species most likely to become extinct as a consequence of this flow reduction. However, public agencies were not able to calculate the cost of each lost species. Therefore, many ingenious techniques for the monetary valuation of environmental goods and services have been devised since the 1940s. These techniques fall into four basic categories.

- **Conventional market approaches.** These approaches, such as the replacement cost technique, use market prices for the environmental service that is affected. For example, degradation of vegetation in developing countries leads to a decrease in available fuelwood. Consequently, animal dung has to be used as a fuel instead of a fertilizer, and farmers must therefore replace dung with chemical fertilizers. By computing the cost of these chemical fertilizers, a monetary value for the degradation of vegetation can then be calculated.

- **Household production functions.** These approaches, such as the travel cost method, use expenditures on commodities that are substitutes or complements for the environmental service that is affected. The travel cost method was first proposed in 1947 by the economist Harold Hotelling, who, in a letter to the director of the US National Park Service, suggested that the actual traveling costs incurred by visitors could be used to develop a measure of the recreation value of the sites visited.

- **Hedonic pricing.** This form of pricing occurs when a price is imputed for an environmental good by examining the effect that its presence has on a relevant market-priced good. For instance, the cost of air and noise pollution is reflected in the price of plots of land that are characterized by different levels of pollution, because people are willing to pay more to build their houses in places with good air quality and little noise. . . .

- **Experimental methods.** These methods include contingent valuation methods, which were devised by the resource economist Siegfried V. Ciriacy-Wantrup. Contingent valuation methods require that individuals express their preferences for some environmental resources by answering questions about hypothetical choices. In particular, respondents to a contingent valuation methods questionnaire will be asked how much they would be willing to pay to ensure a welfare gain from a change in the provision of a nonmarket environmental commodity, or how much they would be willing to accept in compensation to endure a welfare loss from a reduced provision of the commodity.

Among these pricing techniques, the contingent valuation methods approach is the only one that is capable of providing an estimate of existence values, in which biologists have a special interest. Existence value was first defined by Krutilla as the value that individuals may attach to the mere knowledge that rare and diverse species, unique natural environments, or other "goods" exist, even if these individuals do not contemplate ever making active use of or benefiting in a more direct way from them. The name "contingent valuation" comes from the fact that the procedure is contingent on a constructed or simulated market, in which people are asked to manifest, through questionnaires and interviews, their demand function for a certain environmental good (i.e., the price they would pay for one extra unit of the good versus the availability of the good). . . .

The limits of cost–benefit analysis were discussed in the 1960s, after more than two decades of experimentation. In particular, many authors pointed out that cost–benefit analysis encouraged policymakers to focus on things that can be measured and quantified, especially in cash terms, and to disregard problems that are too large to be assessed easily. Therefore, the associated price might not reflect the "true" value of social equity, environmental services, natural capital, or human health. In particular, economists themselves recognize that the increasingly popular contingent valuation methods are undermined by several conceptual problems, such as free-riding, overbidding, and preference reversal.

When it comes to monetary valuation of the goods and services provided by natural ecosystems and landscapes specifically, a number of additional problems undermine the effectiveness of pricing techniques and cost–benefit analysis. These problems include the very definition of "existence" value, the dependence of pricing techniques on the composition of the reference group, and the significance of the simulated market used in contingent valuation.

The definition of "existence" value   A classic example of contingent valuation methods is to ask for the amount of money individuals are willing to pay to ensure the continued existence of a species such as the blue whale. However, the existence value of whales does not take into account potential indirect services and benefits provided by these mammals. It is just the value of the existence of whales for humans, that is, the satisfaction that the existence of blue whales provides to people who want them to continue to exist. Therefore, there is a real risk that species with very low or no aesthetic appeal or whose biological role has not been properly advertised will be given a low value, even if they play a fundamental ecological function. Without adequate information, most people do not understand the extent, importance, and gravity of most environmental problems. As a consequence, people may react emotionally and either underestimate or overestimate risks and effects.

Therefore, it is not surprising that five of the seven guidelines issued by the National Oceanic and Atmospheric Administration [NOAA] about how to conduct contingent valuation discuss how to properly inform and question respondents to produce reliable estimates (e.g., in-person interviews are preferred to telephone surveys to elicit values). Of course, acquisition of reliable and complete information is always possible in theory, but in practice strict adherence to NOAA guidelines makes contingent valuation methods expensive and time consuming.

Difficulties with the reference group for pricing   Pricing techniques such as contingent valuation methods provide information about individual willingness to pay or willingness to accept, which must be summed up in the final balance of cost–benefit analysis. Therefore, the outcome of cost–benefit analysis depends strongly on the group of people that is taken as a reference for valuation—particularly on their income. Van der Straaten noted that the Exxon *Valdez* oil spill in 1989 provides a good example of this dependence. The population of the United States was used as a reference group to calculate the damage to the existence value of the affected species and ecosystems using contingent valuation methods. Exxon was ultimately ordered to pay $5 billion to compensate the people of Alaska for their losses. This huge figure was a consequence of the high income of the US population. If the same accident had occurred in Siberia, where salaries are lower, the outcome would certainly have been different.

This example shows that contingent valuation methods simply provide information about the preferences of a particular group of people but do not necessarily reflect the ecological importance of ecosystem goods and services. Moreover, the outcome of cost–benefit analysis depends on which individual willingness to pay or willingness to accept are included in the cost–benefit analysis. If the quality of the Mississippi River is at issue, should the analysis be restricted to US citizens living close to the river, or should the willingness to pay of Californians and New Yorkers be included too? According to Krutilla's definition of existence value, for many environmental goods and ecological services that may ultimately affect ecosystem integrity at the global level, the preferences of the entire human population should potentially be considered in the analysis. Because practical reasons obviously preclude doing so, contingent valuation methods will inevitably only provide information about the preferences of specific groups of people. For many of the ecological services that may be considered the heritage of humanity, contingent valuation methods analyses performed locally in a particular economic situation should be extrapolated only with great caution to other areas. The process of placing a monetary value on biodiversity and ecosystem functioning through nonuser willingness to pay is performed in the same way as for user willingness to pay, but the identification of people who do not use an environmental good directly and still have a legitimate interest in its preservation is problematic.

Significance of the simulated market   Contingent valuation methods are contingent on a market that is constructed or simulated, not real. It is difficult to believe in the efficiency of what Adam Smith called the "invisible hand" of the market for a process that is the artificial production of economic advisors and does not possess the dynamic feedback that characterizes real competitive markets. Is it even possible to simulate a market where units of biodiversity are bought and sold? As Friend stated, "these contingency evaluation methods (CVM) tend to create an illusion of choice based on psychology (willingness) and ideology (the need to pay) which is supposed, somewhat mysteriously, to reflect an equilibrium between the consumer demand for and producer supply of environmental goods and services."

Many additional criticisms of pricing ecological services are more familiar to biologists. For many ecological services, there is simply no possibility of technological substitution. Moreover, the precise contribution of many species is not known, and it may not be known until the species is close to extinction. . . . In addition, specific ecosystem services, as evaluated by Costanza et al., should not be separated from one another and valued individually because the importance of any piece of biodiversity cannot be determined without considering the value of biodiversity in the aggregate. And finally, the use of marginal value theory may be invalidated by the erratic and catastrophic behavior of many ecological systems, resulting in potentially detrimental effects on the health of humans, the productivity of renewable resources, and the vitality and stability of societies themselves.

Despite the efforts of many economists, we believe that some goods and services, especially those related to ecosystems, cannot reasonably be given a monetary value, although they are of great value to humans. Economists coined the term "intangibles" to define these goods. Cost–benefit analysis cannot easily deal with intangibles. As Nijkamp wrote, more than 20 years ago, "the only reasonable way to take account of intangibles in the traditional cost–benefit analysis seems to be the use of a balance with a debit and a credit side in which all intangible project effects (both positive and negative) are represented in their own (qualitative or quantitative) dimensions" as secondary information. In other words, the result of cost–benefit analysis is primarily a single number, the net monetary benefit that comprises all the effects that can be sensibly converted into monetary returns and costs.

Commensurability of Different Objectives and Multicriteria Analysis

Cost–benefit analysis includes intangibles in the decision-making process only as ancillary information, with the main focus being on those effects that can be converted to monetary value. This approach is not a balanced solution to the problem of making political decisions that are acceptable to a wide number of social groups with a range of legitimate interests. . . .

However, even if the attempt to put a price on everything is abandoned, it is not necessary to give up the attempt to reconcile economic issues with social and environmental ones. Social scientists long ago developed multicriteria techniques to reach a decision in the face of multiple different and structurally incommensurable goals. The most important concept in multicriteria analysis was actually conceived by an Italian economist, Vilfredo Pareto, at the end of the nineteenth century. It is best explained by a simple example. Suppose that a natural area hosting several rare species is a target for the development of a mining activity. Alternative mining projects can have different effects in terms of profits from mining (measured in dollars) and in terms of sustained biodiversity (measured in suitable units, for instance, through the Shannon index). Profit from mining can be corrected using welfare economics to include those environmental and social effects that can be priced (e.g., the benefit of providing jobs to otherwise unemployed people, the cost of treating lung disease of miners, and the cost of the loss of the tourists who used to visit the natural area). . . .

The methods of multicriteria analysis are intended to assist the decision-maker in choosing among . . . alternatives . . . (a task that is particularly difficult when there are several incommensurable objectives, not just two). Nevertheless, the initial step of determining [these] alternatives is of enormous importance, for three reasons. First, [doing so] makes perfect sense even if there is no way of pricing a certain environmental good because each objective can be expressed in its own proper units without reduction to a common scale. Second, the determination of all the feasible alternatives . . . requires the joint effort of a multidisciplinary team that includes, for example, economists, engineers, and biologists and that must predict the effects of alternative decisions on all of the different environmental and social components to which humans are sensitive and which, therefore, deserve consideration. Third, the determination of [feasible alternatives] allows the objective elimination of inadequate alternatives because [they are] independent of the subjective perception of welfare . . . [and] in essence describe the tradeoff between the various incommensurable objectives when every effort is made to achieve the best results in all respects; the attention of the authority that must make the final decision is thus directed toward genuine potential solutions because nonoptimal decisions have already been discarded.

It should be noted that a cost–benefit analysis does not elicit trade-offs between incommensurable goods because it also gives a green light to projects . . . , provided that the benefits that can be converted into a monetary scale exceed the costs. . . . Cost–benefit analysis, however, is not useful for eliciting the tradeoffs between two incommensurable goods, neither of which is monetary. For instance, there might be a conflict between the goals of preserving wildlife within a populated area and minimizing the risk that wild animals are vectors of dangerous diseases. A multicriteria analysis can describe this tradeoff, whereas a cost–benefit analysis cannot.

Another philosophical point concerning the issue of commensurability is the question of implicit pricing. Economists often argue that to make a decision is to put an implicit price on such intangibles as human life or aesthetics and, therefore, to reduce their value to a common scale (as pointed out also by Costanza et al.). . . .

Environmental Impact Assessment and Multiattribute Decision-Making

Because of the flaws of cost–benefit analysis, many countries have taken a different approach to decision-making through the use of environmental impact assessment legislation (e.g., the United States in 1970, with the signing of the National Environmental Policy Act, NEPA; France in 1976, with the act 76/629; the European Union in 1985, with the directive 85/337). Environmental impact assessment procedures, if properly carried out, represent a wiser approach than setting an a priori value of biodiversity and ecosystem services because these procedures explicitly recognize that each situation, and every regulatory decision, responds to different ethical, economic, political, historical, and other conditions and that the final decision must be reached by giving appropriate consideration to several different objectives. As Canter noted, all projects, plans, and policies that are expected to have a significant environmental impact would ideally be subject to environmental impact assessment.

The breadth of goals embraced by environmental impact assessment is much wider than that of cost–benefit analysis. Environmental impact assessment provides a conceptual framework and formal procedures for comparing different alternatives to a proposed project (including the possibilities of not development a site, employing different management rules, or using mitigation measures); for fostering interdisciplinary team formation to investigate all possible environmental, social, and economic consequences of a proposed activity; for enhancing administrative review procedures and coordination among the agencies involved in the process; for producing the necessary documentation to enhance transparency in the decision-making process and the possibility of reviewing all the objective and subjective steps that resulted in a given conclusion; for encouraging broad public participation and the input of different interest groups; and for including monitoring and feedback procedures. Classical multiattribute analysis can be used to rank different alternatives. . . . Ranking usually requires the use of value functions to transform environmental and other indicators (e.g., biological oxygen demand or animal density) to levels of satisfaction on a normalized scale, and the weighting of factors to combine value functions and to rank the alternatives. These weights explicitly reflect the relative importance of the different environmental, social, and economic compartments and indicators.

A wide range of software packages for decision support can assist experts in organizing the collected information; in documenting the various phases of EIA; in guiding the assignment of importance weights; in scaling, rating, and ranking alternatives; and in conducting sensitivity analysis for the overall decision-making process. This last step, of testing the robustness and consistency of multiattribute analysis results, is especially important because it shows how sensitive the final ranking is to small or large changes in the set of weights and value functions, which often reflect different and subjective perspectives. It is important to stress that, although the majority of environmental impact assessments have been conducted on specific projects, such as road construction or the location of chemical plants, there is no conceptual barrier to extending the procedure to evaluation of plans, programs, policies, and regulations. In fact, according to NEPA, the procedure is mandatory for any federal action with an important impact on the environment. The extension of environmental impact assessment to a level higher than a single project is termed "strategic environmental assessment" and has received considerable attention.

Conclusions

An impressive literature is available on environmental impact assessment and multiattribute analysis that documents the experience gained through 30 years of study and application. Nevertheless, these studies seem to be confined to the area of urban planning and are almost completely ignored by present-day economists as well as by many ecologists. Somewhere between the assignment of a zero value to biodiversity (the old-fashioned but still used practice, in which environmental impacts are viewed as externalities to be discarded from the balance sheet) and the assignment of an infinite value (as advocated by some radical environmentalists), lie more sensible methods to assign value to biodiversity than the price tag techniques suggested by the new wave of environmental economists. Rather than collapsing every measure of social and environmental value onto a monetary axis, environmental impact assessment and multiattribute analysis allow for explicit consideration of intangible nonmonetary values along with classical economic assessment, which, of course, remains important. It is, in fact, possible to assess ecosystem values and the ecological impact of human activity without using prices. Concepts such as Odum's eMergy [the available energy of one kind previously required to be used up directly and indirectly to make the product or service] and Rees' ecological footprint [the area of land and water required to support a defined economy or population at a specified standard of living], although perceived by some as naive, may aid both ecologists and economists in addressing this important need.

To summarize our viewpoint, economists should recognize that cost–benefit analysis is only part of the decision-making process and that it lies at the same level as other considerations. Ecologists should accept that monetary valuation of biodiversity and ecosystem services is possible (and even helpful) for part of its value, typically its use value. We contend that the realistic substitute for markets, when they fail, is a transparent decision-making process, not old-style cost–benefit analysis. The idea that, if one could get the price right, the best and most effective decisions at both the individual and public levels would automatically follow is, for many scientists, a sort of Panglossian obsession. In reality, there is no simple solution to complex problems. We fear that putting an a priori monetary value on biodiversity and ecosystem services will prevent humans from valuing the environment other than as a commodity to be exploited, thus reinvigoraing the old economic paradigm that assumes a perfect substitution between natural and human-made capital. As Rees wrote, "for all its theoretical attractiveness, ascribing money values to nature's services is only a partial solution to the present dilemma and, if relied on exclusively, may actually be counterproductive."

POSTSCRIPT

Should a Price Be Put on the Goods and Services Provided by the World's Ecosystems?

In "Can We Put a Price on Nature's Services?" *Report From the Institute for Philosophy and Public Policy* (Summer 1997), Mark Sagoff objects that trying to attach a price to ecosystem services is futile because it legitimizes the accepted cost-benefit approach and thereby undermines efforts to protect the environment from exploitation. The March 1998 issue of *Environment* contains environmental economics professor David Pearce's detailed critique of the 1997 Costanza et al. study. Pearce objects chiefly to the methodology, not the overall goal of attaching economic value to ecosystem services. Costanza et al. reply to Pearce's objections in the same issue. Pearce and Edward B. Barbier have published *Blueprint for a Sustainable Economy* (Earthscan, 2000), in which they discuss how governments worldwide are now applying economics to environmental policy.

Despite the controversy over the worth of assigning economic values to various aspects of nature, researchers continue the effort. Gretchen C. Daily et al., in "The Value of Nature and the Nature of Value," *Science* (July 21, 2000), discuss valuation as an essential step in all decision making and argue that efforts "to capture the value of ecosystem assets . . . can lead to profoundly favorable effects." Daily and Katherine Ellison continue the theme in *The New Economy of Nature: The Quest to Make Conservation Profitable* (Island Press, 2002). In "What Price Biodiversity?" *Ecos* (January 2000), Steve Davidson describes an ambitious program funded by the Commonwealth Scientific and Industrial Research Organization (CSIRO) and the Myer Foundation that is aimed at developing principles and methods for objectively valuing "ecosystem services—the conditions and processes by which natural ecosystems sustain and fulfil human life—and which we too often take for granted. These include such services as flood and erosion control, purification of air and water, pest control, nutrient cycling, climate regulation, pollination, and waste disposal." Jim Morrison, "How Much Is Clean Water Worth?" *National Wildlife* (February/ March 2005), argues that ecosystem services such as cleaning water, controlling floods, and pollinating crops have sufficient economic value to make it profitable to spend millions of dollars to protect natural systems.

Stephen Farber, et al., "Linking Ecology and Economics for Ecosystem Management," *Bioscience* (February 2006), find "the valuation of ecosystem services . . . necessary for the accurate assessment of the trade-offs involved in different management options." Ecosystem valuation is currently being used to justify restoration efforts, "linking the science to human welfare," as shown (for example) in Chungfu Tong, et al., "Ecosystem Service Values and Restoration in the Urban Sanyang Wetland of Wenzhou, China," *Ecological Engineering* (March 2007). See also J. R. Rouquette, et al., "Valuing Nature-Conservation Interests on Agricultural Floodplains," *Journal of Applied Ecology* (April 2009), and Z. M. Chen, et al., "Net Ecosystem Services Value of Wetland: Environmental Economic Account," *Communications in Nonlinear Science & Numerical Simulation* (June 2009). John W. Day, Jr., et al., "Ecology in Times of Scarcity," *Bioscience* (April 2009), report that "In an energy-scarce future, services from natural ecosystems will assume relatively greater importance in supporting the human economy." As a result, the practice of ecology will become both more expensive and more valuable, and ecological engineering (ecoengineering) and restoration will become more common and necessary. Perhaps the connection between ecosystem services and economic value will become less debatable.

Internet References . . .

ECOLEX: A Gateway to Environmental Law

This site, sponsored by the United Nations and the World Conservation Union, is a comprehensive resource for environmental treaties, national legislation, and court decisions.

http://www.ecolex.org/index.php

Environmental Defense

Environment Defense (once The Environmental Defense Fund) is dedicated to "protecting the environmental rights of all people, including future generations." Guided by science, Environmental Defense evaluates environmental problems and works "to create solutions that win lasting economic and social support because they are nonpartisan, cost-efficient, and fair."

http://www.environmentaldefense.org/home.cfm

Arc Ecology Project: Military and the Environment

Arc Ecology is a nonprofit, public interest organization concerned with the ecology of humanity and its place in the global ecology. Among the issues it addresses is the environmental threat posed by military activities.

http://www.arcecology.org/Military.shtml

SourceWatch

SourceWatch is a collaborative project of the Center for Media and Democracy. Its primary focus is on documenting the interconnections and agendas of public relations firms, think tanks, industry-funded organizations and industry-friendly experts that work to influence public opinion and public policy on behalf of corporations, governments, and special interests.

http://www.sourcewatch.org

Principles Versus Politics

*I*n *many environmental issues, it is easy to tell what basic principles apply and therefore determine what is the right thing to do. Ecology is clear on the value of species to ecosystem health and the harmful effects of removing or replacing species. Medicine makes no bones about the ill effects of pollution. But are the environmental problems so bad that we must act immediately? Must damaged ecosystems be restored, and, if so, to their status of how long ago? Should government agencies or businesses receive exemptions from environmental regulations? Should we go slow on environmental regulations for fear of damaging the economy? Such questions arise in connection with most environmental issues, not just the three listed below; but these three will serve to introduce the theme of principles versus politics.*

- Should North America's Landscape Be Restored to Its Pre-Human State?

- Should the Military Be Exempt from Environmental Regulations?

- Will Restricting Carbon Emissions Damage the U.S. Economy?

ISSUE 4

Should North America's Landscape Be Restored to Its Pre-Human State?

YES: C. Josh Donlan, from "Restoring America's Big, Wild Animals," *Scientific American* (June 2007)

NO: Dustin R. Rubenstein, Daniel I. Rubenstein, Paul W. Sherman, and Thomas A. Gavin, from "Pleistocene Park: Does Re-Wilding North America Represent Sound Conservation for the 21st Century?" *Biological Conservation* (vol. 132, 2006)

ISSUE SUMMARY

YES: C. Josh Donlan proposes that because the arrival of humans in the Americas some 13,000 years ago led to the extinction of numerous large animals (including camels, lions, and mammoths) with major effects on local ecosystems, restoring these animals (or their near-relatives from elsewhere in the world) holds the potential to restore health to these ecosystems. There would also be economic and cultural benefits.

NO: Dustin R. Rubenstein, Daniel I. Rubenstein, Paul W. Sherman, and Thomas A. Gavin argue that bringing African and Asian megafauna to North America is unlikely to restore pre-human ecosystem function and may threaten present species and ecosystems. It would be better to focus resources on restoring species where they were only recently extinguished.

As far as we can see into the mists of the past, human actions have affected the environment. Desertification, deforestation, erosion, and soil salinization were happening as soon as agriculture began and the first cities appeared. Paul S. Martin, Emeritus Professor of Geosciences at the Desert Laboratory of the University of Arizona, has long argued that even before that time, humans have been the chief cause of extinctions of large animals as far back as 50,000 years ago (see, e.g., Paul S. Martin and Richard G. Klein, eds., "Prehistoric Overkill," in *Quaternary Extinctions: A Prehistoric Revolution,* University of Arizona Press, 1984). The basic argument is that soon after human beings arrived in areas such as North America or Australia, the large game animals disappeared, either because humans killed (and presumably ate) them or because humans killed their prey. Critics, apparently unwilling to grant "primitive" people armed with stone-tipped spears and arrows enough potency to wipe out whole species, have argued that the disappearance of large animals was just coincidence, or due to diseases brought by humans and their domestic animals, or due to changes in climate. Since Martin first broached the "prehistoric overkill" hypothesis in the 1960s, the evidence in its favor has accumulated, but it has remained controversial. Less controversial is the idea that when the animals went extinct, the ecosystems in which they lived were affected. Large herbivores, for instance, were responsible for distributing the seeds of many trees. When the herbivores vanished, the fruit piled up under the trees and few seeds had the chance to sprout far enough from the parent tree to thrive.

The history of environmentalism has been marked by three major schools of thought. Conservationists want to see nature's resources protected for future human use. Preservationists want to see nature left alone. The third approach, epitomized by Aldo Leopold (*A Sand County Almanac*, 1949; Oxford University Press, 2001), might be called the "reparationist" school; its theme is that where humans have done damage, they have a responsibility to repair it. Its representatives plant dune grasses to fight desertification, plant trees to restore forests, and advocate for restricting grazing animals to help overgrazed land recover. They also argue for removing dams or increasing water releases to restore rivers to health. Their motives may ultimately be either preservationist (repair and leave alone) or conservationist (repair for future human benefit).

Species that vanished thousands of years ago can hardly be preserved or conserved. Can they be "repaired"? They can't be brought back in their original form, of course, but many such species have relatives elsewhere in the world that might fill similar roles in ecosystems (such as seed distribution) if they were transplanted. Should they be transplanted? How far back does the human responsibility to repair human-caused damage to ecosystems extend? In the following selections, C. Josh Donlan proposes that because the arrival of humans in the Americas some 13,000 years ago led to the extinction of numerous large animals (including camels, lions, and mammoths) with major effects on local ecosystems, restoring these animals (or their near-relatives from elsewhere in the world) holds the potential to restore health to these ecosystems. There would also be economic and cultural benefits. Dustin R. Rubenstein, Daniel I. Rubenstein, Paul W. Sherman, and Thomas A. Gavin argue that the ecosystems of 13,000 years ago have evolved, adjusting to the changes imposed at that time. Bringing African and Asian megafauna to North America is unlikely to restore pre-human ecosystem function and may threaten present species and ecosystems. It would be better to focus resources on preventing new extinctions and on restoring species where they were only recently extinguished.

YES

C. Josh Donlan

Restoring America's Big, Wild Animals

In the fall of 2004 a dozen conservation biologists gathered on a ranch in New Mexico to ponder a bold plan. The scientists, trained in a variety of disciplines, ranged from the grand old men of the field to those of us earlier in our careers. The idea we were mulling over was the reintroduction of large vertebrates—megafauna—to North America.

Most of these animals, such as mammoths and cheetahs, died out roughly 13,000 years ago, when humans from Eurasia began migrating to the continent. The theory—propounded 40 years ago by Paul Martin of the University of Arizona—is that overhunting by the new arrivals reduced the numbers of large vertebrates so severely that the populations could not recover. Called Pleistocene overkill, the concept was highly controversial at the time, but the general thesis that humans played a significant role is now widely accepted. Martin was present at the meeting in New Mexico, and his ideas on the loss of these animals, the ecological consequences, and what we should do about it formed the foundation of the proposal that emerged, which we dubbed Pleistocene rewilding.

Although the cheetahs, lions and mammoths that once roamed North America are extinct, the same species or close relatives have survived elsewhere, and our discussions focused on introducing these substitutes to North American ecosystems. We believe that these efforts hold the potential to partially restore important ecological processes, such as predation and browsing, to ecosystems where they have been absent for millennia. The substitutes would also bring economic and cultural benefits. Not surprisingly, the published proposal evoked strong reactions. Those reactions are welcome, because debate about the conservation issues that underlie Pleistocene rewilding merit thorough discussion.

Why Big Animals Are Important

Our approach concentrates on large animals because they exercise a disproportionate effect on the environment. For tens of millions of years, megafauna dominated the globe, strongly interacting and co-evolving with other

From *Scientific American*, vol. 296, issue 6, June 2007, pp. 70–77. Copyright © 2007 by Scientific American, Inc. Reproduced with permission. All rights reserved. www.sciam.com

species and influencing entire ecosystems. Horses, camels, lions, elephants and other large creatures were everywhere: megafauna were the norm. But starting roughly 50,000 years ago, the overwhelming majority went extinct. Today megafauna inhabit less than 10 percent of the globe.

Over the past decade, ecologist John Terborgh of Duke University has observed directly how critical large animals are to the health of ecosystems and how their loss adversely affects the natural world. When a hydroelectric dam flooded thousands of acres in Venezuela, Terborgh saw the water create dozens of islands—a fragmentation akin to the virtual islands created around the world as humans cut down trees, build shopping malls, and sprawl from urban centers. The islands in Venezuela were too small to support the creatures at the top of the food chain—predators such as jaguars, pumas and eagles. Their disappearance sparked a chain of reactions. Animals such as monkeys, leaf-cutter ants and other herbivores, whose populations were no longer kept in check by predation, thrived and subsequently destroyed vegetation—the ecosystems collapsed, with biodiversity being the ultimate loser.

Similar ecological disasters have occurred on other continents. Degraded ecosystems are not only bad for biodiversity; they are bad for human economies. In Central America, for instance, researchers have shown that intact tropical ecosystems are worth at least $60,000 a year to a single coffee farm because of the services they provide, such as the pollination of coffee crops.

Where large predators and herbivores still remain, they play pivotal roles. In Alaska, sea otters maintain kelp forest ecosystems by keeping herbivores that eat kelp, such as sea urchins, in check. In Africa, elephants are keystone players; as they move through an area, their knocking down trees and trampling create a habitat in which certain plants and animals can flourish. Lions and other predators control the populations of African herbivores, which in turn influence the distribution of plants and soil nutrients.

In Pleistocene America, large predators and herbivores played similar roles. Today most of that vital influence is absent. For example, the American cheetah (a relative of the African cheetah) dashed across the grasslands in pursuit of pronghorn antelopes for millions of years. These chases shaped the pronghorn's astounding speed and other biological aspects of one of the fastest animals alive. In the absence of the cheetah, the pronghorn appears "overbuilt" for its environment today.

Pleistocene rewilding is not about recreating exactly some past state. Rather it is about restoring the kinds of species interactions that sustain thriving ecosystems. Giant tortoises, horses, camels, cheetahs, elephants and lions: they were all here, and they helped to shape North American ecosystems. Either the same species or closely related species are available for introduction as proxies, and many are already in captivity in the U.S. In essence, Pleistocene rewilding would help change the underlying premise of conservation biology from limiting extinction to actively restoring natural processes.

At first, our proposal may seem outrageous—lions in Montana? But the plan deserves serious debate for several reasons. First, nowhere on Earth is pristine, at least in terms of being substantially free of human influence. Our demographics, chemicals, economics and politics pervade every part of the planet. Even in our largest national parks, species go extinct without active intervention. And human encroachment shows alarming signs of worsening. Bold actions, rather than business as usual, will be needed to reverse such negative influences. Second, since conservation biology emerged as a discipline more than three decades ago, it has been mainly a business of doom and gloom, a struggle merely to slow the loss of biodiversity. But conservation need not be only reactive. A proactive approach would include restoring natural processes, starting with ones we know are disproportionately important, such as those influenced by megafauna.

Third, land in North America is available for the reintroduction of megafauna. Although the patterns of human land use are always shifting, in some areas, such as parts of the Great Plains and the Southwest, large private and public lands with low or declining human population densities might be used for the project. Fourth, bringing megafauna back to America would also bring tourist and other dollars into nearby communities and enhance the public's appreciation of the natural world. More than 1.5 million people visit San Diego's Wild Animal Park every year to catch a glimpse of large mammals. Only a handful of U.S. national parks receive that many visitors. Last, the loss of some of the remaining species of megafauna in Africa and Asia within this century seems likely—Pleistocene rewilding could help reverse that.

How It Might Be Done

We are not talking about backing up a van and kicking some cheetahs out into your backyard. Nor are we talking about doing it tomorrow. We conceive of Pleistocene rewilding as a series of staged, carefully managed ecosystem manipulations. What we are offering here is a vision—not a blueprint—of how this might be accomplished. And by no means are we suggesting that rewilding should be a priority over current conservation programs in North America or Africa. Pleistocene rewilding could proceed alongside such conservation efforts, and it would likely generate conservation dollars from new funding sources, rather than competing for funds with existing conservation efforts.

The long-term vision includes a vast, securely fenced ecological history park, encompassing thousands of square miles, where horses, camels, elephants and large carnivores would roam. As happens now in Africa and regions surrounding some U.S. national parks, the ecological history park would not only attract ecotourists but would also provide jobs related both to park management and to tourism.

To get to that distant point, we would need to start modestly, with relatively small-scale experiments that assess the impacts of megafauna on North American landscapes. These controlled experiments, guided by sound science and by the fossil record, which indicates what animals actually lived here, could occur first on donated or purchased private lands and could begin immediately. They will be critical in answering the many questions about the reintroductions and would help lay out the costs and benefits of rewilding.

One of these experiments is already under way. Spurred by our 2004 meeting, biologists recently reintroduced Bolson tortoises to a private ranch in New

Mexico. Bolson tortoises, some weighing more than 100 pounds, once grazed parts of the southwestern U.S. before disappearing around 10,000 years ago, victims of human hunting. This endangered tortoise now clings to survival, restricted to a single small area in central Mexico. Thus, the reintroduction not only repatriates the tortoise to the U.S., it increases the species' chance for survival. Similar experiments are also occurring outside North America.

The reintroduction of wild horses and camels would be a logical part of these early experiments. Horses and camels originated on this continent, and many species were present in the late Pleistocene. Today's feral horses and asses that live in some areas throughout the West are plausible substitutes for extinct American species. Because most of the surviving Eurasian and African species are now critically endangered, establishing Asian asses and Przewalski's horse in North America might help prevent the extinction of these animals. Bactrian camels, which are critically endangered in the Gobi Desert, could provide a modern proxy for Camelops, a late Pleistocene camel. Camels, introduced from captive or domesticated populations, might benefit U.S. ecosystems by browsing on woody plants that today are overtaking arid grasslands in the Southwest, an ecosystem that is increasingly endangered.

Another prong of the project would likely be more controversial but could also begin immediately. It would establish small numbers of elephants, cheetahs and lions on private property.

Introducing elephants could prove valuable to nearby human populations by attracting tourists and maintaining grasslands useful to ranchers (elephants could suppress the woody plants that threaten southwestern grasslands). In the late Pleistocene, at least four elephant species lived in North America. Under a scientific framework, captive elephants in the U.S. could be introduced as proxies for these extinct animals. The biggest cost involved would be fencing, which has helped reduce conflict between elephants and humans in Africa.

Many cheetahs are already in captivity in the U.S. The greatest challenge would be to provide them with large, securely fenced areas that have appropriate habitat and prey animals. Offsetting these costs are benefits—restoring what must have been strong interactions with pronghorn, facilitating ecotourism as an economic alternative for ranchers, many of whom are struggling financially, and helping to save the world's fastest carnivore from extinction.

Lions are increasingly threatened, with populations in Asia and some parts of Africa critically endangered. Bringing back lions, which are the same species that once lived in North America, presents daunting challenges as well as many potential benefits. But private reserves in southern Africa where lions and other large animals have been successfully reintroduced offer a model—and these reserves are smaller than some private ranches in the Southwest.

If these early experiments with large herbivores and predators show promising results, more could be undertaken, moving toward the long-term goal of a huge ecological history park. What we need now are panels of experts who, for each species, could assess, advise and cautiously lead efforts in restoring megafauna to North America.

A real-world example of how the reintroduction of a top predator might work comes from the wolves of Yellowstone National Park [see "Lessons from the Wolf," by Jim Robbins; *Scientific American*, June 2004]. The gray wolf became extinct in and around Yellowstone during the 1920s. The loss led to increases in their prey—moose and elk—which in turn reduced the distribution of aspens and other trees they eat. Lack of vegetation destroyed habitat for migratory birds and for beavers. Thus, the disappearance of the wolves propagated a trophic cascade from predators to herbivores to plants to birds and beavers. Scientists have started to document the ecosystem changes as reintroduced wolves regain the ecological role they played in Yellowstone for millennia. An additional insight researchers are learning from putting wolves back into Yellowstone is that they may be helping the park cope with climate change. As winters grow milder, fewer elk die, which means less carrion for scavengers such as coyotes, ravens and bald eagles. Wolves provide carcasses throughout the winter for the scavengers to feed on, bestowing a certain degree of stability.

The Challenges Ahead

As our group on the ranch in New Mexico discussed how Pleistocene rewilding might work, we foresaw many challenges that would have to be addressed and overcome. These include the possibility that introduced animals could bring novel diseases with them or that they might be unusually susceptible to diseases already present in the ecosystem; the fact that habitats have changed over the millennia and that reintroduced animals might not fare well in these altered environments; and the likelihood of unanticipated ecological consequences and unexpected reactions from neighboring human communities. Establishing programs that monitor the interactions among species and their consequences for the well-being of the ecosystem will require patience and expertise. And, of course, it will not be easy to convince the public to accept predation as an important natural process that actually nourishes the land and enables ecosystems to thrive. Other colleagues have raised additional concerns, albeit none that seems fatal.

Many people will claim that the concept of Pleistocene rewilding is simply not feasible in the world we live in today. I urge these people to look to Africa for inspiration. The year after the creation of Kruger National Park was announced, the site was hardly the celebrated mainstay of southern African biodiversity it is today. In 1903 zero elephants, nine lions, eight buffalo and very few cheetahs lived within its boundaries. Thanks to the vision and dedication of African conservationists, 7,300 elephants, 2,300 lions, 28,000 buffalo and 250 cheetahs roamed Kruger 100 years later—as did 700,000 tourists, bringing with them tens of millions of dollars.

In the coming century, humanity will decide, by default or design, the extent to which it will tolerate other species and thus how much biodiversity will endure. Pleistocene rewilding is not about trying to go back to the past; it is about using the past to inform society about how to maintain the functional fabric of nature. The potential scientific, conservation and cultural benefits of restoring mega-fauna are clear, as are the costs. Although sound science can help mitigate the potential costs, these ideas will make many uneasy. Yet given the apparent dysfunction of North American ecosystems and Earth's overall state, inaction carries risks as well. In the face of tremendous uncertainty, science and society must weigh the costs and benefits of bold, aggressive actions like Pleistocene rewilding against those of business as usual, which has risks, uncertainties and costs that are often unacknowledged. We have a tendency to think that if we maintain the status quo, things will be fine. All the available information suggests the opposite. . . .

Dustin R. Rubenstein, Daniel I. Rubenstein, Paul W. Sherman, and Thomas A. Gavin **NO**

Pleistocene Park: Does Re-Wilding North America Represent Sound Conservation for the 21st Century?

Introduction

Ancestors of elephants and lions once roamed much of North America. Recently, a diverse group of conservation biologists has proposed to create a facsimile of this bygone era by reintroducing charismatic African and Asian megafauna to western North America to replace species that disappeared during the Pleistocene extinctions, some 13,000 years ago. Arguing that their vision is justified on "ecological, evolutionary, economic, aesthetic and ethical grounds", Donlan et al. believe that modern "Pleistocene Parks" would provide refuges for species that are themselves threatened or endangered, and that repopulating the American west with these large mammals would improve local landscapes, restore ecological and evolutionary potential, and make amends for the ecological excesses of our ancestors.

To understand the uniqueness of this proposal, a terminological clarification is necessary. The "re-wilding" of ecosystems is the practice of reintroducing extant species (captive-bred or wild caught) back to places from which they were extirpated in historical times (i.e., in the past several hundred years). Because re-wilding deals with recently extirpated species and short evolutionary time scales, it is reasonable to assume that there have been minimal evolutionary changes in the target species and their native habitats. Re-wilding of ecosystems is not a new conservation practice and, indeed, it has become a standard management tool.

By contrast, "Pleistocene re-wilding" of ecosystems is a revolutionary idea that would involve introducing to present-day habitats either (1) extant species that are descended from species that occurred in those habitats during the Pleistocene, but that went extinct about 13,000 years ago, or (2) modern-day ecological proxies for extinct Pleistocene species. Pleistocene re-wilding is thus a novel plan for ecological restoration on a more grandiose temporal and spatial scale than is re-wilding.

Pleistocene re-wilding has been discussed for many years, and in 1989, it was attempted in Siberia, Russia, when mega-herbivores including wood bison (*Bison bison athabascae*), Yakutian horses (*Equus* sp.), and muskoxen (*Ovibos moschatus*) were introduced in an effort to recreate the grassland ecosystem of the Pleistocene. However, the North American Pleistocene re-wilding proposal

From *Biological Conservation*, vol. 132, issue 2, October 2006, pp. 232–238. Copyright © 2006 by Elsevier Science Ltd. Reprinted by permission via Rightslink.

of Donlan et al. is far more ambitious than this because it aims to reconstruct an ancient ecosystem by translocating a more diverse array of African and Asian megafauna to geographical regions and plant communities that have evolved without such creatures since the Pleistocene. Species targeted for introduction span several trophic levels and include predators such as African cheetahs (*Acinonyx jubatus*) and lions (*Panthera leo*), and large herbivores like African (*Loxodonta africana*) and Asian (*Elephas maximus*) elephants, various equids (*Equus* spp.), and Bactrian camels (*Camelus bactranus*). This plan includes animals that are both descendant species of extinct taxa and ecological proxies for extinct species.

Pleistocene re-wilding of North America has two principal goals: (1) to restore some of the evolutionary and ecological potential that was lost from North America 13,000 years ago; and (2) to help prevent the extinction of some of the world's existing megafauna by creating new, and presumably better protected, populations in North America. Discussion of the proposal is just beginning and, although some initial concerns have been raised, supporters and detractors agree that Pleistocene re-wilding is a bold and innovative idea, deserving of careful consideration. This paper was developed with the intent of extending healthy and fruitful scientific debate about Pleistocene re-wilding of North America.

The Ecology and Evolution of Pleistocene Re-Wilding: Restoring Ecological Potential to North American Ecosystems

Although an ethical desire to redress the excesses of our ancestors might serve as an initial justification for Pleistocene re-wilding, the ecological and evolutionary merits of such a plan must be considered carefully. Pleistocene re-wilding of North America would involve a monumental introduction of large mammals into areas where they have been extinct for millennia, and into habitats that have existed without such creatures for similarly long periods of evolutionary time. The potential negative ecological effects of transplanting exotic species to non-native habitats are well-known. The results of Pleistocene re-wilding in North America are unknown and might well be catastrophic; ecosystem functioning could be disrupted, native flora and fauna, including species of conservation value, could be negatively impacted, and a host of other unanticipated ecological problems could arise.

Pleistocene re-wilders believe that it is possible to enhance ecological potential, that is, to recreate evolutionarily-relevant mammalian species assemblages and restore ecosystem functioning to Pleistocene levels, because they believe that the flora of North American ecosystems is essentially unchanged since the Pleistocene. However, plant communities are dynamic and constantly in flux, genotypically and phenotypically, and there has been over 13,000 years for grassland and shrub-steppe communities to evolve and plant assemblages to change in the absence of the full suite of Pleistocene mega-herbivores. When managers discuss restoring ecological potential, or simply ecosystem restoration, it is important for them to be clear about what they are trying to restore and to what level of restoration they are trying to reach. Whereas Pleistocene re-wilding could potentially increase the ecological potential of some of North America's ecosystems by reintroducing predators on species like pronghorn or bighorn sheep (and thus, indirectly restoring the evolutionary potential of these prey species), or by restoring herbivorous keystone species like elephants to the temperate grasslands, it is questionable whether it would restore ecological potential to Pleistocene levels.

Indeed, rather than restoring our "contemporary" wild ecosystems to the "historic" wild ecosystems of the Pleistocene and their original levels of ecosystem functioning, which are unknown, Pleistocene re-wilding could instead result in "re-wilded" novel, or emerging, ecosystems with unique species compositions and new or altered levels of ecosystem functioning. Biogeographic assemblages and evolutionary lineages would be co-mingled in novel ways; new parasites and diseases could be introduced; and food chains would be disrupted. Moreover, without really knowing how Pleistocene ecosystems functioned, there will be no way to determine whether Pleistocene re-wilding restored ancient ecosystems or disrupted contemporary ones.

While the reintroduction of large grazers can, in some cases, shape and restore grassland ecosystems, this will depend on whether the grazers are indigenous or exotic. Modern introductions of exotic feral horses have dramatically altered vegetation in marsh and grassland ecosystems throughout the New World, and these changes have had direct impacts on a variety of native animal species, some positive, but some negative. Moreover, exotic grazers, such as the one-humped camel (*Camelus dromedarius*), have wreaked havoc upon desert ecosystems in Australia by selectively eating rare plant species. Similarly, the reintroduction of large predators can also have unexpected results on populations of prey species. For instance, wolves reintroduced to Yellowstone National Park, USA preyed upon elk more, and other species of ungulates less, than what was predicted prior to reintroduction.

Of course, it might be argued that these problems would quickly become apparent if Pleistocene re-wilding were first attempted on a small-scale, experimental basis. However, experiments of this nature cannot be done quickly and may take decades and generations to play out. For instance, the Siberian Pleistocene Park experiment began in 1989, and as of yet, few of the results have been published. Moreover, it may not be possible to conduct adequate, meaningful experiments on small spatial scales because many of these species have large home ranges or migrate great distances. For instance, African cheetahs can have home ranges of nearly 200 km^2, and African elephants can migrate distances of up to nearly 150 km or more.

Despite the potential dangers to ecosystem functioning, the reintroductions proposed by Donlan et al. would place many of the animals in temperate grasslands and shrub-steppe habitats, which are among the most threatened, but least protected, ecosystems in the world. If the reintroduction of exotic megafauna could help preserve these ecosystems, conservationists must weigh the possibility of preserving disrupted or novel North American ecosystems against the possibility of losing those ecosystems altogether.

The Ethics and Aesthetics of Pleistocene Re-Wilding: Protecting and Restoring the Evolutionary Potential of Threatened Megafauna

Humans were at least partly responsible for exterminating some species of Pleistocene megafauna and, today, anthropogenic impacts continue to contribute to the extinction of the world's remaining megafauna. Donlan et al. argue that humans bear an ethical responsibility to prevent future megafaunal extinctions and redress past losses. They suggest that introducing large Asian and African vertebrates to North America will not only ensure their long-term survival, but also restore their evolutionary potential (i.e., increase the number of individuals worldwide to allow them greater chances to radiate and generate new phenotypic and genotypic variants). Although this plan is certainly well-intentioned, the underlying reasoning is flawed. In essence, it is an attempt to preserve charismatic African and Asian species that are being driven to extinction by humans in their native habitats by refocusing efforts in places where those species have never occurred and where humans drove their distant ancestors extinct. Although Donlan et al. do not advocate giving up on conserving megafauna in developing nations, diverting attention from some of the world's most economically poor, but most biologically rich, countries to make amends for the ecological excesses of our North American ancestors could cripple, rather than assist, the conservation movement worldwide.

The human population is growing and natural habitats are declining in extent and diversity everywhere. Couple this with the political and economic strife that is occurring in many developing nations and it is not difficult to see why native megafauna, especially large mammals, are declining in numbers worldwide. Despite this dire situation, Pleistocene re-wilding of North America is not the only viable solution to preserve the world's megafauna. In the developing world, new conservation models are being implemented that go hand-in-hand with human development as wildlife must pay for itself by generating economic benefits for local citizenry to help alleviate poverty. Although there are many challenges in developing such programs, there is much to be gained by overcoming them because most of the native megafauna in developing regions inhabit private, often unprotected, lands outside of parks. For instance, across Africa, 84% of African elephant habitat is outside of protected areas, and in Kenya, 70% of the wildlife lives outside of protected areas for at least part of the year.

Conserving African and Asian megafauna does not require relocating them to North America. However, it will require new conservation plans that ensure local citizenry receive economic benefits from wildlife. Available human and financial resources might be better expended on preserving land, promoting ecotourism, building fences in areas of high human–wildlife conflict, and establishing educational and research programs in areas of Africa and Asia where indigenous megafauna are most at risk, rather than on introducing those same large, exotic species to North America.

In addition, the question of how the Pleistocene re-wilding plan would affect existing conservation efforts in North America must be considered. Conservationists often struggle with local opposition to re-wilding with native predators, and even the reintroduction of relatively benign large mammals (e.g., moose) meets resistance. The introduction of modern relatives of extinct predators will be opposed even more strongly by state governments and locally-affected citizens. And, with good reason: escapes are inevitable, resulting in human–wildlife conflict as often occurs near protected areas in Africa and Asia. . . . It is difficult enough for North American conservationists to address the real concerns of local citizens about attacks by mountain lions (native predators) on joggers. One can only imagine the anti-conservation backlash that would be generated by news coverage of farmers coping with crop destruction by herds of elephants, or lions and cheetahs attacking cattle, or even children.

While Pleistocene re-wilding may help maintain the evolutionary potential of modern, extant species, it cannot restore the evolutionary potential of extinct species that no longer exist. And even attempting to restore evolutionary potential of endangered species using modern-day species from foreign continents as proxies for creatures that went extinct in North America is controversial. Donlan et al. highlighted the peregrine falcon (*Falco peregrinus*) to illustrate how using similar, but not genetically identical, (sub-)species can indeed serve as proxies for nearly-extinct taxa. Moreover, for nearly two decades, conservation biologists have proposed introducing closely related proxy species for extinct birds on New Zealand and other Pacific islands. However, species that went extinct some 13,000 years ago are probably more genetically different from their modern-day proxies, who have continued to evolve for millennia, than are two sub-species of modern falcons or modern Pacific island birds. For instance, although recent molecular data suggest that the common horse (*Equus caballus*) is genetically similar to its evolutionary ancestral species, modern elephants, cheetahs, and lions are quite genetically distinct from their extinct Pleistocene relatives.

Rather than use modern-day species from foreign continents as proxies for creatures that went extinct in North America, conservation efforts should focus on re-wilding native species into their historical ranges throughout North America to restore ecosystems and increase the evolutionary potential of indigenous species. For instance, native herbivores like bison (*Bison bison*), pronghorn (*Antilcapra americana*), elk (*Cervus elaphus*), jack rabbits (*Lepus townsendii*), and various ground-dwelling squirrels (*Spermophilus* spp.) and prairie dogs (*Cynomys* spp.), as well as native predators like black-footed ferrets (*Mustela nigripes*), bobcats (*Lynx rufus*), badgers (*Taxidea taxus*), and swift foxes (*Vulpes velox*) are likely candidates for reintroduction to geographic regions from which they were extirpated in the past several hundred years.

Donlan et al. (2005) suggested that another appropriate candidate for reintroduction is the Bolson tortoise (*Gopherus avomarginatus*). Because this animal once lived throughout the southwestern United States and still persists in small areas of similar habitat in Mexico, it may not differ greatly from its ancestral form and, therefore, it might be a reasonable candidate for reintroduction. However, before attempting such a reintroduction, one would also have to consider how much the plant and animal communities in the tortoise's native geographic habitats have changed (evolved) since this reptile went locally extinct.

If reintroducing charismatic megafauna is an important goal of Pleistocene re-wilding because of its possible galvanizing effect on public support for conservation, then one might consider expanding reintroductions of some of North America's own megafauna like wolves (*Canis lupus*) or grizzly bears (*Urus arctos*) to other portions of their known recent (i.e., historical) ranges. And if more predators are deemed necessary, an even better candidate for re-wilding would be the puma (*Puma concolor*), because it is more genetically similar to the long-extinct American cheetah (*Miracinonyx trumani*) than the African cheetah is to the American cheetah. Moreover, the puma is a native mammalian predator that barely survived the Pleistocene extinctions 13,000 years ago, and still remains threatened throughout much of its North American range.

The Economics and Politics of Pleistocene Re-Wilding: Uncertainty and Tradeoffs

The political and economic ramifications of Pleistocene re-wilding of North America are unclear. Certainly, it will be expensive because land acquisition and preparation, translocation, monitoring, protection, and containment require considerable human and financial resources. Moreover, all of these efforts would likely cost proportionally more in North America than they would in Africa or Asia, given the higher prices of salaries and supplies. Because conservation funding is limited, Pleistocene re-wilding may compete for resources that might otherwise have gone to local conservation efforts. Although it is possible that introducing charismatic African and Asian megafauna to North America could ignite public and political support, ultimately leading to an overall increase in funding for conservation projects worldwide, other new ideas might increase the pool of resources with less risk to North American ecosystems and conservation efforts worldwide.

Donlan et al. are careful to point out that the initial steps of Pleistocene re-wilding can occur without large-scale translocations of proxy species because many of these animals are already in captivity in the United States. This would potentially reduce costs, as well as avoid potential political problems between the United States and developing nations, the ultimate "sources" for the animals. However, reintroductions from wild populations have been more successful than those from captive populations, and Pleistocene re-wilding inevitably would involve translocating animals from Asia and Africa to North America, either to increase population numbers or to improve population viability by augmenting genetic diversity. Such translocations of megafauna into areas where they were recently extirpated occur routinely throughout Africa and Asia, and learning from these examples could shed some light on the practicality of Pleistocene re-wilding of North America.

The Practicality of Pleistocene Re-Wilding: The Reality of Reintroductions

One of the goals of Pleistocene re-wilding of North America is to ". . . restore equid species to their evolutionary homeland" and, indeed, some of the best known and most successful reintroductions of endangered species to their historical ranges involve equids. For example, the Tahki, or Przewalski's horse (*Equus ferus przewalskii*), which is endemic to Mongolia and China, was considered extinct in the wild by the end of the 1960s and fewer than 400 Tahki remained in captivity in 1979. By the beginning of the 1990s, however, efforts began to reestablish populations in the wild and two reintroduction sites were chosen in Mongolia. By 2000, the population at one site had declined only slightly, while that at the other had increased by 50%. These encouraging trends led to a third introduction in 2005 and suggest that the re-wilding of the Tahki's historic range is likely to succeed.

In an attempt to repopulate Israel with recently extirpated biblical animals, onagers, a race of Asian asses (*Equus hemionus*), were translocated from Israel's Hai Bar breeding reserve to a nearby erosional crater in the Negev desert. Between 1968 and 1993 multiple reintroductions of 50 individuals took place. It was not until the end of the 1990s, however, before the population started to expand numerically and spatially. Low fertility of translocated adult females relative to that of their wild-born daughters and male-biased sex ratios among the progeny limited recruitment. These unanticipated biological constraints suggest that even reintroductions of native species to their historical habitats are not assured of succeeding.

Repopulating the historic range of the endangered Grevy's zebra (*Equus grevyi*) in east Africa is viewed as critically important to saving the species from extinction. Fewer then 2000 Grevy's zebras remain in small areas of Ethiopia and northern Kenya, whereas only 35 years ago, over 20,000 individuals inhabited areas all the way to the horn of Africa. Efforts to repopulate areas of the Grevy's historic range have involved capturing and moving small groups of appropriate sex ratios and age structures to holding areas before subsequent release. While two such reintroductions in Kenya, one to Tsavo National Park and one to Meru National Park, began successfully, neither has led to expanding populations. In fact, in Meru, differences in the composition and abundance of mammal species in the Grevy's zebras' new range have, at times, led to rapid declines in their numbers. Therefore, even reintroductions within the natural geographic regions of a species are often fraught with surprises due to diseases, unexpected differences in environmental conditions, and naïveté toward predators. . . .

Another Jurassic Park?

We all remember "Jurassic Park", Crichton's fictional account of re-wilding an isolated island with extinct dinosaurs recreated from ancient DNA. Pleistocene re-wilding of North America is only a slightly less sensational proposal. It is a little like proposing that two wrongs somehow will make a right: both the modern-day proxy species are "wrong" (i.e., different genetically from the species that occurred in North America during the Pleistocene), and the ecosystems into which they are to be reintroduced are "wrong" (i.e., different in composition from the Pleistocene ecosystems, as well as from those in which the modern-day proxy species evolved). Pleistocene re-wilding of North America will not restore evolutionary potential of North America's extinct megafauna because the species in question are evolutionarily distinct, nor will it restore ecological potential of North America's modern ecosystems because they have continued to evolve over the past 13,000 years. In addition, there is a third and potentially greater "wrong" proposed: adding these exotic species to current ecological communities could potentially devastate populations of indigenous, native animals and plants.

Although Donlan et al. argued that Pleistocene re-wilding of North America is justified for ecological, evolutionary, economic, aesthetic, and ethical reasons, there are clearly numerous ecological and evolutionary concerns. On the one hand, the plan might help conserve and maintain the evolutionary potential of some endangered African and Asian megafauna, as well as indirectly enhance the evolutionary potential of native North American prey species that have lacked appropriate predators since the Pleistocene. On the other hand, the plan cannot restore the evolutionary potential of extinct species and it is unlikely to restore the ecological potential of western North America's grassland and shrub-steppe communities. Instead, it may irreparably disrupt current ecosystems and species assemblages. Moreover, there are many potential practical limitations to Donlan et al.'s plan. Reintroduced camels did not survive for long in the deserts of the American West. Could African megafauna, especially large carnivores, really populate the same areas? Would the genetically depauperate cheetah succumb to novel diseases? Would elephants survive the harsh prairie winters, lacking the thick coats of their mastodon ancestors?

Answering these questions and accomplishing Pleistocene re-wilding of North America would require a massive effort and infusion of funds and could take more time to experimentally test than some of these critically endangered species have left to survive in their existing native habitats. If financial and physical resources were available on this scale, they would be better spent on developing and field-testing new ways to manage and conserve indigenous populations of African, Asian, and North American wildlife in their historically-populated native habitats, on conducting ecological, behavioral, and demographic studies of these organisms in the environments in which they evolved, and on educating the public on each continent about the wonders of their own dwindling flora and fauna.

POSTSCRIPT

Should North America's Landscape Be Restored to Its Pre-Human State?

Rewilding as described by C. Josh Donlan in the present essay and in C. Josh Donlan, Joel Berger, Carl E. Bock, Jane H. Bock, David A. Burney, James A. Estes, Dave Foreman, Paul S. Martin, Gary W. Roemer, Felisa A. Smith, Michael E. Soulé, and Harry W. Greene, "Pleistocene Rewilding: An Optimistic Agenda for Twenty-First Century Conservation," *American Naturalist* (November 2006), is probably not likely to happen. But as Rubenstein notes, the basic idea has been around for a while. One major presentation of the idea is Dave Foreman's *Rewilding North America* (Island Press, 2004), which calls for the construction of networks of connected protected areas, rather than isolated protected areas as is more common today. In essence, networks give threatened species more room, an important consideration especially for predators that may need large home ranges in order to be able to find sufficient food or suitable den sites. The basic idea was developed while Foreman was with the Wildlands Project, of which he was a cofounder in 1991. Today the Wildlands Project works to develop such networks using both public and private lands. Foreman left the Project in 2003 to found the Rewilding Institute, which "looks at North America in shades of landscape permeability: the degree to which the land is open and safe for the movement of large carnivores and other wide-ranging and sensitive species between large core habitats." To enhance permeability, it works to create "megalinkages," large networks of protected areas that provide "habitat corridors through a hostile sea" of cities, suburbs, and farms. Foreman's status as a coauthor of the second Donlan paper mentioned above suggests that his thinking is evolving from building networks and megalinkages to protect contemporary threatened wildlife to restoring wildlife that vanished long, long ago.

On a more local level, centered on New England, there is RESTORE: The North Woods (http://www.restore.org), whose aim is "America's first restored landscape: a place of vast recovered wilderness, of forests where the wolf and caribou roam free, of clear waters alive with salmon and trout, of people once again living in harmony with nature." RESTORE's efforts are directed toward creating a large national park in northern Maine and restoring "extirpated and imperiled wildlife, including the eastern timber wolf, Canada lynx, and Atlantic salmon." In the American Southwest, the Turner Endangered Species Fund has initiated a project to restore the endangered Bolson tortoise to a portion of its late Pleistocene range; see Joe Truett and Mike Phillips, "Beyond Historic Baselines: Restoring Bolson Tortoises to Pleistocene Range," *Ecological Restoration*

(June 2009). In a similar spirit, beaver are being reintroduced inScotland; see Christopher Werth, "Unleash the Critters," *Newsweek* (April 20, 2009).

Peter Taylor, *Beyond Conservation: A Wildland Strategy* (Earthscan, 2005), calls rewilding "putting a new soul in the landscape," although his focus, like RESTORE's, is on restoring only what has been lost in the last few centuries. See also J. C. Hallman, "Pleistocene Dreams: A Radical Conservation," *Science & Spirit* (May/June 2008). L. Martin, *Twilight of the Mammoths: Ice Age Extinctions and the Rewilding of America* (University of California Press, 2005), notes that if the Pleistocene overkill hypothesis is true, humans bear a moral responsibility to repair the damage they have caused. This puts him on Donlan's side of the debate, but that debate is hardly over. In Eric Jaffe's "Brave Old World," *Science News* (November 11, 2006), Paul Sherman is quoted as saying that the aim of the papers by Rubenstein et al. is to stimulate further discussion. However, Rubenstein's group (including Sherman) published "Rewilding Rebuttal," *Scientific American* (October 2007), a few months after the Donlan paper, saying that if the debate is to go forward, Donlan et al. must "abandon sensationalism."

Is it "sensationalism" or "idealism" to speak of undoing damage done thousands of years ago? Whatever one calls it, it may not be very realistic. As conservationists must admit, fighting even to protect contemporary wildlife in isolated areas is difficult. Building networks and megalinkages is even more challenging. The reason is simply that modern civilization imposes a great many competing demands for space and resources. In a Special Section on "The Rise of Restoration Ecology," /Science/ (July 31, 2009), Stephen T. Jackson and Richard J. Hobbs, "Ecological Restoration in the Light of Ecological History," note that a historical viewpoint is essential to understanding ecosystems, but "many historical restoration targets will be unsustainable in the coming decades."

ISSUE 5

Should the Military Be Exempt from Environmental Regulations?

YES: Benedict S. Cohen, "Impact of Military Training on the Environment," Testimony before the Senate Committee on Environment and Public Works (April 2, 2003).

NO: Jamie Clark, "Impact of Military Training on the Environment," Testimony before the Senate Committee on Environment and Public Works (April 2, 2003).

ISSUE SUMMARY

YES: Benedict S. Cohen argues that environmental regulations interfere with military training and other "readiness" activities, and that though the U.S. Department of Defense will continue "to provide exemplary stewardship of the lands and natural resources in our trust" those regulations must be revised to permit the military to do its job without interference.

NO: Jamie Clark argues that reducing the Department of Defense's environmental obligations is dangerous because both people and wildlife would be threatened with serious, irreversible, and unnecessary harm.

Most of us have heard of "scorched earth" wars, in which an army destroys forests and farms in order to deny the enemy their benefit. We have surely seen the images of a Europe laid waste by World War II. More recently, we may recall, in the Gulf War we saw oil deliberately released to flood desert sands and the waters of the Persian Gulf. Enough smoke poured from burning oil wells to threaten both local climate change and human health. See, for example, Randy Thomas, "Eco War," *Earth Island Journal* (Spring 1991); B. Ruben, "Gulf Smoke Screens," *Environmental Action* (July/August 1991); and Jeffrey L. Lange, David A. Schwartz, Bradley N. Doebbeling, Jack M. Heller, and Peter S. Thorne, "Exposures to the Kuwait Oil Fires and Their Association with Asthma and Bronchitis among Gulf War Veterans," *Environmental Health Perspectives* (November 2002). Weaponry can have environmental effects by destroying dams, by physically destroying plants and animals, by causing erosion, and by disseminating toxic materials; see Henryk Bem and Firyal Bou-Rabee, "Environmental and Health Consequences of Depleted Uranium Use in the 1991 Gulf War," *Environment International* (March 2004).

The environmental impact of war would seem impossible to deny. But even after the Gulf War, the U.S. Department of Defense tried to suppress satellite photos showing the extent of the damage; see Shirley Johnston, "Gagged on Smoke," *Earth Island Journal* (Summer 1991). And for many years, it insisted that nuclear war was survivable, until researchers made it clear that even a small nuclear war would produce a "nuclear winter" that would probably destroy civilization, if not the human species. See T. Rueter and T. Kalil, "Nuclear Strategy and Nuclear Winter," *World Politics* (July 1991), and Carl Sagan and Richard Turco, *A Path Where No Man Thought: Nuclear Winter and Its Implications* (Random House, 1990).

Preparations for war may also have serious environmental impacts. Puerto Rico's island of Vieques was long a bomb depot and bombing range for the U.S. Navy. Local residents protested vigorously and documented heavy-metal contamination of the local ecosystem. After the Navy left the island, it "has continued to deny that it has been anything but an excellent environmental steward in Vieques." See Shane DuBow and Scott S. Warren, "Vieques on the Verge," *Smithsonian* (January 2004).

In 2002, the U.S. Congress, through the Readiness and Range Preservation Initiative, granted the Department of Defense a temporary exemption to the Migratory Bird Treaty Act that allowed the "incidental taking" of endangered birds during bombing and other training on military lands. In 2003, the Department of Defense asked Congress for additional exemptions from environmental regulations, specifically the Clean Air Act, Marine Mammal Protection Act, Endangered Species Act, Migratory Bird Treaty Act, and federal toxic waste laws. Paul Mayberry, deputy undersecretary of defense for readiness, said the exemptions were justified because many environmental restrictions were putting the nation's military readiness at stake. See "Pentagon Seeks Clarity in Environmental Laws Affecting Ranges," Agency Group 09, FDCH Regulatory Intelligence Database (March 21, 2003). The U.S. Senate Committee on Environment and Public Works held a hearing on "The Impact of Military Training on the Environment" on April 2, 2003. In the following selections, Benedict S. Cohen, Deputy General Counsel for Environment and Installations, Department of Defense, argues in his testimony before the committee that environmental regulations interfere with military training and other "readiness" activities, and that though the Department of Defense will continue "to provide exemplary stewardship of the lands and natural resources in our trust" those regulations must be revised to permit the military to do its job without interference. Jamie Clark, Senior Vice President for Conservation Programs, National Wildlife Federation, argues that reducing the Department of Defense's environmental obligations is dangerous for two reasons. First, both people and wildlife will be threatened with serious, irreversible, and unnecessary harm. Second, other federal agencies and industry sectors with important missions, using the same logic as used here by the Department of Defense, would demand similar exemptions from environmental laws.

YES

Benedict S. Cohen

Impact of Military Training on the Environment

Mr. Chairman and distinguished members of this Committee, I appreciate the opportunity to discuss with you the very important issue of sustaining our test and training capabilities, and the legislative proposal that the Administration has put forward in support of that objective. In these remarks I would like particularly to address some of the comments and criticisms offered concerning these legislative proposals.

Addressing Encroachment

We have only recently begun to realize that a broad array of encroachment pressures at our operational ranges are increasingly constraining our ability to conduct the testing and training that we must do to maintain our technological superiority and combat readiness. Given world events today, we know that our forces and our weaponry must be more diverse and flexible than ever before. Unfortunately, this comes at the same time that our ranges are under escalating demands to sustain the diverse operations required today, and that will be increasingly required in the future.

    This current predicament has come about as a cumulative result of a slow but steady process involving many factors. Because external pressures are increasing, the adverse impacts to readiness are growing. Yet future testing and training needs will only further exacerbate these issues, as the speed and range of our weaponry and the number of training scenarios increase in response to real-world situations our forces will face when deployed. We must therefore begin to address these issues in a much more comprehensive and systematic fashion and understand that they will not be resolved overnight, but will require a sustained effort.

Environmental Stewardship

Before I address our comprehensive strategy, let me first emphasize our position concerning environmental stewardship. Congress has set aside 25 million acres of land—some 1.1% of the total land area in the United States. These lands were entrusted to the Department of Defense (DoD) to use efficiently and to care for properly. In executing these responsibilities we are committed

From Testimony before the Senate Committee on Environment and Public Works, April 2, 2003.

to more than just compliance with the applicable laws and regulations. We are committed to protecting, preserving, and, when required, restoring, and enhancing the quality of the environment.

- We are investing in pollution prevention technologies to minimize or reduce pollution in the first place. Cleanup is far more costly than prevention.
- We are managing endangered and threatened species, and all of our natural resources, through integrated natural resource planning.
- We are cleaning up contamination from past practices on our installations and are building a whole new program to address unexploded ordnance on our closed, transferring, and transferred ranges.

Balance

The American people have entrusted these 25 million acres to our care. Yet, in many cases, these lands that were once "in the middle of nowhere" are now surrounded by homes, industrial parks, retail malls, and interstate highways.

On a daily basis our installation and range managers are confronted with a myriad of challenges—urban sprawl, noise, air quality, air space, frequency spectrum, endangered species, marine mammals, and unexploded ordnance. Incompatible development outside our fence-lines is changing military flight paths for approaches and take-offs to patterns that are not militarily realistic—results that lead to negative training and potential harm to our pilots. With over 300 threatened and endangered species on DoD lands, nearly every major military installation and range has one or more endangered species, and for many species, these DoD lands are often the last refuge. Critical habitat designations for an ever increasing number of threatened or endangered species limit our access to and use of thousands of acres at many of our training and test ranges. The long-term prognosis is for this problem to intensify as new species are continually added to the threatened and endangered list.

Much too often these many encroachment challenges bring about unintended consequences to our readiness mission. This issue of encroachment is not going away. Nor is our responsibility to "train as we fight."

2003 Readiness and Range Preservation Initiative (Rrpi)

Overview

DoD's primary mission is maintaining our Nation's military readiness, today and into the future. DoD is also fully committed to high-quality environmental stewardship and the protection of natural resources on its lands. However, expanding restrictions on training and test ranges are limiting realistic preparations for combat and therefore our ability to maintain the readiness of America's military forces.

Last year, the Administration submitted to Congress an eight-provision legislative package, the Readiness and Range Preservation Initiative (RRPI).

Congress enacted three of those provisions as part of the National Defense Authorization Act for Fiscal Year 2003. Two of the enacted provisions allow us to cooperate more effectively with local and State governments, as well as private entities, to plan for growth surrounding our training ranges by allowing us to work toward preserving habitat for imperiled species and assuring development and land uses that are compatible with our training and testing activities on our installations.

Under the third provision, Congress provided the Department a regulatory exemption under the Migratory Bird Treaty Act for the incidental taking of migratory birds during military readiness activities. We are grateful to Congress for these provisions, and especially for addressing the serious readiness concerns raised by recent judicial expansion of the prohibitions under the Migratory Bird Treaty Act. I am pleased to inform this Committee that as a direct result of your legislation, Air Force B-1 and B-52 bombers, forward deployed to Anderson Air Force Base, Guam, are performing dry run training exercises over the Navy's Bombing Range at Farallon de Medinilla in the Commonwealth of the Northern Mariana Islands.

Last year, Congress also began consideration of the other five elements of our Readiness and Range Preservation Initiative. These five proposals remain essential to range sustainment and are as important this year as they were last year—maybe more so. The five provisions submitted this year reaffirm the principle that military lands, marine areas, and airspace exist to ensure military preparedness, while ensuring that the Department of Defense remains fully committed to its stewardship responsibilities. These [. . .] remaining provisions:

- Authorize use of Integrated Natural Resource Management Plans in appropriate circumstances as a substitute for critical habitat designation;
- Reform obsolete and unscientific elements of the Marine Mammal Protection Act, such as the definition of "harassment," and add a national security exemption to that statute;
- Modestly extend the allowable time for military readiness activities like bed-down of new weapons systems to comply with Clean Air Act; and
- Limit regulation of munitions on operational ranges under the Comprehensive Environmental Response, Compensation, and Liability Act (CERCLA) and Resource Conservation and Recovery Act (RCRA), if and only if those munitions and their associated constituents remain there, and only while the range remains operational.

Before discussing the specific elements of our proposal, I would like to address some overarching issues. A consistent theme in criticisms of our proposal is that it would bestow a sweeping or blanket exemption for the Defense Department from the Nation's environmental laws. No element of this allegation is accurate.

First, our initiative would apply only to military readiness activities, not to closed ranges or ranges that close in the future, and not to "the routine operation of installation operating support functions, such as administrative offices, military exchanges, commissaries, water treatment facilities, storage, schools, housing, motor pools . . . nor the operation of industrial activities, or the construction or demolition of such facilities." Our initiative thus is not applicable to the Defense Department activities that have traditionally been of greatest concern to state and federal regulators. It does address only uniquely military activities—what DoD does that is unlike any other governmental or private activity. DoD is, and will remain, subject to precisely the same regulatory requirements as the private sector when we perform the same types of activities as the private sector. We seek alternative forms of regulation only for the things we do that have no private-sector analogue: military readiness activities.

Moreover, our initiative largely affects environmental regulations that don't apply to the private sector or that disproportionately impact DoD:

- Endangered Species Act "critical habitat" designation has limited regulatory consequences on private lands, but can have crippling legal consequences for military bases.
- Under the Marine Mammal Protection Act, the private sector's Incidental Take Reduction Plans give commercial fisheries the flexibility to take significant numbers of marine mammal each year, but are unavailable to DoD—whose critical defense activities are being halted despite far fewer marine mammal deaths or injuries a year.
- The Clean Air Act's "conformity" requirement applies only to federal agencies, not the private sector.

Our proposals therefore are of the same nature as the relief Congress afforded us last year under the Migratory Bird Treaty Act, which environmental groups are unable to enforce against private parties but, as a result of a 2000 circuit court decision were able and willing to enforce, in wartime, against vital military readiness activities of the Department of Defense.

Nor does our initiative "exempt" even our readiness activities from the environmental laws; rather, it clarifies and confirms existing regulatory policies that recognize the unique nature of our activities. It codifies and extends EPA's existing Military Munitions Rule; confirms the prior Administration's policy on Integrated Natural Resource Management Plans and critical habitat; codifies the prior Administration's policy on "harassment" under the Marine Mammal Protection Act; ratifies longstanding state and federal policy concerning regulation under RCRA and CERCLA of our operational ranges; and gives states and DoD temporary flexibility under the Clean Air Act. Our proposals are, again, of the same nature as the relief Congress provided us under the Migratory Bird Treaty Act last year, which codified the prior Administration's position on DoD's obligations under the Migratory Bird Treaty Act.

Ironically, the alternative proposed by many of our critics—invocation of existing statutory emergency authority—would fully exempt DoD from the waived statutory requirements for however long the exemption lasted, a more far-reaching solution than the alternative forms of regulation we propose.

Accordingly, our proposals are neither sweeping nor exemptive; to the contrary, it is our critics who urge us to rely on wholesale, repeated use of emergency exemptions for routine, ongoing readiness activities that could easily be accommodated by minor clarifications and changes to existing law.

Existing Emergency Authorities

As noted above, many of our critics state that existing exemptions in the environmental laws and the consultative process in 10 USC 2014 render the Defense Department's initiative unnecessary.

Although existing exemptions are a valuable hedge against unexpected future emergencies, they cannot provide the legal basis for the Nation's everyday military readiness activities.

- The Marine Mammal Protection Act, like the Migratory Bird Treaty Act the Congress amended last year, has no national security exemption.
- 10 USC 2014, which allows a delay of at most five days in regulatory actions significantly affecting military readiness, is a valuable insurance policy for certain circumstances, but allows insufficient time to resolve disputes of any complexity. The Marine Corps' negotiations with the Fish and Wildlife Service over excluding portions of Camp Pendleton from designation as critical habitat took months. More to the point, Section 2014 merely codifies the inherent ability of cabinet members to consult with each other and appeal to the President. Since it does not address the underlying statutes giving rise to the dispute, it does nothing for readiness in circumstances where the underlying statute itself—not an agency's exercise of discretion—is the source of the readiness problem. This is particularly relevant to our RRPI proposal because none of the five amendments we propose have been occasioned by the actions of state or federal regulators. Four of the five proposed amendments (RCRA, CERCLA, MMPA, and ESA), like the MBTA amendment Congress passed last year, were occasioned by private litigants seeking to overturn federal regulatory policy and compel federal regulators to impose crippling restrictions on our readiness activities. The fifth, our Clean Air Act amendment, was proposed because DoD and EPA concluded that the Act's "general conformity" provision unnecessarily restricted the flexibility of DoD, state, and federal regulators to accommodate military readiness activities into applicable air pollution control schemes. Section 2014, therefore, although useful in some circumstances, would be of no use in addressing the critical readiness issues that our five RRPI initiatives address.
- Most of the environmental statutes with emergency exemptions clearly envisage that they will be used in rare circumstances, as a last resort, and only for brief periods.
- Under these statutes, the decision to grant an exemption is vested in the President, under the highest possible standard: "the paramount interest of the United States," a standard understood to involve exceptionally grave threats to national survival. The exemptions are also usually limited to renewable periods of a year (or in some cases as much as three years for certain requirements).
- The ESA's section 7(j) exemption process, which differs significantly from typical emergency exemptions, allows the Secretary of Defense to direct the Endangered Species Committee to exempt agency actions

in the interest of national security. However, the Endangered Species Committee process has given rise to procedural litigation in the past, potentially limiting its usefulness—especially in exigent circumstances. In addition, because it applies only to agency actions rather than to ranges themselves, any exemption secured by the Department would be of limited duration and benefit: because military testing and training evolve continuously, such an exemption would lose its usefulness over time as the nature of DoD actions on the range evolved.

- The exemption authorities do not work well in addressing those degradations in readiness that result from the cumulative, incremental effects of many different regulatory requirements and actions over time (as opposed to a single major action).
- Moreover, readiness is maintained by thousands of discrete test and training activities at hundreds of locations. Many of these are being adversely affected by environmental provisions. Maintaining military readiness through use of emergency exemptions would therefore involve issuing and renewing scores or even hundreds of Presidential certifications annually.
- And although a discrete activity (e.g., a particular carrier battle group exercise) might only rarely rise to the extraordinary level of a "paramount national interest," it is clearly intolerable to allow all activities that do not individually rise to that level to be compromised or ended by overregulation.
- Finally, to allow continued unchecked degradation of readiness until an external event like Pearl Harbor or September 11 caused the President to invoke the exemption would mean that our military forces would go into battle having received degraded training, with weapons that had received degraded testing and evaluation. Only the testing and training that occurred after the emergency exemption was granted would be fully realistic and effective.

The Defense Department believes that it is unacceptable as a matter of public policy for indispensable readiness activities to require repeated invocation of emergency authority—particularly when narrow clarifications of the underlying regulatory statutes would enable both essential readiness activities and the protection of the environment to continue. Congress would never tolerate a situation in which another activity vital to the Nation, like the practice of medicine, was only permitted to go forward through the repeated use of emergency exemptions.

That having been said, I should make clear that the Department of Defense is in no way philosophically opposed to the use of national security waivers or exemptions where necessary. We believe that every environmental statute should have a well-crafted exemption, as an insurance policy, though we continue to hope that we will seldom be required to have recourse to them. [. . .]

Specific Proposals

This year's proposals do include some clarifications and modifications based on events since last year. Of the five, the Endangered Species Act (ESA) and Clean Air Act provisions are unchanged. Let me address the changed provisions first.

RCRA and CERCLA

The legislation would codify and confirm the longstanding regulatory policy of EPA and every state concerning regulation of munitions use on operational ranges under RCRA and CERCLA. It would confirm that military munitions are subject to EPA's 1997 Military Munitions Rule while on range, and that cleanup of operational ranges is not required so long as material stays on the range. If such material moves off range, it still must be addressed promptly under existing environmental laws. Moreover, if munitions constituents cause an imminent and substantial endangerment on range, EPA will retain its current authority to address it on range under CERCLA section 106. (Our legislation explicitly reaffirms EPA's section 106 authority.) The legislation similarly does not modify the overlapping protections of the Safe Drinking Water Act, NEPA, and the ESA against environmentally harmful activities at operational military bases. The legislation has no effect whatsoever on DoD's cleanup obligations under RCRA or CERCLA at Formerly Used Defense Sites, closed ranges, ranges that close in the future, or waste management practices involving munitions even on operational ranges (such as so-called OB/OD activities).

The core of our concern is to protect against litigation the longstanding, uniform regulatory policy that (1) use of munitions for testing and training on an operational range is not a waste management activity or the trigger for cleanup requirements, and (2) that the appropriate trigger for DoD to address the environmental consequences of such routine test and training uses involving discharge of munitions is (a) when the range closes, (b) when munitions or their elements migrate or threaten to migrate off-range, or (c) when munitions or their elements create an imminent and substantial endangerment on-range. [. . .]

This legislation is needed because of RCRA's broad definition of "solid waste," and because states possess broad authority to adopt more stringent RCRA regulations than EPA (enforceable both by the states and by environmental plaintiffs). EPA therefore has quite limited ability to afford DoD regulatory relief under RCRA. Similarly, the broad statutory definition of "release" under CERCLA may also limit EPA's ability to afford DoD regulatory relief. And the President's site-specific, annually renewable waiver (under a paramount national interest standard in RCRA and a national security standard in CERCLA) is inapt for the reasons discussed above. [. . .]

Marine Mammal Protection Act

Although I realize this Committee is not centrally concerned with the Marine Mammal Protection Act (MMPA), I would like to take a moment to discuss it for purposes of completeness. This year's MMPA proposal includes some new provisions. This year's proposal, like last year's, would amend the term "harassment" in the MMPA, which currently focuses on the mere "potential" to injure or disturb marine mammals.

Our initiative adopts verbatim a reform proposal developed during the prior Administration by the Commerce, Interior, and Defense Departments and applies it to military readiness activities. That proposal espoused a recommendation by the National Research Council (NRC) that the currently overbroad definition of "harassment" of marine mammals—which includes "annoyance" or "potential to disturb"—be focused on biologically significant effects. As recently as 1999, the National Marine Fisheries Service (NMFS) asserted that under the sweeping language of the existing statutory definition harassment "is presumed to occur when marine mammals react to the generated sounds or visual cues"—in other words, whenever a marine mammal notices and reacts to an activity, no matter how transient or benign the reaction. As the NRC study found, "If [this] interpretation of the law for level B harassment (detectable changes in behavior) were applied to shipping as strenuously as it is applied to scientific and naval activities, the result would be crippling regulation of nearly every motorized vessel operating in U.S. waters."

Under the prior Administration, NMFS subsequently began applying the NRC's more scientific, effects-based definition. But environmental groups have challenged this regulatory construction as inconsistent with the statute. As you may know, the Navy and the National Oceanic and Atmospheric Administration suffered an important setback last year involving a vital anti-submarine warfare sensor—SURTASS LFA, a towed array emitting low-frequency sonar that is critical in detecting ultra-quiet diesel-electric submarines while they are still at a safe distance from our vessels. In the SURTASS LFA litigation environmental groups successfully challenged the new policy as inconsistent with the sweeping statutory standard, putting at risk NMFS' regulatory policy, clearly substantiating the need to clarify the existing statutory definition of harassment that we identified in our legislative package last year. [. . .]

The last change we are proposing, a national security exemption process, also derives from feedback the Defense Department received from environmental advocates last year after we submitted our proposal, as I discussed above. Although DoD continues to believe that predicating essential military training, testing, and operations on repeated invocations of emergency authority is unacceptable as a matter of public policy, we do believe that every environmental statute should have such authority as an insurance policy. The comments we received last year highlighted the fact that the MMPA does not currently contain such emergency authority, so this year's submission does include a waiver mechanism. Like the Endangered Species Act, our proposal would allow the Secretary of Defense, after conferring with the Secretaries of Commerce or Interior, as appropriate, to waive MMPA provisions for actions or categories of actions when required by national security. This provision is not a substitute for the other clarifications we have proposed to the MMPA, but rather a failsafe mechanism in the event of emergency.

The only substantive changes are those described above. The reason that the text is so much more extensive than last year's version is that last year's version was drafted as a freestanding part of title 10—the Defense Department title—rather than an amendment to the text of the MMPA itself. This year, because we were making several changes, we concluded that as a drafting matter we should include our changes in the MMPA itself. That necessitated a lot more language, largely just reciting existing MMPA language that we are not otherwise modifying.

The environmental impacts of our proposed reforms would be minimal. Although our initiative would exclude transient, biologically insignificant effects from regulation, the MMPA would remain in full effect for biologically significant effects—not only death or injury but also disruption of significant activities. The Defense Department could neither harm marine mammals nor disrupt their biologically significant activities without obtaining authorization from FWS or NMFS, as appropriate.

Nor does our initiative depart from the precautionary premise of the MMPA. The Precautionary Principle holds that regulators should proceed conservatively in the face of scientific uncertainty over environmental effects. But our initiative embodies a conservative, science-based approach validated by the National Research Council. By defining as "harassment" any readiness activities that "injure or have the significant potential to injure," or "disturb or are likely to disturb," our initiative includes a margin of safety fully consistent with the Precautionary Principle. The alternative is the existing grossly overbroad, unscientific definition of harassment, which sweeps in any activity having the "potential to disturb." As the National Research Council found, such sweeping overbreadth is unscientific and not mandated by the Precautionary Principle. [. . .]

The Defense Department already exercises extraordinary care in its maritime programs: all DoD activities worldwide result in fewer than 10 deaths or injuries annually (as opposed to 4800 deaths annually from commercial fishing activities). And DoD currently funds much of the most significant research on marine mammals, and will continue this research in future.

Although the environmental effects of our MMPA reforms will be negligible, their readiness implications are profound. Application of the current hair-trigger definition of "harassment" has profoundly affected both vital R&D efforts and training. Navy operations are expeditionary in nature, which means world events often require planning exercises on short notice. To date, the Navy has been able to avoid the delay and burden of applying for a take permit only by curtailing and/or dumbing down training and research/testing. For six years, the Navy has been working on research to develop a suite of new sensors and tactics (the Littoral Advanced Warfare Development Program, or LWAD) to reduce the threat to the fleet posed by ultraquiet diesel submarines operating in the littorals and shallow seas like the Persian Gulf, the Straits of Hormuz, the South China Sea, and the Taiwan Strait. These submarines are widely distributed in the world's navies, including "Axis of Evil" countries such as Iran and North Korea and potentially hostile great powers. In the 6 years that the program has operated, over 75% of the tests have been impacted by environmental considerations. In the last 3 years, 9 of 10 tests have been affected. One was cancelled entirely, and 17 different projects have been scaled back.

Endangered Species Act

Our Endangered Species Act provision is unchanged from last year. The legislation would confirm the prior Administration's decision that an Integrated Natural Resources Management Plan (INRMP) may in appropriate circumstances obviate the need to designate critical habitat on military installations. These plans for conserving natural resources on military property, required by the Sikes Act, are developed in cooperation with state wildlife agencies, the U.S. Fish and Wildlife Service, and the public. In most cases they offer comparable or better protection for the species because they consider the base's environment holistically, rather than using a species-by-species analysis. The prior Administration's decision that INRMPs may adequately provide for appropriate endangered species habitat management is being challenged in court by environmental groups, who cite Ninth Circuit caselaw suggesting that other habitat management programs provided an insufficient basis for the Fish and Wildlife Service to avoid designating Critical Habitat. These groups claim that no INRMP, no matter how protective, can ever substitute for critical habitat designation. This legislation would confirm and insulate the Fish and Wildlife Service's policy from such challenges.

Both the prior and current Administrations have affirmed the use of INRMPs as a basis for possible exclusion from critical habitat. Such plans are required to provide for fish and wildlife management, land management, forest management, and fish and wildlife-oriented recreation; fish and wildlife habitat enhancement; wetland protection, enhancement, and restoration; establishment of specific natural resource management goals, objectives, and timeframes; and enforcement of natural resource laws and regulations. And unlike the process for designation of critical habitat, INRMPs assure a role for state regulators. Furthermore, INRMPs must be reviewed by the parties on a regular basis, but not less than every five years, providing a continuing opportunity for FWS input.

By contrast, in 1999, the Fish and Wildlife Service stated in a Notice of Proposed Rulemaking that "we have long believed that, in most circumstances, the designation of 'official' critical habitat is of little additional value for most listed species, yet it consumes large amounts of conservation resources. [W]e have long believed that separate protection of critical habitat is duplicative for most species."

Our provision does not automatically eliminate critical habitat designation, precisely because under the Sikes Act, the statute giving rise to INRMPs, the Fish & Wildlife Service is given approval authority over those elements of the INRMP under its jurisdiction. This authority guarantees the Fish & Wildlife Service the authority to make a case-by-case determination concerning the adequacy of our INRMPs as a substitute for critical habitat designation. And if the Fish & Wildlife Service does not approve the INRMP, our provision will not apply to protect the base from critical habitat designation.

Our legislation explicitly requires that the Defense Department continue to consult with the Fish and Wildlife Service and the National Marine Fisheries Service under Section 7 of the Endangered Species Act (ESA); the other provisions of the ESA, as well as other environmental statutes such as the National Environmental Policy Act, would continue to apply, as well.

The Defense Department's proposal has vital implications for readiness. Absent this policy, courts, based on complaints filed by environmental litigants, compelled the Fish and Wildlife Service to re-evaluate "not prudent" findings for many critical habitat determinations, and as a result FWS proposed to designate over 50% of the 12,000-acre Marine Corps Air Station (MCAS) Miramar and over 56% of the 125,000-acre Marine Corps Base (MCB) Camp Pendleton. Prior to adoption of this policy, 72% of Fort Lewis and 40% of the Chocolate Mountains Aerial Gunnery Range were designated as critical habitat for various species, and analogous habitat restrictions were imposed on 33% of Fort Hood. These are vital installations.

Unlike Sikes Act INRMPs, critical habitat designation can impose rigid limitations on military use of bases, denying commanders the flexibility to manage their lands for the benefit of both readiness and endangered species.

Clean Air Act General Conformity Amendment

Our Clean Air Act amendment is unchanged since last year. The legislation would provide more flexibility for the Defense Department in ensuring that emissions from its military training and testing are consistent with State Implementation Plans under the Clean Air Act by allowing DoD and the states a slightly longer period to accommodate or offset emissions from military readiness activities.

The Clean Air Act's "general conformity" requirement, applicable only to federal agencies, has repeatedly threatened deployment of new weapons systems and base closure/realignment despite the fact that relatively minor levels of emissions were involved. [. . .]

Conclusion

In closing Mr. Chairman, let me emphasize that modern warfare is a "come as you are" affair. There is no time to get ready. We must be prepared to defend our country wherever and whenever necessary. While we want to train as we fight, in reality our soldiers, sailors, airmen and Marines fight as they train. The consequences for them, and therefore for all of us, could not be more momentous.

DoD is committed to sustaining U.S. test and training capabilities in a manner that fully satisfies that military readiness mission while also continuing to provide exemplary stewardship of the lands and natural resources in our trust. [. . .]

 **NO**

Impact of Military Training on the Environment

Prior to arriving at the National Wildlife Federation in 2001, I served for 13 years at the U.S. Fish and Wildlife Service, with the last 4 years as the Director of the agency. Prior to that, I served as Fish and Wildlife Administrator for the Department of the Army, Natural and Cultural Resources Program Manager for the National Guard Bureau, and Research Biologist for U.S. Army Medical Research Institute. I am the daughter of a U.S. Army Colonel, and lived on or near military bases throughout my entire childhood.

Based on this experience, I am very familiar with the Defense Department's long history of leadership in wildlife conservation. On many occasions during my tenures at FWS and the Defense Department, DOD rolled up its sleeves and worked with wildlife agency experts to find a way to comply with environmental laws and conserve imperiled wildlife while achieving military preparedness objectives.

The Administration now proposes in its Readiness and Range Preservation Initiative that Congress scale back DOD's responsibilities to conserve wildlife and to protect people from the hazardous pollution that DOD generates. This proposal is both unjustified and dangerous. It is unjustified because DOD's longstanding approach of working through compliance issues on an installation-by-installation basis works. As DOD itself has acknowledged, our armed forces are as prepared today as they ever have been in their history, and this has been achieved without broad exemptions from environmental laws.

The DOD proposal is dangerous because, if Congress were to broadly exempt DOD from its environmental protection responsibilities, both people and wildlife would be threatened with serious, irreversible and unnecessary harm. Moreover, other federal agencies and industry sectors with important missions, using the same logic as used here by DOD, would line up for their own exemptions from environmental laws.

My expertise is in the Endangered Species Act (ESA), so I would like to focus my testimony on why exempting the Defense Department from key provisions of the ESA would be a serious mistake. I will rely on my fellow witnesses to explain why the proposed exemptions from other environmental and public health and safety laws is similarly unwise.

From Testimony before the Senate Committee on Environment and Public Works, April 2, 2003.

Concerns with the ESA Exemption

The Defense Department's proposed ESA exemption suffers from three basic flaws: it would severely weaken this nation's efforts to conserve imperiled species and the ecosystems on which all of us depend; it is unnecessary for maintaining military readiness; and it ignores the Defense Department's own record of success in balancing readiness and conservation objectives under existing law.

1. Section 2017 Removes a Key Species Conservation Tool

Section 2017 of the Administration's Readiness and Range Preservation Initiative would preclude designations of critical habitat on any lands owned or controlled by DOD if DOD has prepared an Integrated Natural Resources Management Plan (INRMP) pursuant to the Sikes Act and has provided "special management consideration or protection" of listed species pursuant to Section 3(5)(A) of the ESA.

This proposal would effectively eliminate critical habitat designations on DOD lands, thereby removing an essential tool for protecting and recovering species listed under the ESA. Of the various ESA protections, the critical habitat provision is the only one that specifically calls for protection of habitat needed for recovery of listed species. It is a fundamental tenet of biology that habitat must be protected if we ever hope to achieve the recovery of imperiled fish, wildlife and plant species.

Section 2017 would replace this crucial habitat protection with management plans developed pursuant to the Sikes Act. The Sikes Act does not require the protection of listed species or their habitats; it simply directs DOD to prepare INRMPs that protect wildlife "to the extent appropriate." Moreover, the Sikes Act provides no guaranteed funding for INRMPs and the annual appropriations process is highly uncertain. Even the best-laid management plans can go awry when the anticipated funding fails to come through. Yet, under Section 2017, even poorly designed INRMPs that allow destruction of essential habitat and put fish, wildlife or plant species at serious risk of extinction would be substituted for critical habitat protections.

Section 2017 contains one minor limitation on the substitution of INRMPs for critical habitat designations: such a substitution is allowed only where the INRMP provides "special management consideration or protection" within the meaning of Section 3(5)(A) of the ESA. Unfortunately, this limitation does nothing to ensure that INRMPs truly conserve listed species.

The term "special management consideration or protection" was never intended to provide a biological threshold that land managers must achieve in order to satisfy the ESA. The term is found in Section 3(5) of the ESA, which sets forth a two-part definition of critical habitat. Section 3(5)(A) states that critical habitat includes areas occupied by a listed species that are "essential for the conservation of the species" and "which may require special management consideration or protection." Section 3(5)(B) states that critical habitat also includes areas not currently occupied by a listed species that are simply "essential for the conservation of the species."

As this language makes clear, an ESA §3(5) finding by the U.S. Fish and Wildlife Service or National Marine Fisheries Service (Services) that a parcel of land "may require special management consideration or protection" is not the same as finding that it is already receiving adequate protection. Such a finding simply highlights the importance of a parcel of land to a species, and it should lead to designation of that land as critical habitat. See *Center for Biological Diversity* v. *Norton,* 240 F. Supp. 2d 1090 (D. Ariz. 2003) (rejecting, as contrary to plain meaning of ESA, defendant's interpretation of "special management consideration or protection" as providing a basis for substituting a U.S. Forest Service management plan for critical habitat protection). By allowing DOD to substitute INRMPs for critical habitat designations whenever it unilaterally makes a finding of "special management consideration or protection," Section 2017 significantly weakens the ESA.

Section 2017 is also problematic because it would eliminate many of the ESA Section 7 consultations that have stimulated DOD to "look before it leaps" into a potentially harmful training exercise. As a result of Section 7 consultations, DOD and the Services have routinely developed what is known as "work-arounds," strategies for avoiding or minimizing harm to listed species and their habitats while still providing a rigorous training regimen.

Section 2017 purports to retain Section 7 consultations. However, the duty to consult only arises when a proposed federal action would potentially jeopardize a listed species or adversely modify or destroy its critical habitat. By removing critical habitat designations on lands owned or controlled by DOD, Section 2017 would eliminate one of the two possible justifications for initiating a consultation, reducing the likelihood that consultations will take place. This would mean that DOD and the Services would pay less attention to species concerns and would be less effective in conserving imperiled species and maintaining the sustainability of the land.

The reductions in species protection proposed by DOD would have major implications for our nation's rich natural heritage. DOD manages approximately 25 million acres of land on more than 425 major military installations. These lands are home to at least 300 federally listed species. Without the refuge provided by these bases, many of these species would slide rapidly toward extinction. These installations have played a crucial role in species conservation and must continue to do so.

2. The ESA Exemption Is Not Necessary to Maintain Military Readiness

The ESA already has the flexibility needed for the Defense Department to balance military readiness and species conservation objectives. Three key provisions provide this flexibility. First, under the consultation provision of Section 7(a)(2) of the Act, DOD is provided with the opportunity to develop solutions in tandem with the Services to avoid unnecessary harm to listed species from military activities. Typically, the Services conclude, after informal consultation, that the proposed action will not adversely affect a listed species or its designated critical habitat or, after formal consultation, that it will not likely jeopardize a listed species or destroy or adversely modify its critical habitat. See, e.g., U.S. Army Environmental Center, Installation Summaries from the FY 2001 Survey of Threatened and Endangered Species on Army Lands (August 2002) at 9 (noting successful conclusion of 282 informal consultations and 36 formal consultations, with no "jeopardy" biological opinions). In both informal and formal consultations, the Services either will recommend that the action go forward without changes, or it will work with DOD to design "work arounds" for avoiding and minimizing harm to the species and its habitat. In either case, DOD accomplishes its readiness objectives while achieving ESA compliance.

Second, under Section 4(b)(2) of the ESA, the Services are authorized to exclude any area from critical habitat designation if they determine that the benefits of exclusion outweigh the benefits of specifying the area. (An exception is made for when the Services find that failure to designate an area as critical habitat will result in the extinction of a species—a finding that the Services have never made.) In making this decision, the Services must consider "the economic impact, and any other relevant impact" of the critical habitat designation. DOD has recently availed itself of this provision to convince the U.S. Fish and Wildlife Service to exclude virtually all of the habitat at Camp Pendleton—habitat deemed critical to five listed species in proposed rulemakings—from final critical habitat designations. Thus, for situations where the Section 7(a)(2) consultation procedures place undue burdens on readiness activities, DOD already has a tool for working with the Services on excluding land from critical habitat designation. Attached to my testimony is a factsheet that shows how the Services have worked cooperatively with DOD on these exclusions, and another factsheet showing the importance of maintaining the Services' role in evaluating proposed exclusions.

Third, under Section 7(j) of the ESA an exemption "shall" be granted for an activity if the Secretary of Defense finds the exemption is necessary for reasons of national security. To this date, DOD has never sought an exemption under Section 7(j), highlighting the fact that other provisions of the ESA have provided DOD with all the flexibility it needs to reconcile training needs with species conservation objectives.

Where there are site-specific conflicts between training needs and species conservation needs, the ESA provides these three mechanisms for resolving them in a manner that allows DOD to achieve its readiness objectives. Granting DOD a nationwide ESA exemption, which would apply in many places where no irreconcilable conflicts between training needs and conservation needs have arisen, would be harmful to imperiled species and totally unnecessary to achieve readiness objectives.

a. DOD Has Misstated the Law Regarding Its Ability to Continue with a Cooperative, Case-by-Case Approach to Critical Habitat Designations

DOD has stated that the ESA exemption is necessary because a recent court ruling in Arizona would prevent DOD from taking the cooperative, case-by-case approach to critical habitat designations that was developed when I served as

Director of the Fish and Wildlife Service. This description of the court ruling is inaccurate—the ruling clearly allows DOD to continue the cooperative, case-by-case approach if it wishes.

The court ruling at issue is entitled *Center for Biological Diversity* v. *Norton,* 240 F. Supp. 2d 1090 (D. Ariz. 2003). In this case, FWS excluded San Carlos Apache tribal lands from a critical habitat designation pursuant to ESA §4(b)(2) because the tribal land management plan was adequate and the benefits of exclusion outweighed the benefits of inclusion. The federal district court upheld the exclusion as within FWS's broad authority under ESA §4(b)(2). At the same time, the court held that lands could not legitimately be excluded from a critical habitat designation on the basis of the "special management" language in ESA §3(5).

Under the court's reasoning, FWS continues to have the broad flexibility to exclude DOD lands from a critical habitat designation on the basis of a satisfactory INRMP and the benefits to military training that the exclusion would provide. The ruling simply clarifies that such exclusions must be carried out pursuant to ESA §4(b)(2) rather than ESA §3(5). Thus, DOD's assertion that the Center for Biological Diversity ruling prevents it from working with FWS to secure exclusions of DOD lands from critical habitat designations is inaccurate.

b. DOD's Anecdotes Do Not Demonstrate That the ESA Has Reduced Readiness

The DOD has offered a series of misleading anecdotes describing difficulties it has encountered in balancing military readiness and conservation objectives. Before Congress moves forward with any exemption legislation, the appropriate Congressional committees should get a more complete picture of what is really happening at DOD installations.

Some of DOD's anecdotes are simply unpersuasive on their face, such as DOD's repeated assertion that environmental laws have prevented the armed services from learning how to dig foxholes and that troops abroad have been put at greater risk as a result. There is simply no evidence that environmental laws have ever prevented foxhole digging. Moreover, given its vast and varied landholdings and the many management options available, the Defense Department certainly can find places on which troops can learn to dig foxholes without encountering endangered species or other environmental issues.

Other anecdotes have simply disregarded the truth. For example, DOD and its allies have repeatedly argued that more than 50 percent of Camp Pendleton may not be available for training due to critical habitat designations. In fact, only five species have been proposed for critical habitat designations at Camp Pendleton. In each of these five instances, DOD raised concerns about impacts to military readiness, and in each instance, FWS worked closely with DOD to craft a solution. FWS ultimately excluded virtually all of the habitats for the five listed species on Camp Pendleton from critical habitat designations—even though FWS had earlier found that these habitats were essential to the conservation of the species. As a result of FWS's exclusion decisions, less than one percent of the training land at Camp Pendleton, and less than 4 percent of all of Camp Pendleton, is designated critical habitat. (Most of the critical habitat designated at Camp Pendleton is non-training land leased to San Onofre State Park, agricultural operations, and others. DOD's repeated suggestion that more than 50 percent of Camp Pendleton is at risk of being rendered off-limits to training due to critical habitat is simply inaccurate.

DOD also has argued that training opportunities and expansion plans at Fort Irwin have been thwarted by the desert tortoise. Yet just two weeks ago this official line was contradicted by the reality on the ground. In an article dated March 21, 2003, Fort Irwin spokesman Army Maj. Michael Lawhorn told the Barstow Desert Dispatch that he is unaware of any environmental regulations that interfere with troops' ability to train there. He also said there isn't any environmental law that hinders the expansion. [. . .]

These examples of misleading anecdotes highlight the need for Congress to look behind the reasons that are being put forward by DOD as the basis for weakening environmental laws. DOD uses the anecdotes in an attempt to demonstrate that conflicts between military readiness and species conservation objectives are irreconcilable. However, solutions to these conflicts are within reach if DOD is willing to invest sufficient time and energy into finding them. DOD has vast acres of land on which to train and vast stores of creativity and expertise among its land managers. With careful inventorying and planning, DOD can find a proper balance.

Has DOD made the necessary effort to inventory and plan for its training needs? In June 2002, the General Accounting Office issued a report entitled "Military Training: DOD Lacks a Comprehensive Plan to Manage Encroachment on Training Ranges," suggesting that the answer is no. The GAO found:

- DOD has not fully defined its training range requirements and lacks information on training resources available to the Services to meet those requirements, and that problems at individual installations may therefore be overstated.
- The Armed Services have never assessed the overall impacts of encroachment on training.
- DOD's readiness reports show high levels of training readiness for most units. In those few instances of when units reported lower training readiness, DOD officials rarely cited lack of adequate training ranges, areas or airspace as the cause.
- DOD officials themselves admit that population growth around military installations is responsible for past and present encroachment problems.
- The Armed Services' own readiness data do not show that environmental laws have significantly affected training readiness.

Ten months after the issuance of the GAO report, DOD still has not produced evidence that environmental laws are at fault for any of the minor gaps in readiness that may exist. EPA Administrator Whitman confirmed this much at a recent hearing. At a February 26, 2003, Senate Environment and Public Works Committee hearing on EPA's budget, EPA Administrator Whitman stated that she was "not aware of any particular area where environmental protection regulations are preventing the desired training."

To this date, DOD has not provided Congress with the most basic facts about the impacts of ESA critical habitat requirements on its readiness activities. Out of DOD's 25 million acres of training land, how many acres are designated critical habitat? At which installations? Which species? In what ways have the critical habitat designations limited readiness activities? What efforts did DOD make to alert FWS to these problems and to negotiate resolutions? Without answers to these most basic questions, Congress cannot fairly conclude that the ESA is at fault for any readiness gaps or that a sweeping ESA exemption is warranted.

3. DOD has Worked Successfully with the Services to Balance Readiness and Species Conservation Objectives

The third reason why enacting DOD's proposed ESA changes would be a mistake is because the current approach—developing solutions at the local level, rather than relying on broad, national exemptions—has worked. My experience at both FWS and DOD has shown me that solutions developed at the local level are sometimes difficult to arrive at, but they are almost always more intelligent and long-lasting than one-size-fits-all solutions developed at the national level.

Allow me to provide a few brief examples. At the Marine Corps Base at Camp Lejeune in North Carolina, every colony tree of the endangered red-cockaded woodpecker is marked on a map, and Marines are trained to operate their vehicles as if those mapped locations are land mines. Here is the lesson that Major General David M. Mize, the Commanding General at Camp Lejeune, has drawn from this experience:

> "Returning to the old myth that military training and conservation are mutually exclusive; this notion has been repeatedly and demonstrably debunked. In the overwhelming majority of cases, with a good plan along with common sense and flexibility, military training and the conservation and recovery of endangered species can very successfully coexist."
>
> "Military installations in the southeast are contributing to red-cockaded woodpecker recovery while sustaining our primary mission of national military readiness."
>
> "I can say with confidence that the efforts of our natural resource managers and the training community have produced an environment in which endangered species management and military training are no longer considered mutually exclusive, but are compatible."

These sentiments, which I share, were relayed by Major General Mize just eight weeks ago at a National Defense University symposium sponsored by the U.S. Army Forces Command (FORSCOM) and others. At that symposium, representatives of Camp Lejeune Marine Corps Base, Eglin Air Force Base, Fort Bragg Army Base, Fort Stewart Army Base, Camp Blanding Training Center in Florida, the U.S. Army Environmental Center, and other Defense facilities—some of the most heavily utilized training bases in the country—heralded the success that Defense Department installations have had in furthering endangered species conservation while maintaining military readiness.

On the Mokapu Peninsula of Marine Corps Base Hawaii, the growth of non-native plants, which can decrease the reproductive success of endangered waterbirds, is controlled through annual "mud-ops" maneuvers by Marine Corps Assault Vehicles. Just before the onset of nesting season, these 26 ton vehicles are deployed in plow-like maneuvers that break the thick mats of invasive plants, improving nesting and feeding opportunities while also giving drivers valuable practice in unusual terrain. [. . .]

These success stories highlight a major trend that I believe has been missed by those promoting the DOD exemptions. In recent years, DOD has increasingly recognized the importance of sustainability because it meets several importance objectives at once. Sustainable use of the land helps DOD achieve not only compliance with environmental laws, but also long-term military readiness and cost-effectiveness goals. For example, by operating tanks so that they avoid the threatened desert tortoise, DOD prevents erosion, a problem that is extremely difficult and costly to remedy. If DOD abandons its commitment to environmental compliance, it will incur greater long-term costs for environmental remediation and will sacrifice land health and military readiness.

A November 2002 policy guidance issued by the then-Secretary of the Navy to the Chief of Naval Operations and the Commandant of the Marine Corps suggests that certain members of DOD's leadership are indeed willing to abandon the sustainability goal. The policy guidance on its face seems fairly innocuous—it purports to centralize at the Pentagon all decision making on proposed critical habitat designations and other ESA actions. However, the Navy Secretary's cover memo makes clear that its purpose is also to discourage any negotiation of solutions to species conservation challenges by Marines or Navy personnel in the field, lest these locally-developed "win-win" solutions undercut DOD's arguments on Capitol Hill that the ESA is broken. According to paragraph 2 of the cover memo, "concessions [. . .] could run counter to the legislative relief that we are continuing to pursue with Congress."

Similar sentiments were voiced by Deputy Defense Secretary Paul Wolfowitz in his March 7, 2003, memo to the chiefs of the Army, Navy and Air Force. Deputy Secretary Wolfowitz argued that "it is time for us to give greater consideration to requesting exemptions" from environmental laws and pleaded for specific examples of instances in which environmental regulations hamper training. The implicit message is that efforts at the installation level to resolve conflicts between conservation and training objectives should be suspended, and that such conflicts instead should be reported to the Pentagon, where environmental protections will simply be overridden.

These messages to military personnel in the field mark a very unfortunate abdication of DOD's leadership in wildlife conservation. To maintain its leadership role as steward of this nation's endangered wildlife, DOD must encourage its personnel to continue developing innovative solutions and not thwart those efforts.

Conclusion

With the Iraq war ongoing and terrorism threats always present, no one can dismiss the importance of military readiness. However, there is no justification for the Defense Department to retreat from its environmental stewardship commitments at home. As base commanders have been telling us, protecting endangered species and other important natural resources is compatible with maintaining military readiness.

Surveys show that the American people today want environmental protection from the federal government, including the Defense Department, as much as ever. According to an April 2002 Zogby Poll, 85% of registered voters believe that the Defense Department should be required to follow America's environmental and public health laws and not be exempt. Americans believe that no one, including the Defense Department, should be above the law.

Congress should reject the proposed environmental exemptions in the Administration's defense authorization package. This proposal, along with the parallel proposal in the Administration's FY04 budget request that Congress cut spending on DOD's environmental programs by $400 million, are a step in the wrong direction.

DOD has a long and impressive record of balancing readiness activities with wildlife conservation. The high quality of wildlife habitats at many DOD installations provides tangible evidence of DOD's positive contribution to the nation's conservation goals. At a time when environmental challenges are growing, DOD should be challenged to move forward with this successful model and not to sacrifice any of the progress that has been made. [. . .]

POSTSCRIPT

Should the Military Be Exempt from Environmental Regulations?

After the April 2003 Senate committee hearing, this issue received some press attention as environmentalists and congressional Democrats prepared to oppose the Bush Administration. See "War on the Environment," *The Ecologist* (May 2003), and John Stanton, "Activists, Democrats Brace for Defense Environment Showdown," *CongressDaily AM* (May 13, 2003). The Senate voted for the exemptions, and early in 2004, the issue came before the House of Representatives. Dan Miller, First Assistant Attorney General, Natural Resources and Environmental Section, Colorado Department of Law, testified against them before the House Committee on Energy and Commerce Subcommittee on Energy and Air Quality, saying that "Even read in the narrowest possible fashion, the [proposed exemptions] would hamstring state and EPA cleanup authorities at over 24 million acres of 'operational ranges,' an area the size of Maryland, Massachusetts, New Jersey, Hawaii, Connecticut and Rhode Island combined. As a practical matter, environmental regulators would likely be precluded from using RCRA, CERCLA, and related state authorities to require any investigation or cleanup of groundwater contamination on these ranges, even if the contamination had migrated off-range, polluted drinking or irrigation water supplies, and even if it posed an imminent and substantial endangerment to human health. And it is likely that DOD's amendments would be construed more broadly to exempt even more contamination from state and EPA oversight. . . . If we have learned anything in the past thirty years of environmental regulation, it is that relying on federal agencies to 'voluntarily' address environmental contamination is often fruitless. One need look no further than the approximately 130 DOD facilities on the Superfund National Priorities List, or DOD's poor record of compliance with state and federal environmental laws to see that independent, legally enforceable state oversight of federal agencies is required to achieve effective results." In May, the House Committees on Armed Services Readiness and Energy and Commerce announced that they were not about to consider the exemption. See "Hefley: No Plans to Exempt Military from Enviro Laws," *CongressDaily AM* (May 6, 2004).

But the issue is not about to go away. The Government Accounting Office prepared a background paper, "Military Training: DOD Approach to Managing Encroachment on Training Ranges Still Evolving," GAO-03-621T (April 2, 2003), delivered by Barry W. Holman, Director, Defense Infrastructure Issues, as testimony before the Senate Committee on Environment and Public Works. It discussed eight encroachment issues, including urban growth around military bases, air and noise pollution, unexploded ordnance and other munitions, endangered species habitat, and protected marine resources. Since urban growth is not likely to cease and the number of endangered species and protected marine resources is sure to increase, encroachment is not about to diminish. The Department of Defense, says the GAO, must better document the impact of the encroachment on training and costs; it has not yet produced required reports to Congress. So far, "workarounds" have been enough to deal with the problem, but that may not remain sufficient. E. G. Willard, Tom Zimmerman, and Eric Bee, "Environmental Law and National Security: Can Existing Exemptions in Environmental Laws Preserve DOD Training and Operational Prerogatives without New Legislation?" *Air Force Law Review* (2004), conclude that existing exemptions are not enough to support military readiness and say that more are needed. "The bottom line is that we must be able to train the way we fight, and we must be able to operate to defend the country and its interests." Paul D. Thacker, "Are Environmental Exemptions for the U.S. Military Justified?" *Environmental Science & Technology* (October 15, 2004), noted that "many critics of the administration say that the campaign is more about undermining environmental laws than protecting military readiness."

Recent incidents involving beached whales, apparently due to use of sonar by Navy ships, have returned the issue to the public eye. E. C. M. Parsons, et al., "Navy Sonar and Cetaceans: Just How Much Does the Gun Need to Smoke Before We Act?" *Marine Pollution Bulletin* (June 2008), note that such strandings are not uncommon and that they often occur near where sonar is being used. Parsons, et al., also charge "senior government officials" with obstructing measures to prevent harm to the whales by limiting sonar use. According to Joan Biskupic, "Justices to Debate Whether Navy Sonar Harms Whales," *USA Today* (June 24, 2008), after the Natural Resources Defense Council sued, the courts restricted the use of sonar. When the Bush Administration appealed, saying that the restrictions jeopardize "the Navy's ability to train sailors and Marines for wartime deployment [which is] essential to the national security," the Supreme Court agreed to hear the case. The Supreme Court removed the restrictions in November 2008; Joan Biskupic, "High Court OKs Sonar Training off Calif. Coast," *USA Today* (November 13, 2008). See also "Can the U.S. Navy Proceed with Its Plan for Sonar Training Exercises off the Coast of Southern California?" *Supreme Court Debates* (November 2008).

ISSUE 6

Will Restricting Carbon Emissions Damage the U.S. Economy?

YES: Paul Cicio, from "Competitiveness and Climate Policy: Avoiding Leakage of Jobs and Emissions," testimony before the House Committee on Energy and Commerce Subcommittee on Energy and Environment (March 18, 2009).

NO: Eileen Claussen, from "Competitiveness and Climate Policy: Avoiding Leakage of Jobs and Emissions," testimony before the House Committee on Energy and Commerce Subcommittee on Energy and Environment (March 18, 2009).

ISSUE SUMMARY

YES: Paul Cicio argues that lacking global agreements, capping greenhouse gas emissions of the industrial sector will make domestic production less competitive in the global market, drive investment and jobs offshore, increase exports, and damage the economy. The real greenhouse gas problem lies with other sectors of the economy, and that is where attention should be focused.

NO: Eileen Claussen argues that environmental regulations have little impact on trade patterns and that though controlling greenhouse gas emissions will affect industrial production, most of the impact will come from a decline in consumption.

$\mathbf{F}$ollowing World War II the United States and other developed nations experienced an explosive period of industrialization accompanied by an enormous increase in the use of fossil fuel energy sources and a rapid growth in the manufacture and use of new synthetic chemicals. In response to growing public concern about the pollution and other forms of environmental deterioration resulting from this largely unregulated activity, the U.S. Congress passed the National Environmental Policy Act of 1969. This legislation included a commitment on the part of the government to take an active and aggressive role in protecting the environment. The next year the Environmental Protection Agency (EPA) was established to coordinate and oversee this effort. During the next two decades an unprecedented series of legislative acts and administrative rules were promulgated, placing numerous restrictions on industrial and commercial activities that might result in the pollution, degradation, or contamination of land, air, water, food, and the workplace.

Such forms of regulatory control have always been opposed by the affected industrial corporations and developers as well as by advocates of a free-market policy. More moderate critics of the government's regulatory program recognize that adequate environmental protection will not result from completely voluntary policies. They suggest that a new set of strategies is needed. Arguing that "top down, federal, command and control legislation" is not an appropriate or effective means of preventing ecological degradation, they propose a wide range of alternative tactics, many of which are designed to operate through the economic marketplace. The first significant congressional response to these proposals was the incorporation of tradable pollution emission rights into the 1990 Clean Air Act amendments as a means for achieving the set goals for reducing acid rain-causing sulfur dioxide emissions. More recently, the 1997 international negotiations on controlling global warming in Kyoto, Japan, resulted in a protocol that includes emissions trading as one of the key elements in the plan to limit the atmospheric buildup of greenhouse gases.

Charles W. Schmidt, "The Market for Pollution," *Environmental Health Perspectives* (August 2001), argues that emissions trading schemes represent "the most significant developments" in the use of economic incentives to motivate corporations to reduce pollution. . . . Cap-and-trade systems work by setting a limit (the cap) on how much of a pollutant can be emitted per year. Permits to emit a portion of the pollutant are then made available—either free or as the result of an auction process—to businesses that emit that pollutant. Businesses that do not emit as much as their permits allow can then sell their unused permits. This provides an incentive to reduce emissions by spending money to improve efficiency. But a cap-and-trade system does require spending money for initial permits. This means added expenses, with an impact on profitability. Many businesses are concerned that this will affect their competitive position and even their ability to stay in business. A crucial question thus becomes whether environmental or economic protection should come first.

In "A Low-Cost Way to Control Climate Change," *Issues in Science and Technology* (Spring 1998), Byron Swift argues that the "cap-and-trade" feature of the U.S. Acid Rain Program has been so successful that a similar system for implementing the Kyoto Protocol's emissions trading mandate as a cost-effective means of controlling greenhouse gases should work.

In the following selections, Paul Cicio argues that lacking global agreements, capping greenhouse gas emissions of the industrial sector will make domestic production less competitive in the global market, drive investment and jobs offshore, increase exports, and damage the economy. The real greenhouse gas problem lies with other sectors of the economy, and that is where attention should be focused. Eileen Claussen agrees that global agreements are essential for effective control of greenhouse gas emissions but argues that environmental regulations have little impact on trade patterns and that though controlling greenhouse gas emissions will affect industrial production, most of the impact will come from a decline in consumption. Industrial impacts will be modest and can easily be managed with a number of policy measures.

YES

Paul Cicio

Competitiveness and Climate Policy: Avoiding Leakage of Jobs and Emissions

Key Points

Capping the greenhouse gas (GHG) emissions of the industrial sector will drive investment and jobs offshore and increase imports. It will not bring major developing countries to the table but they will benefit through increased exports to the U.S. Even the third phase of the EU Emissions Trading Scheme (ETS) contains a provision to ensure their trade exposed industries receive compensation in order to prevent job loss and emissions leakage. Regulating the U.S. industrial sector "before" negotiating an international agreement undermines our ability to achieve a fair and effective GHG reduction agreement for U.S. industry.

For the industrial sector, climate policy is also trade, energy, economic and employment policy. They are all intrinsically linked and inseparable. It is for this reason that regulating GHG emissions for the industrial sector be negotiated with both developed and developing countries in the context of a fair trade and productivity.

The U.S. industrial sector is not the problem. In the U.S., the industrial sector's GHG emissions have risen only 2.6% above 1990 levels while emissions from the residential sector are up 29%, commercial up 39%, transportation up 27% and electricity generation up 29%.

The industrial sector competes globally and requires a global GHG policy solution that is based on productivity, something that the developing countries industrial sector can potentially agree to. A GHG cap is an unacceptable policy alternative for them and for us.

The U.S. cannot grow the economy without using more volume of our products. The only question is whether the product will be supplied from domestic sources or imports. In fact a cap limits economic efficiency because it even limits the ability to maximize production from existing facilities that are not running at installed capacity. Since 2000, U.S. manufacturing has been losing ground. From 2000 to 2008, imports are up 29% and manufacturing unemployment fell 22%, losing 3.8 million jobs, a direct statistical correlation.

U.S. Senate, March 18, 2009.

The use of energy by the industrial sector is value-added. Our products enable GHG emission reductions. Lifecycle studies show that they save much more energy and GHG emissions than what is used/emitted in their production. Raising energy costs raises the cost of these valuable products.

The industrial sector already has a price signal for GHG emissions, it is called global competition and because we are energy intensive, we either drive down our energy costs or go out of business.

Under cap and trade, the industrial sector pays twice. Through the additional cost of carbon embedded in energy purchases and through the higher cost of natural gas and electricity. Higher demand for natural gas will result in higher prices for all consumers. Since natural gas power generation sets the marginal price of electricity, higher natural gas prices will mean higher electricity prices for all consumers.

A cap will damage the ability of the U.S. industrial sector to take back market share from imports and increase exports.

Cap and trade does not address our country's fundamental need to significantly increase the availability, affordability and reliability of low carbon sources of supply.

Carbon trading and market manipulation is of great concern. The U.S. government has proven unable to prevent market manipulation for mature energy and food commodities and credit default swaps—carbon markets will be much harder to regulate.

If the U.S. proceeds to cap GHGs, it must provide to industry free allowances equal to the resulting increased direct and indirect costs due to GHG regulation until major competing developing countries have similar cost increases.

Congressional Justification for Not Capping GHG Emissions of the Industrial Sector

Congress has a choice to make and it is a decision it cannot afford to make incorrectly. It must decide whether to maintain and possibly increase U.S. manufacturing jobs by not capping GHG emissions on the industrial sector— or create jobs in foreign countries by importing manufacturing products to supply the needs of our economy.

The Industrial Energy Consumers of America is an association of leading manufacturing companies with $510 billion in annual sales and with more than 850,000 employees nationwide. It is an organization created to promote the interests of manufacturing companies. IECA membership represents a diverse set of industries including: plastics, cement, paper, food processing, brick, chemicals, fertilizer, insulation, steel, glass, industrial gases, pharmaceuticals, aluminum and brewing.

The decision should not be hard because there is very sound economic and environmental justification for Congress to not act in the short term to cap GHG reductions on the industrial sector but to forge a different policy path that will provide sustained GHG reductions globally by harnessing real market forces called competition.

The industrial sector needs a globally level playing field that lets the best companies win. Adding costs by unilateral action helps "all" of our competitors in other countries take our business and our jobs. **We need U.S. leadership to forge a global effort to address industrial sector GHG emission reductions that is focused on "fair trade" and "productivity." This is the only way to potentially bring developing nations to the table.**

Productivity is a language that all manufacturers understand and fundamental to competition. We believe that all governments want increased productivity by their industrial sector. We urge you to take action in this more realistic direction.

The world in which the industrial manufacturing company operates is diverse and business is often won or lost on the difference between pennies per unit of product. Competitiveness is everything. Some segments of industry, such as the power producers, may support cap and trade, but that's because they don't compete globally and they simply pass through their increased costs, we don't have that luxury.

Unlike that vision that many Americans have of China building coal-fired power plants using antiquated technology, it is vitally important that the Congress understand that a great number of companies that we compete with from developing countries are top-in-class competitors. They are utilizing the latest, world class technology. Some of these facilities are state owned or supported. Many also have subsidized energy costs. Energy costs most often determine our competitiveness and it can be our largest non-controllable cost.

The congress can act in the public interest to consider both the cost and benefits of not imposing the cap on the industrial sector. The benefits of not imposing a GHG cap include good paying jobs, exports that reduce our balance of payments and the domestic production of products that are solutions to our climate challenges.

So far, only the environmental costs have been debated. We caution you to consider that your policy decision can lead to a further acceleration of the loss of the industrial sector. Just look at the facts. Due to the loss of competitiveness since 2000, the manufacturing sector has lost 3.8 million jobs thru 2008. During this same time period, imports rose 29%, a direct statistical correlation.

President Obama rightfully points to the disappearing middle class as troubling. We agree. The U.S. began to lose the middle class when the industrial sector began to lose competitiveness along with our high paying jobs that most often pay benefits. The timing is consistent. We encourage the president and Congress to work with us to put new industrial policies in place that will increase competitiveness and grow the industrial sector and greatly restore the middle class.

To their credit, Representative Inslee and Doyle have rightfully recognized the need to protect manufacturing competitiveness. They are well intentioned but their solution is not really a solution for an industry that competes globally. We will still be burdened with costs and uncertainty. Most importantly, it does not do anything to bring the industrial sectors of developing countries into a climate agreement. Instead, a global solution is warranted that puts us on equal footing with our competitors. The international agreement should be negotiated first, not second. Regulating the U.S. industrial sector in advance of negotiations completely removes our negotiation leverage.

The global reality is that developing nations place a significant priority on their manufacturing sector for both domestic economic growth and exports. They have a long history of providing all types of subsidies that include energy and trade credits. If they subsidize energy costs for their manufacturers, why wouldn't they also subsidize the cost of GHG reductions to enable exports to the U.S.? U.S. industry needs a level playing field—and then let us compete.

The justification is obvious and in the best interests of the country. The industrial sector's absolute GHG emissions are only 2.6% above 1990 levels and the rate of change has been flat due to energy efficiency improvements and a declining manufacturing presence. In contrast, according to the EPA, the transportation sector emissions are up 27%, residential up 29%, commercial up 39% and power generation up 29%. The point is that the industrial sector is *not* a contributor to growing GHG emissions and should not be a high priority for GHG reduction mandates.

Secondly, the products we produce are essential for economic growth of the country and a vibrant opportunity to create new high paying jobs. As the economy rebounds, our country will require significant volumes of the products that we produce such as cement, steel, aluminum, chemicals, plastics, paper, glass, and fertilizer which are all energy intensive. You can't produce renewable energy without our products. The question Congress must answer is whether it wants these products to be supplied by production facilities in the U.S. or imported from foreign countries.

If Congress places a declining GHG cap on the industrial sector, you can be pretty confident that U.S. companies will "not" invest their capital nor create jobs in the U.S. The reason is obvious. There is a lack of confidence that other countries will place a GHG cap on their manufacturers any time soon which would place U.S. industry at a significant competitive disadvantage. Setting a starting date of 2012 for a GHG cap will result in industrial companies making preemptive capital decisions on where to locate and increase the production of their products that anticipates these assumptions.

Third, products from the manufacturing sector provide the "enabling solutions" to the challenges of climate change and it is important that GHG regulation does not increase the cost of these products to deter consumer purchases.

It takes energy to save energy. Insulation can be made from glass, plastic or paper, all of which are energy intensive. Double pane windows use twice the amount of glass but save an enormous amount of energy over the life of a building. Reducing the weight of autos, trucks and aircraft is an essential solution but requires greater use of aluminum, composite plastics and different grades of steel. More steel and plastics are needed for wind turbines. The production of solar silicon used to make solar panels is energy intensive. There are literally a thousand examples of how manufacturing products contribute to the climate solution and it is important to keep the cost of these products low.

The industrial sector is the "green sector." Manufacturing has a remarkable track record of reducing energy while continuing to increase the output of product. They predominantly use natural gas as a fuel versus coal. They are the largest consumer of biomass that is used for making paper and as a fuel for producing energy efficient steam and power. They utilize combined heat and power extensively and substantial quantities of recycled steel, aluminum, glass and paper which is extraordinarily energy efficient.

Fourth, placing a GHG cap on manufacturing makes it much more difficult for our sector to reclaim domestic market share and increase exports. The U.S. has a significant trade deficit in part due to declining manufacturing product exports that accelerated in 2000 as U.S. natural gas prices rose and imports increased.

A lot of these imports are from China, a country that values its manufacturing sector. And now, the U.S. is dependent upon China to finance its burgeoning debt. Improving the competitive health of our manufacturing sector can help reduce this dependency. Increasing competitiveness of the industrial sector and increasing exports is an important matter of public policy that needs addressed.

The decision is yours to make. Company CEOs have a responsibility to their shareholders to protect the company's interest and they will. The manufacturing sector is agile and mobile to survive and thrive—it is just a question of where.

Climate Policy and Manufacturing Competitiveness

IECA has not attempted to gain consensus by the industrial sector on what is the best way to regulate GHG emissions for the U.S. economy or for the manufacturing sector. However, there is little question how the majority view policy options.

Every discussion begins and ends with "competitiveness." Manufacturers compete globally and for many, the cost of energy and carbon will determine whether they will successfully compete in domestic and global markets.

The "absolute" cost of energy and carbon does not matter so long as all of our competitors around the world have the same increased costs. What matters to manufacturers is the "relative" cost of energy and carbon compared to our major global competitors regardless of whether they are in Europe or a developing country.

For that reason, U.S. climate policy must not increase our relative costs. This means that manufacturing competitiveness must also be dealt with at the international level. While this presents a challenge for policy makers, it also provides a wonderful opportunity.

Those of us from the industry believe that more GHG emission reductions can be achieved globally when industrial climate policy instruments are focused on productivity that is, increasing production while reducing energy consumption. It's a win-win and recognizes that all players can only manage the energy use inside their plant and often have little control on the type of energy available.

There is general agreement by U.S. manufacturers that other countries will not knowingly sacrifice their manufacturing jobs in response to climate policy. Since China tends to be a policy lynchpin, it is important to note that they especially will not sacrifice their manufacturing competitiveness to address climate change.

It is China's manufacturing sector that has raised its status to a world power by creating jobs and exports that have provided a significant and unequaled trade surplus. Now, the U.S. is dependent upon them to buy our treasury bills and finance our debt. This is not an enviable position for the U.S. nor is it necessary.

To its credit, the Chinese government has a history of emphasizing the importance of the manufacturing sector which is in contrast to the U.S. government. China has also provided export tax credits, subsidies for energy costs and manages its currency. Some U.S. government officials claim that currency control gives China a 40% competitive advantage over U.S. manufacturers. Whether it's the currency or not, China's manufacturing sector is winning and U.S. manufacturing is losing.

Any U.S. climate policy option must hold manufacturing harmless until major competitors in both developed and countries in transition have comparable energy and carbon cost increases. Comparable reduction requirements do not meet the test. Without this protection, U.S. manufacturers will protect their shareholders and move production facilities to countries that offer a competitive environment.

Well intentioned members of Congress have proposed a cap and trade system that would provide manufacturers with "some" free allowances that would decline over time and would cover "some" of the resulting higher energy costs. While appreciated, these provisions are not adequate to allow the industrial sector to compete, grow domestic production and exports. Many U.S. industries have been working on energy efficiency for decades and simply don't have technology available to make step changes needed to meet these ratcheting targets.

Under these provisions we will still have a declining GHG cap that reduces our production; unpredictable costs for energy, carbon and transaction costs; and unnecessary cost increases. It also does not do anything to help our domestic customers who will be asked to absorb higher costs for our products.

Economy-wide cap and trade is simply the wrong policy platform for the manufacturing sector. IECA wants a climate policy that will allow U.S. manufacturing to: invest in the U.S.; does not create winners and losers; does not penalize those who have already invested in energy efficiency; and transparency so that the system cannot be manipulated or gamed.

Relatively few manufacturers in the industry support cap and trade. The ones that do have either inherent special circumstances that allow them to gain a relative competitive advantage; have already moved their energy intensive manufacturing offshore; will significantly benefit from increased product sales or are simply not energy intensive and are not measurably impacted.

We do not know any manufacturing companies who support carbon cap and trade with auction. This is completely understandable because the manufacturing sector needs predictability over long time horizons for capital investment. The auction of carbon allowances does not give price certainty plus manufacturers are disadvantaged in competing for the auctioned carbon with regulated utilities who can afford to pay any price and then pass the cost on to consumers to pay.

If the government lets Wall Street participate, the auction option gets even worse. In general, manufacturers believe that only companies who are required to reduce GHG emissions should be allowed to purchase carbon allowances or offsets. This leaves Wall Street out.

Auctioning is the quickest way to lose manufacturing jobs and they will go silently, one at a time and without an announcement. Each manufacturing production unit has a cost break-even that varies significantly from plant to plant and from company to company. As the cost of carbon rises, the manufacturer will not have any choice but to shut it down.

Very few companies support cap and trade even if allowances are initially provided free of charge because they recognize that these temporary allowances are not a safety net and their economic viability is in jeopardy long term. The engineering limitations of their manufacturing facilities leave little room for imagination—just realism.

A carbon tax is better than a cap and trade program because it does not constrict our ability to increase the volume of product produced, it is superior in transparency, and more easily adjusted at the border. Nonetheless, it is a cost that is not welcomed and unnecessary for the industrial sector to reduce carbon intensity. Clearly, a high carbon tax will be just as effective of putting us out of business.

There are about 350,000 manufacturing facilities in the U.S. It is estimated that about 7,800 facilities would emit 10,000 tons of CO_2 per year. By itself, regulating the industrial sector presents a significant regulatory challenge for the federal government. While only 7,800 would be regulated, the other 342,200 facilities and the American consuming public would be asked to absorb higher resulting product costs. [. . .]

Carbon Trading: Take Action to Prevent Market Manipulation and Fraud

We offer a simple question. If the U.S. government cannot prevent market manipulation, market power, fraud and excessive speculation in mature commodities like oil, natural gas and food commodities, not to mention loan derivatives, why would the government believe it can do so with carbon?

The reality is that preventing market manipulation and fraud in the carbon market will be much harder because all reductions are "project by project." Mature commodities like energy or food commodities have physically deliverable products. Carbon reductions are a response to doing capital projects that reduce carbon and the level or rate of reduction can change at any time. Some reductions will be for compliance reasons and some to generate carbon offsets, both are the underlying value or asset.

The national and international economic failures we are experiencing are the result of the financial industry's creation of highly leveraged instruments called credit default swaps and excessive commodity speculation during the first half of 2008.

Financial companies issued a significant number of credit default swaps that are insurance like contacts that other companies bought as protection against the default of mortgage backed securities. They reaped huge profits until the underlying asset values fell. When the mortgage market values began to fall, banks that had purchased the swaps demanded collateral from insurance companies which they could not pay. The house of cards crumbled.

From January to July of 2008, that same financial industry (Wall Street trading houses, hedge funds, sovereign funds and managers of passive index funds) drove the price of energy and food commodities to record levels. Experts now admit that with only a small exception, supply and demand fundamentals had little to do with the run up.

The natural gas market provides an excellent example. The price of natural gas about doubled from January to August of 2008. In that same time period, domestic supply of natural gas rose by 8 percent, national inventories were comfortably within their five year averages and demand was almost identical to the previous year. There was no supply versus demand reason for the doubling of the price. IECA estimates excessive speculation during that time period cost consumers around $40 billion.

Some people respond that we can learn from those lessons and that we will not make the same mistakes as it applies to carbon markets. This does not give us comfort. On all counts, the Congress has failed to act to fix the regulatory oversight shortfalls that have cost consumers billions several times over.

Even after Enron manipulated the market that cost consumers billions, Congress did not act to close the Enron Loophole. After the collapse of the Amaranth, the giant hedge fund, it was discovered that it had successfully controlled almost 60 percent of the U.S. natural gas market contracts and the Commodity Futures Trading Commission did not even know it. Congress did not act to fix it.

Then came last year's excessive speculation of the energy and food commodity market. A year has passed and Congress has not passed any laws to close multiple loopholes that allow speculators unlimited speculation nor have they addressed the long-only index funds. Lastly, Congress has not acted to change the laws to prevent new credit default swaps.

Trading carbon can and will suffer from both problems and more easily. The underlying value of carbon projects can change dramatically without warning leaving the purchaser with little recourse. Traders from around the world view carbon as their next great windfall profit. Just look at the EU market to see how carbon is traded, not for its underlying cost of abatement, but traded as an energy commodity.

Carbon Offsets

The key thing to remember about offsets is that it represents a capital investment. Where ever the capital is invested will create new jobs. IECA companies want to create jobs in the U.S. But, as stated earlier, under cap and trade, companies will have no choice but to protect their shareholders and invest, if necessary in foreign countries to create offsets to stay in business.

In general, manufacturing companies would rather invest in projects to reduce GHG emissions and increase energy efficiency in their domestic facilities than buy carbon offsets from potentially our competitors in countries like China. The United Nation's "Clean Development Mechanism" (CDM) has approved projects in the manufacturing sector. For the last several years a large number of European countries have purchased CDM and Joint Implementation offsets to help meet their EU reduction requirements. We feel confident the U.S. tax payer is not going to do the same for us.

Countries like China have turned the CDM into a money maker by adding a substantial tax to CDM credits and some companies have turned generation of CDM credits into increased sale of products.

A Trade Issue, a WTO Issue, Equals Uncertainty and Competitiveness Risk

Because of this multiple exposure reality of the industrial sector, the congress and the industrial sector must evaluate any proposed carbon policy through the filters of both international trade competitiveness impact and cost impact.

Depending on the sector involved, this can lead to different answers on what type of policy is best suited to reducing greenhouse gas emissions and protecting domestic jobs and competitiveness. This is why a one size fits all cap and trade program is problematic for the manufacturing sector as a whole and why the industrial sector attitudes toward every alternative policy must be nuanced. Moreover, we are very concerned that the entire climate policy debate may become confused and tangled with a larger public finance debate.

As such, it is imperative that any legislative approach to dealing with the greenhouse gas issue include a strong and effective border mechanism to ensure that imports face the same costs and burdens as domestic production. Regardless of what Congress may do in terms of allocating allowances or otherwise reducing costs for trade-sensitive industries (which is critical), it is inevitable that such industries will face higher (and likely growing) costs associated with climate legislation. If we do nothing to ensure that foreign firms selling in this market bear these same costs, the result will simply be more imports from countries without similar environmental measures—a catastrophic result not only for our industries, but for the environment as well.

No one is suggesting a border mechanism that penalizes foreign production. Imports should be subject to the same costs of carbon that are imposed on domestic producers—no more and no less. This is essential to level the playing field until there is a uniform, global approach in place to address the climate issue. While a number of the bills that have been introduced in the House and Senate include border provisions, they have unfortunately included any number of loopholes and deficiencies that would undermine their effectiveness. The worst thing we could do is to put in place some type of "fig leaf" to purportedly address the problem without actually resolving it.

There have been a lot of questions about whether a border provision in the context of climate legislation would be compatible with WTO rules. The truth is that nobody knows for certain how WTO rules will be applied in this area because there is simply no binding precedent. Several points are clear, however.

First, there are very strong arguments that we can impose equivalent burdens on both domestic production and imports, so long as imports are treated no worse than domestic producers.

Second, given that any border mechanism is almost certain to be the subject of examination at the WTO it makes no sense to put in place an ineffective provision. We should enact a meaningful mechanism that will truly impose equivalent burdens on imports and domestic production, and then see how the issues are resolved internationally.

Third, if it turns out that WTO rules are interpreted so as *not* to permit an effective border provision, that information will be critical to Congress as it considers climate policy. The fact is that no climate measure can or will be successful if it cannot ensure that imports bear the same burdens as domestic production. [. . .]

President Obama's Cap and Trade Budget Proposal

President Obama's budget blueprint would establish a 100% auction based system, the revenues from which have been promised to an assortment of uses— some related to achieving climate policy objectives and some completely unrelated. We strongly encourage the Congress to not use climate policy as a federal revenue raiser.

Our interpretation of President Obama's budget proposal would mean that only the industrial and commercial sector would pay for the higher energy/carbon compliance costs. The electric and natural gas utilities will be able to pass the costs onto consumers under state utility regulation in states that are regulated. In those states, the electric utility sector will experience an increase in the average cost of producing electricity. In deregulated states there will be an increase in the cost of production for the marginal generation unit which clears the market. The Obama plan would provide rebates to some retail consumers to cover their increased costs. If this is correct, this means that only a small portion of the economy will bear the costs. This is not sound climate policy. [. . .]

Eileen Claussen

 **NO**

Competitiveness and Climate Policy: Avoiding Leakage of Jobs and Emissions

Addressing global climate change presents policy challenges at both the domestic and the international levels, and the issue of competitiveness underscores the very close nexus between the two. The immediate task before this subcommittee, and before the Congress, is developing and enacting a comprehensive domestic program to limit and reduce U.S. greenhouse gas (GHG) emissions. Moving forward with a mandatory program to reduce U.S. emissions in advance of a comprehensive international agreement presents both risks and opportunities. On the one hand, domestic GHG limits may lead to a shift of some energy-intensive production to countries without climate constraints, resulting in "emissions leakage" and posing competitiveness concerns for some domestic industries. On the other hand, a mandatory domestic program in the United States is an essential step towards the development of an effective global climate agreement.

In the long term, a strong multilateral framework ensuring that all major economies contribute their fair share to the global climate effort is, I believe, the most effective means of addressing competitiveness concerns. Achieving such an agreement must be a fundamental objective of U.S. climate policy. In designing a domestic climate program, the question before Congress is what to do in the interim—until an effective global agreement is in place. In considering this question, it is important to distinguish two distinct but closely related policy challenges: how best to encourage strong climate action by other countries, and in particular, by the major emerging economies; and how best to minimize potential competitiveness impacts on U.S. industry. I believe that each of these two objectives is most effectively addressed through a different set of policy responses, and it is important to ensure that our efforts to address one do not undermine the other.

I will focus today primarily on the second of these challenges: designing transitional policies to minimize potential competitiveness impacts on U.S. industry. Our analysis of the underlying issues leads us to conclude that the potential competitiveness impacts of domestic climate policy are modest and are manageable. In my testimony, I will: 1) present our analysis of the nature and potential magnitude of the competitiveness challenge; 2) discuss a range of options for addressing competitiveness concerns; and 3) outline what we believe

U.S. Senate, March 18, 2009.

would be the most effective approach. This approach would employ output-based emission allocations to vulnerable industries, phased out over time, and other transition assistance to affected workers and communities.

Understanding Competitiveness Concerns

A first step in considering options to address competitiveness is assessing the potential scope and magnitude of potential competitiveness impacts. It is not the competitiveness of the U.S. economy as a whole that is at issue. (According to an MIT analysis of the Lieberman-Warner Climate Security Act of 2007, the cost of meeting the bill's emission reduction targets in 2050, by when the U.S. economy is projected to triple in size, would result in GDP being 1% less than would otherwise be the case.) Rather, the concern centers on a relatively narrow segment of the U.S. economy: energy-intensive industries whose goods are traded globally, such as steel, aluminum, cement, paper, glass, and chemicals. As heavy users of energy, these industries will face higher costs as a result of domestic GHG constraints; however, as the prices of their goods are set globally, their ability to pass along these price increases is limited.

Competitiveness impacts can be experienced as a loss in market share to foreign producers, a shift in new investment, or, in extreme cases, the relocation of manufacturing facilities overseas. In assessing the economic consequences of past environmental regulation in the United States, most analyses find little evidence of significant competitive harm to U.S. firms. Many studies conclude that other factors—such as labor costs, the availability of capital, and proximity to raw materials and markets—weigh far more heavily in firms' location decisions. One comprehensive review—synthesizing dozens of studies of the impact of U.S. environmental regulation on a range of sectors—concluded that while new environmental rules imposed significant costs on regulated industries, they did not appreciably affect patterns of trade.

In the case of GHG regulation, the additional cost to firms could include the compliance cost of purchasing allowances to cover direct emissions; indirect compliance costs embedded in higher fuel or electricity prices; further demand-driven price increases for lower-GHG fuels such as natural gas; and the costs of equipment and process changes to abate emissions or reduce energy use.

In gauging the potential impacts of GHG regulation, it is important to distinguish the "competitiveness" effect from the broader economic impact on a given industry or firm. A mandatory climate policy will present costs for U.S. firms regardless of what action is taken by other countries. In the case of energy-intensive industries, one potential impact of pricing carbon could be a decline in demand for their products as consumers substitute less GHG-intensive products. This is distinct, however, from the international "competitiveness" impact of GHG regulation, which is only that portion of the total impact on a firm resulting from an imbalance between stronger GHG constraints within, and weaker GHG constraints outside, the United States.

To empirically quantify the potential magnitude of this competitiveness impact, the Pew Center commissioned an analysis by economists at the

Resources for the Future. This work, which we will be publishing shortly, analyzes 20 years of data in order to discern the historical relationship between electricity prices and production, consumption, and employment in more than 400 U.S. manufacturing industries. On that basis, the analysis then projects the potential competitiveness impacts of a U.S. carbon price, assuming no comparable action in other countries. (The analysis assumes a CO_2 price of $15 per ton. The Energy Information Administration's core case analysis of the Lieberman-Warner cap-and-trade bill estimated a 2012 allowance price of $16.88 per ton CO_2.)

The analysis finds an average production decline of 1.3 percent across U.S. manufacturing, but also a 0.6 percent decline in consumption, suggesting a competitiveness effect of just 0.7 percent. For energy-intensive industries (those whose energy costs exceed 10 percent of shipment value), the analysis projects that average U.S. output declines about 4 percent. However, consumption declines 3 percent, so that only a 1 percent decline in production (or one-fourth of the total decline) can be attributed to an increase in imports, or a loss of competitiveness. For specific energy-intensive industries, including chemicals, paper, iron and steel, aluminum, cement, and bulk glass, the analysis projects a competitiveness impact ranging from 0.6 percent to 0.9 percent, although within certain subsectors, the impact could be higher. What this analysis demonstrates very clearly is that most of the projected decline in production stems from a reduction in domestic demand, not an increase in imports. In other words, most of the projected economic impact on energy-intensive industries reflects a move toward less emissions-intensive products—as would be expected from an effective climate change policy—not a movement of jobs and production overseas. At the price level studied, the projected competitiveness impacts, as well as the broader economic effects on energy-intensive industries, are modest and, in our view, can be readily managed with a range of policy instruments.

Policy Options

In the design of a domestic cap-and-trade system, competitiveness concerns can be addressed in part through a variety of cost-containment measures, such as banking and borrowing and the use of offsets, which can help reduce the costs to all firms, including energy-intensive, trade-exposed industries. However, other transitional policies may be needed to directly address competitiveness concerns in the period preceding the establishment of an effective international framework. Options include: fully or partially exempting potentially vulnerable firms from the cap-and trade system; compensating firms for the costs of GHG regulation through allowance allocation or tax rebates; transition assistance to help firms adopt lower-GHG technologies, and to help communities and workers adjust to changing labor markets; and border measures such as taxes on energy-intensive imports from countries without GHG controls. In addition, a domestic policy could be designed to encourage and anticipate international sectoral agreements establishing the respective obligations of major producing companies within given sectors.

Exclusion from Coverage

One option is to fully or partially exclude vulnerable sectors or industries from coverage under the cap-and-trade program. For instance, under the Lieberman-Warner Climate Security Act of 2008, the direct "process" emissions of many energy-intensive industries would not be subject to GHG limits. Exclusions would relieve trade-exposed industries of any requirement to hold emission allowances and thereby eliminate direct regulatory costs, shielding them not only from competitiveness impacts but also from some of the broader economic effects of pricing carbon. However, by limiting the scope of the cap-and-trade system, exclusions would undermine the goal of reducing GHG emissions economy-wide, and would reduce the economic efficiency of a national GHG reduction program. They also would give exempted industries an economic advantage over nonexempt domestic firms and sectors, including competitors. Moreover, firms whose emissions are exempted would still face the indirect costs of higher energy prices.

Compensation for the Costs of GHG Regulation

Another option is to include these sectors in the cap-and-trade system but compensate them for the costs of GHG regulation. Key design considerations include the scope, form, and means of calculating such compensation, and whether and how it should be phased out.

As noted earlier, firms covered by the cap-and-trade system face both direct and indirect costs of regulation. The direct, or compliance, cost is the cost of purchasing any allowances needed to cover direct emissions regulated under the cap. Indirect costs include higher prices for electricity and natural gas (reflecting an embedded carbon price and, in the case of natural gas, rising demand for this less GHG-intensive fuel), and the costs of equipment and process changes to abate emissions or reduce energy use. For energy-intensive industries, the indirect cost of higher energy prices represents a significant portion of the total potential cost.

One form of compensation is providing free emission allowances. In the case of direct emissions, allowances could be granted on the basis of historic emissions ("grandfathering") and energy-intensive sectors could receive a more generous allocation than other emitters. For instance, energy-intensive industries could receive a full free allocation while others receive allocations for 80 percent of their historic emissions. Over time, the energy-intensive sectors could continue to be treated more generously—for instance, continuing to receive a higher proportion of free allowances as the allocation system transitions to fuller auctioning. Because free allocation provides the same economic incentive to reduce emissions as does an auction, keeping energy-intensive sectors under the cap, but providing free allowances, provides for greater environmental effectiveness and economic efficiency than excluding them.

Additional allowances could be provided to compensate for indirect costs. However, as future energy prices cannot be predicted, there is no way of determining in advance whether this allocation matches the firms' actual costs.

Another form of compensation for direct and/or indirect costs could be tax credits or rebates. One potential source of revenue for such measures is proceeds from the auction of emission allowances. A tax rebate would be a direct payment to compensate a firm for GHG regulatory costs; a tax credit could alternatively offset those costs by reducing a non-GHG burden such as corporate or payroll taxes, or healthcare or retirement costs.

Whatever form the compensation takes, one critical issue is the basis for calculating the appropriate level. In the case of direct compliance costs, granting allowances on the basis of historical emissions can effectively penalize early action and reward relatively heavier emitters within an industry. In addition, it does not necessarily guard against emissions leakage or a loss of jobs, as a firm could choose to maximize profits by selling its free allowances and reducing production. There is also the risk that firms will be over-compensated and realize windfall profits.

Alternatively, compensation could be "output-based," pegged to actual production levels and/or energy consumption. Firms could be compensated in full for direct or indirect costs; or an output-based approach could apply a performance standard (i.e., emissions or energy use per unit of production) to encourage and reward lower GHG intensity production. The Inslee-Doyle Carbon Leakage Prevention Act introduced in the 110th Congress would have allocated allowances to compensate for both direct and indirect costs based on a facility's level of output, adjusted by an "efficiency factor" which could be adjusted over time to provide firms an ongoing incentive to switch to lower-GHG processes and energy sources. The compensation would shield them from regulatory costs, lowering the risk of emissions leakage and competitiveness impacts, while maintaining an incentive for improved environmental performance and continued operation.

As with the exclusion of trade-exposed sectors from the cap, the remedy provided by these compensation approaches extends beyond any actual competitiveness effect. Whether based on output or historical emissions, most of the proposals offered to date aim to compensate firms for most or all of the increased costs associated with GHG regulation, not just for the impacts they may face due to the asymmetry between GHG constraints within and outside the United States. To limit compensation to competitiveness impacts alone would require in-depth financial knowledge of each firm and/or complex calculations that could be reliably performed only once the impacts have occurred. A drawback of a compensation approach is that the financial resources required—whether drawn from auction revenue or other sources—are not available for other climate- or non-climate-related purposes.

If compensation is provided, one important consideration is how long it should be maintained and at what level. Phasing out the compensation would give firms additional incentive to improve their GHG performance but would also make them more vulnerable to competitiveness impacts. A mandatory program could provide for periodic review of any allowances or other compensation to vulnerable sectors to consider adjusting them on the basis of new information. For instance, if the legislation establishes a specific timetable for moving from free allocation to auctioning, this transition might be slowed for specific industries if there are clear indications of competitiveness impacts. Alternatively, compensation could be phased out or ended if other countries take stronger action or new international agreements are reached. The review could focus narrowly on the issue of trade-related impacts or it could be a broad-based review also looking at new science, technology, and economic data.

Transition Assistance

Another option is to provide transition assistance to vulnerable firms to help them adopt lower-GHG technologies, and to communities and workers affected by competitiveness impacts. In the case of firms, measures could include tax incentives such as accelerated depreciation to encourage the replacement of inefficient technologies, or tax credits for the development or adoption of lower-GHG alternatives. Firms could also be incentivized to switch to low carbon energy sources, for example through subsidies for purchases or generation of renewable energy.

Where competitiveness impacts are unavoidable, assistance can be provided to both workers and communities. Previous government efforts to help communities adjust to economic changes resulting from national policies provide lessons for shaping similar efforts as part of climate change policy. At the level of individual workers, policies such as the Workforce Investment Act providing income support and retraining to help move workers into new jobs can provide a blueprint for transition programs to assist workers adversely affected by competitiveness imbalances under a climate policy.

Border Adjustment Measures

Another strategy is to try to equalize GHG-related costs for U.S. and foreign producers by imposing a cost or other requirement on energy-intensive imports from countries with weaker or no GHG constraints. One option is a border tax based on an import's "embedded" emissions (equal to the compliance costs for a domestic producer of an equivalent good). An alternative approach, described by proponents as more likely to withstand challenge under international trade rules, would instead require that imports be accompanied by allowances for their associated emissions. The Lieberman-Warner bill would have required allowances for energy-intensive imports from countries not determined by an appointed commission to be undertaking "comparable" action to reduce emissions. To avoid driving up allowance prices for U.S. firms, importers would buy from an unallocated pool of "reserve allowances" at a price set by the government. In the 110th Congress, the Bingaman-Specter bill, the Dingell-Boucher discussion draft, and Chairman Markey's ICAP bill all adopted variations of this approach.

One major shortcoming of this approach is its limited effectiveness in reducing competitiveness impacts. As the border adjustment measures would apply only to imports to the United States, they would not help "level the playing field" in the larger global market where U.S. producers may face greater competition from foreign producers.

Among the other issues raised by unilateral border measures is their consistency with World Trade Organization (WTO) rules. The legality of a given measure would depend in part on its specific design and on the types of climate policies in place domestically. As such approaches have not been previously employed, there are no definitive rulings, and experts differ in their interpretation of relevant WTO precedents. The legal uncertainties ultimately would be resolved only through the adjudication of a WTO challenge, a likely prospect if unilateral border measures were to be applied by the United States or another country.

Trade measures also present significant administrative challenges—in particular, calculating the GHG intensity of imported goods. Would the imported good's GHG intensity be calculated at the sector, firm, or plant level? Would such an assessment rely on data from the exporting country? In addition, criteria are needed to determine whether a country is meeting a "comparability" or other standard. Under the Lieberman-Warner bill, "comparable action" would have been defined as either a) a percentage reduction in GHGs equivalent to that achieved by the United States, or b) as determined by the commission, "tak[ing] into consideration . . . the extent to which" a country has implemented measures and deployed state-of-the-art technologies to reduce emissions. A literal application of a "comparability" standard to developing countries—particularly if border requirements are imposed upon or very soon after mandatory domestic limits are put in place—would likely be viewed internationally as inconsistent with the principle of "common but differentiated responsibilities" agreed to by the United States in the UN Framework Convention on Climate Change (UNFCCC).

Another important consideration is the potential impact on trade and international relations. If the United States were to impose border requirements, there is a greater likelihood that it would become the target of similar measures. European policymakers also are weighing the use of border measures and have argued that the emission targets under consideration in the United States are not comparable to those adopted by the European Union. U.S. trade officials and others also have voiced strong concern about the potential for retaliatory trade measures by targeted countries, leading to escalating trade conflicts. Proponents argue that the threat of unilateral trade measures would give the United States greater leverage in international climate negotiations. However, there is a significant risk that they would engender more conflict than cooperation, in the end making it more difficult to reach agreements that could more effectively address competitiveness concerns.

International Sectoral Agreements

All of the preceding options are measures that would be implemented domestically. Another approach that would help reduce emissions within and outside the United States, while addressing competitiveness concerns, is to negotiate international agreements setting GHG standards or other measures within energy-intensive globally-traded sectors. For example, major steel-producing countries could agree on standards limiting GHGs per ton of steel, which could be differentiated initially according to national circumstances and converge over time. Sectoral agreements could take a number of forms, depending on the specific sectors, and could be negotiated as stand-alone agreements or as part of a comprehensive climate framework.

Within the domestic context, a purely sector-by-sector approach would sacrifice the broad coverage and economic efficiency of an economy-wide cap-and-trade program. However, sectoral agreements could exist alongside a cap-and-trade program, and the system could be designed to encourage U.S. producers to work toward their establishment. One option would be to provide for a sector's exclusion from the cap once an international agreement of comparable stringency is in place (although, as noted, diminishing the scope of the cap-and-trade system by exempting one or more sectors would limit its economic efficiency). An alternative is to keep the sectors under the cap but align their obligations under the domestic program and the international sectoral agreement. For instance, a firm's emissions allowance under the trading system could be based on the GHG standard that is agreed internationally.

In keeping with the principle of "common but differentiated responsibilities," an international sectoral agreement may not set fully equivalent requirements for all countries, particularly at the outset. In that event, compensation for energy-intensive industries could be maintained at some level and phased out as the requirements for other countries rise to those borne by the United States.

Recommendations: An Allowance-Based Approach

Based on our assessment of the available options, the Pew Center believes that Congress should seek to address competitiveness concerns by: 1) strongly encouraging the executive branch to negotiate a new multilateral climate agreement establishing strong, equitable, and verifiable commitments by all major economies; 2) including in domestic legislation incentives for such an agreement, including support for stronger action by major developing countries; and 3) including in cap-and-trade legislation transitional measures to cushion the impact of mandatory GHG limits on energy-intensive trade-exposed industries and the workers and communities they support. These transitional measures should be structured as follows:

- In the initial phase of a cap-and-trade program, free allowances should be granted to vulnerable industries to compensate them for the costs of GHG regulation. For direct costs, allocations should be based on actual production levels. For indirect costs, allowances should reflect the emitter's production-based energy consumption, taking into account the GHG intensity of its energy supplies.
- Based on an analysis of GHG performance within a given sector, allocations should be set initially so that producers with average GHG performance are fully compensated for regulatory costs, while those

performing above or below the norm receive allowances whose value is greater or less than the their costs, respectively. This factor should be adjusted over time as an incentive to producers to continually improve their GHG performance.

- Allowance levels should decline over time, gradually transitioning to full auctioning, although at a slower rate than for other sectors.
- A review should be conducted periodically to assess whether sectors are experiencing competitiveness impacts and, if warranted, to adjust allowance levels and/or the rate of transition to full auctioning.
- A portion of allowance auction revenue should be earmarked for programs to assist workers and communities in cases where GHG constraints are demonstrated to have caused dislocation.
- Transition assistance should be curtailed for a given sector upon entry into force of a multilateral or sectoral agreement establishing reasonable obligations for foreign producers, or upon a Presidential determination that such measures have been instituted domestically.

We believe this approach addresses the transitional competitiveness concerns likely to arise under a mandatory cap-and-trade program, while maintaining the environmental integrity of the program and providing an ongoing incentive for producers to improve their GHG performance. We commend the subcommittee for focusing the attention of Congress on this critical issue, and would be happy to work with you as you develop legislation to address this and other dimensions of the climate challenge.

POSTSCRIPT

Will Restricting Carbon Emissions Damage the U.S. Economy?

The threat of global warming from continuing emissions of greenhouse gases has prompted the extension of emissions trading to carbon dioxide. Europe is implementing its Greenhouse Gas Emissions Trading Scheme, although so far its effectiveness is in doubt; Marianne Lavelle, "The Carbon Market Has a Dirty Little Secret," *U.S. News and World Report* (May 14, 2007), reports that in Europe the value of tradable emissions allowances has fallen so low, partly because too many allowances were issued, that it has become cheaper to burn more fossil fuel and emit more carbon than to burn and emit less. Future trading schemes will need to be designed to avoid the problem, and the U.S. Congress is actively considering ways to address the issue (see "Support Grows for Capping and Trading Carbon Emissions," *Issues in Science and Technology,* Summer 2007). Meanwhile, there is great interest in what is known as "carbon offsets," by which corporations, governments, and even individuals compensate for carbon dioxide emissions by investing in activities that remove carbon dioxide from the air or reduce emissions from a different source. See Anja Kollmuss, "Carbon Offsets 101," *World Watch* (July/August 2007). Unfortunately, present carbon-offset schemes contain loopholes that mean they may do little to reduce overall emissions; see Madhusree Mukerjee, "A Mechanism of Hot Air," *Scientific American* (June 2009).

On May 21, 2009, the House Energy and Commerce Committee approved H.R. 2454, "The American Clean Energy and Security Act." The goal of the Act, said Committee Chair Henry A. Waxman (D-CA), is to "break our dependence on foreign oil, make our nation the world leader in clean energy jobs and technology, and cut global warming pollution. I am grateful to my colleagues who supported this legislation and to President Obama for his outstanding leadership on these critical issues." Among other things, the Act establishes Title VII of the Clean Air Act to provide a declining limit on global warming pollution (a "cap" as in "cap-and-trade") and to hold industries accountable for pollution reduction under the limit. The aim is to cut global warming pollution by 17% compared to 2005 levels in 2020, by 42% in 2030, and by 83% in 2050. (See the summary of the Act at http://energycommerce.house.gov/Press_111/20090515/hr2454_summary.pdf). In June 2009, the House of Representatives passed the bill, which also called for utilities to use more renewable energy sources. However, the bill had not yet been introduced in the Senate.

Many people feel that it is about time that the U.S. took such action. According to the Global Humanitarian Forum's "Human Impact Report: Climate Change—The Anatomy of a Silent Crisis" (May 29, 2009; see http://www.ghf-ge.org/programmes/human_impact_report/index.cfm) global warming is already affecting over 300 million people and is responsible for 300,000 deaths per year. Action now is clearly appropriate, even though it does seem inevitable that this deadly impact of global warming must grow worse for many years before it can be stopped. However the debate over the proper actions to take is by no means over. Some analysts argue that a carbon tax would be more effective; see Bettina B. F. Wittneben, "Exxon Is Right: Let Us Re-Examine Our Choice for a Cap-and-Trade System over a Carbon Tax," *Energy Policy* (June 2009).

Internet References . . .

350.org

350.org is an international campaign dedicated to building a movement to unite the world around solutions to the climate crisis—the solutions that justice demands. Its goal is to inspire the world to reduce the level of carbon dioxide in the atmosphere to 350 parts per million, the level scientists have identified as the safe upper limit for CO_2 in our atmosphere.

http://www.350.org/

Intergovernmental Panel on Climate Change

The Intergovernmental Panel on Climate Change (IPCC) was formed by the World Meteorological Organization (WMO) and the United Nations Environment Programme (UNEP) to assess the scientific, technical, and socio-economic information relevant for the understanding of the risk of human-induced climate change.

http://www.ipcc.ch/

Climate Change

The United Nations Environmental Program maintains this site as a central source for substantive work and information resources with regard to climate change.

http://www.unep.org/themes/climatechange/

CAFE Overview

The U.S. National Highway Traffic Safety Administration provides an overview of existing CAFE standards here.

http://www.nhtsa.dot.gov/cars/rules/cafe/overview.htm

The National Renewable Energy Laboratory

The National Renewable Energy Laboratory is the nation's primary laboratory for renewable energy and energy efficiency research and development. Among other things, it works on wind power and biofuels.

http://www.nrel.gov/learning/re_basics.html

Nuclear Energy

The U.S. Department of Energy's Office of Nuclear Energy leads U.S. efforts to develop new nuclear energy generation technologies; to develop advanced, proliferation-resistant nuclear fuel technologies that maximize energy from nuclear fuel; and to maintain and enhance the national nuclear technology infrastructure.

http://www.ne.doe.gov/

Energy Issues

*H*umans *cannot live and society cannot exist without producing environmental impacts. The reason is very simple: Humans cannot live and society cannot exist without using resources (e.g., soil, water, ore, wood, space, plants, animals, oil, sunlight), and those resources come from the environment. Many of these resources (e.g., wood, oil, coal, water, wind, sunlight, uranium) have to do with energy. The environmental impacts come from what must be done to obtain these resources and what must be done to dispose of the wastes generated in the process of obtaining and using them. The issues that arise are whether and how we should obtain these resources, whether and how we should deal with the wastes, and whether alternative answers to these questions may be preferable to the answers that experts think they already have.*

In 2007, the Intergovernmental Panel on Climate Change released its fourth assessment report, summarizing the scientific consensus as the climate is warming, human activities are responsible for it, and the impact on human well-being and ecosystems will be severe. This brought energy issues to the fore with unprecedented urgency. The six issues presented here are by no means the only issues related to energy, but they will serve to demonstrate the vigor and variety of the current energy debates.

- Should We Drill for Offshore Oil?
- Is Carbon Capture Technology Ready to Limit Carbon Emissions?
- Is It Time to Put Geothermal Energy Development on the Fast Track?
- Should Cars Be More Efficient?
- Are Biofuels Responsible for Rising Food Prices?
- Is It Time to Revive Nuclear Power?

ISSUE 7

Should We Drill for Offshore Oil?

YES: **Stephen L. Baird,** from "Offshore Oil Drilling: Buying Energy Independence or Buying Time?" *The Technology Teacher* (November 2008).

NO: **Mary Annette Rose,** from "The Environmental Impacts of Offshore Oil Drilling," *The Technology Teacher* (February 2009).

ISSUE SUMMARY

YES: Stephen L. Baird argues that the demand for oil will continue even as we develop alternative energy sources. Drilling for offshore oil will not give the United States energy independence, but the nation cannot afford to ignore energy sources essential to maintaining its economy and standard of living.

NO: Mary Annette Rose argues that the environmental impacts of exploiting offshore oil—including toxic pollution, ocean acidification, and global warming—are so complex and far-reaching that any decision to expand U.S. oil drilling must be based on more than public opinion driven by consumer demands for cheap energy, economic trade imbalances, and politics.

Petroleum was once known as "black gold" for the wealth it delivered to those who found rich deposits. Initially those deposits were located on land, in places such as Pennsylvania, Texas, Oklahoma, California, and Saudi Arabia. As demand for oil rose, so did the search for more deposits, and it was not long before they were being found under the waters of the North Sea and the Gulf of Mexico, and even off the beaches of California.

In 1969, a drilling rig off Santa Barbara, California, suffered a blowout, releasing more than three million gallons of oil and fouling 35 miles of the coast with tarry goo. John Bratland, "Externalities, Conflict, and Offshore Lands," *Independent Review* (Spring 2004), calls this incident the origin of the modern conflict over offshore drilling for oil. He notes that since then most accidental oil releases have been related to transportation (as when an oil tanker runs into rocks; the *Exxon Valdez* spill was a striking example; see John Terry, "Oil on the Rocks—the 1989 Alaskan Oil Spill," *Journal of Biological Education,* Winter 1991). Underwater oil releases have largely been prevented by the development of blowout-prevention technology. But Santa Barbara has not forgotten, and residents do not trust blow-out prevention technology. See William M. Welch, "Calif.'s Memories of 1969 Oil Disaster Far from Faded," *USA Today* (July 14, 2008).

Some people do not seem disturbed by the prospect of oil blowouts or spills. Ted Falgout, director of the port at Port Fourchon, Louisiana, looks at the forest of oil rigs in the Gulf of Mexico and "sees green: the color of money that comes from the nation's busiest haven of offshore drilling. 'It's OK to have an ugly spot in your backyard,' Falgout says, 'if that spot has oil coming out of it.'" See Rick Jervis, William M. Welch, and Richard Wolf, "Worth the Risk? Debate on Offshore Drilling Heats Up," *USA Today* (July 13, 2008).

In 2008, oil and gasoline prices reached record highs. Many people were concerned that prices would continue to rise, with the result being rapid investment in alternative energy sources such as wind. At the same time, those who favored increased drilling, both on and offshore, began to call for the government to open up more land for exploration (see Fred Barnes, "Let's Drill," *The Weekly Standard,* May 26, 2008). In its last few months in office, the Bush Administration issued leases for lands near national parks and monuments in Utah and lifted an executive order banning offshore drilling. Both measures were considered justified because they would reduce dependence on foreign sources of oil, ease a growing balance of payments problem, and ensure a continuing supply of oil. Critics pointed out that any oil from new wells, on land or at sea, would not reach the market for a decade or more. They also stressed the risks to the environment and called for more attention to alternative energy technologies.

In the following selections, Stephen L. Baird argues that the demand for oil will continue even as we develop alternative energy sources. Drilling for offshore oil will not give the United States energy independence, but the nation cannot afford to ignore energy sources essential to maintaining its economy and its standard of living. He claims the environmental objections just do not add up. In a direct response to Baird's essay, Mary Annette Rose argues that the environmental impacts of exploiting offshore oil—including toxic pollution, ocean acidification, and global warming—are so complex and far-reaching that any decision to expand U.S. oil drilling must be based on more than public opinion driven by consumer demands for cheap energy, economic trade imbalances, and politics.

YES

Stephen L. Baird

Offshore Oil Drilling: Buying Energy Independence or Buying Time?

Skyrocketing fuel prices, unprecedented home foreclosures, rising unemployment, escalating food prices, increasing climate disasters, and the continued war on two fronts have prompted greater public support for renewed offshore drilling for oil. A Gallup poll conducted in May of 2008 found that 57 percent of respondents favored such drilling, while 41 percent were opposed. [. . .] The political landscape is also being changed in favor of offshore drilling, with the results of a Zogby poll (Zogby International has been tracking public opinion since 1984) showing that three in four likely voters—74 percent—support offshore drilling for oil in U.S. coastal waters, and more than half (59 percent) also favor drilling for oil in the Alaska National Wildlife Refuge. [. . .] The tide is turning in favor of offshore drilling, with environmental concerns given less thought because of the increasing financial strain being realized by a majority of the American public. The debate on offshore drilling has captured headlines in newspapers, stirred debate on talk radio, and has been at the forefront on the nightly news.

The rising tide for support of offshore drilling recently gathered momentum when, on July 14, 2008, President George W. Bush lifted a 1990 executive order by the first President Bush banning offshore drilling, while at the same time calling for drilling in the Arctic National Wildlife Refuge. As of August 2008, however, a 1982 congressional ban is still in place, making Bush's action a symbolic gesture, and now the congressional ban is being debated in terms of both environmental issues and U.S. energy independence. In an almost complete reversal of policy, on July 30, 2008, the U.S. Department of the Interior released a news report saying that the nation's energy situation has dramatically changed in the past year. Secretary of the Interior, Dirk Kempthorne, said, "Areas that were considered too expensive to develop a year ago are no longer necessarily out of reach based on improvements to technology and safety." Kempthorne went on to say that, "The American people and the President want action, and a new initiative (the development of a new oil and natural gas leasing program for the U.S. Outer Continental Shelf) can accelerate an offshore exploration and development program that would increase production from additional domestic energy resources." President Bush is urging Congress

From *The Technology Teacher,* November 2008, pp. 13–17. Copyright © 2008 by International Technology Education Association. Reprinted by permission.

to enact legislation that would allow states to have a say regarding operations off their shores and to share in the resulting revenues. [. . .] Shortly after the Interior Department released plans for jumpstarting new offshore oil exploration, on August 16, 2008 the Speaker of The House, Nancy Pelosi, dropped her opposition to a vote on coastal oil exploration and expanded offshore drilling (with appropriate safeguards and without taxpayer subsidies to big oil) as part of broad energy legislation to be addressed when Congress returned in September. [. . .] Today, with the high price of oil and a widening gap between U.S. energy consumption and supply, the ban on offshore oil drilling is being rethought by the general public, politicians, and the oil industry.

The energy stalemate between environmentalists and industry that has inhibited U.S. offshore oil production since the late 1960s is being broken, environmental arguments no longer add up, and working Americans are now taking energy policy inaction personally. According to a Pew Research Center poll conducted in July 2008, 60 percent of respondents considered energy supplies more important than environmental protection, and a majority of young Americans, 18–29, now consider energy exploration more important than conservation. [. . .]

Addressing Environmental and Safety Concerns

Though offshore drilling conjures up fears of catastrophic spills, (such as the 80,000 barrels that spilled six miles off Santa Barbara, California, inundating beaches and aquatic life in January 1969), the petroleum industry rightly argues that safety measures have improved considerably in recent years. According to the U.S. Minerals Management Service, since 1975, 101,997 barrels spilled from among the 11.855 billion barrels of American oil extracted offshore. This is a 0.001 percent pollution rate. That equates to 99.999 percent clean—compare that with Mother Nature herself, as 620,500 barrels of oil ooze organically from North America's ocean floors each year. [. . .]

The United States has been a leader in the creation of the modern offshore oil industry and has pioneered many new safety technologies, ranging from blowout preventers to computer-controlled well data designed to help oil companies' efforts to prevent disasters. Sensors and other instruments now help platform workers monitor and handle the temperatures and pressures of subsea oil, even as drilling is occurring. Hurricanes have become manageable, with oil lines now being capped at or beneath the ocean floor. Even if oil platforms snap loose and blow away, industrial seals restrain potentially destructive petroleum leaks from hundreds or even thousands of feet below the ocean's surface. In August and September of 2005, the 3,050 offshore oil structures endured the wrath of Hurricanes Katrina and Rita without damaging petroleum spills. While 168 platforms and 55 rigs were destroyed or seriously damaged, the oil they pumped remained safely encased, thanks to heavy underwater machinery. The U.S. Minerals Management Service concluded, "Due to the prompt evacuation and shut-in preparations made by operating and service personnel, there was no loss of life and no major oil spills attributed to either storm." [. . .] If it can be done in an environmentally friendly fashion—and with oil companies themselves footing the bill—increasing opportunities for new offshore drilling might be worthwhile.

Offshore territories and public lands like the Alaska National Wildlife Refuge (ANWR) that don't allow drilling have been estimated to contain up to 86 billion barrels of oil according to the U.S. government's Energy Information Administration. Although analysts say that amount of oil will not greatly affect the price of oil, and that renewed offshore drilling would have little impact on gas prices anytime soon, in the short term, oil prices could go down slightly if Congress lifts its moratorium on new offshore drilling because the market would factor in the prospect of additional oil supplies later on. A spokeswoman for the American Petroleum Institute said that, "If we had new territory, we could hypothetically make a big find." [. . .] Offshore drilling might not be the end-all solution to our oil dependence, but any serious energy proposal has to be comprehensive and should include more oil supply and production from the outer continental shelf.

How Dependent Are We on Foreign Oil?

Although the United States is the third largest oil producer (the U.S. produces 10 percent of the world's oil and consumes 24 percent), most of the oil we use is imported. The U.S. imported about 60 percent of the oil consumed in 2006. [. . .] About half of the oil we import comes from the Western Hemisphere (North, South, Central America, and the Caribbean including U.S. territories). [. . .] We imported only 16 percent of our crude oil and petroleum products from the Persian Gulf countries of Bahrain, Iraq, Kuwait, Qatar, Saudi Arabia, and the United Arab Emirates. During 2006, our five biggest suppliers of crude oil were: Canada (17.2%), Mexico (12.4%), Saudi Arabia (10.7%), Venezuela (10.4%), and Nigeria (8.1%). It is usually impossible to tell whether the petroleum products that you use came from domestic or imported sources of oil once they are refined. [. . .] According to the United States Energy Information Administration, the United States spends more than $20 billion, on average, per month to purchase oil, gasoline, and diesel fuel from abroad.

The negative aspects of this dependency are fairly obvious, and they have been well documented. First, oil imports contribute heavily to the United States' trade deficit, which is at record levels. Second, the United States is forced to make political decisions that it might not make otherwise (invading Iraq, cooperating with hostile governments such as Venezuela and Nigeria, looking the other way at Saudi Arabia's reactionary regime, etc.) because it needs their oil. Third, up to now the availability of oil at a fairly reasonable price has left the United States to continue down a path of using more and more energy. [. . .] From Nixon to now, every sitting President has promised to make sure that we wouldn't have a future energy problem . . . though we certainly do now.

Richard Nixon, 1974: "We will lay the foundation for our future capacity to meet America's energy needs from America's own resources."

> Gerald Ford, 1975: "I am proposing a program, which will begin to restore our country's surplus capacity in total energy. In this way, we will be able to assure ourselves reliable and adequate energy and help foster a new world energy stability for other major consuming nations."
>
> Jimmy Carter, 1980: "We must take whatever actions are necessary to reduce our dependence on foreign oil—and at the same time reduce inflation."
>
> Ronald Reagan, 1982: "We will ensure that our people and our economy are never again held hostage by the whim of any country or oil cartel."
>
> George H. W. Bush, 1990: "The Congress should, this month, enact measures to increase domestic energy production and energy conservation in order to reduce dependence on foreign oil."
>
> George W. Bush, 2008: "And here we have a serious problem, America is addicted to oil." [. . .]

It is somewhat misleading when politicians talk about "America's addiction to oil" because there are some mitigating factors that make this situation a lot less dire than it might seem. The two largest foreign suppliers of oil to the United States are friendly to us: Canada and Mexico. These countries have increased their exports to the United States for the past decade and are well-positioned to continue doing so. Thus, fears that the United States will be dependent on "enemy" regimes are overblown, and the price of oil is a world-market price, so the United States is not being gouged. The United States can buy oil from anywhere (except where it imposes sanctions, like Iran), and it doesn't really matter if the oil comes from internal sources or imports. In fact, the United States exports some oil from Alaska, because it is more efficient to send that oil to Japan than it is to send it down to refineries in California. [. . .] The world oil markets are very competitive, and the locating, capturing, refining, and selling of oil is a very complex system. Saying that the United States is too dependent on foreign oil and that this will spell disaster in the near future is not an accurate statement. A more accurate statement would be to say that we have many wasteful energy habits, and that we need to focus on how to reduce our energy use and to expand alternative energy sources without drastically affecting our lifestyles and our economy. But oil is essential to our country's normal functioning, and therefore more American oil must be part of an American energy solution.

Why Drill Offshore for New Oil?

Is more drilling for American oil an essential part of lowering energy costs and freeing us from dependence on foreign sources of energy? Opening up new areas for exploration in the Outer Continental Shelf and the Alaska National Wildlife Refuge in the United States, even if new supplies won't actually reach our gas tanks for several years, would immediately impact the amount of upward speculation on long-term commodity investment in oil. Oil speculators would see a greater supply ahead and that the future of oil would be less constrained on the supply side. Also, fears of Middle Eastern turmoil or South American unrest that could disrupt supply shipments would be much less of a reason to drive up the price of crude if a stable United States could supply additional millions of barrels of oil.

Today, oil drilling is prohibited in all offshore regions along the North Atlantic coast, most of the Pacific coast, parts of the Alaska coast, and most of the eastern Gulf of Mexico. The central and western portions of the Gulf of Mexico therefore account for almost all current domestic offshore oil production, providing 27 percent of the United States' domestic oil production. The areas under the congressional ban contain an estimated additional 18 billion barrels of oil. This estimate is considered conservative since little exploration has been conducted in most of those areas during the past quarter of a century due to the congressional ban. Estimates tend to increase dramatically as technology improves and exploration activities occur. [. . .] Major advances in seismic technology and deep-water drilling techniques have already led the Interior Department's Minerals Management Service to increase its original estimate of untapped Gulf of Mexico oil from 9 billion barrels to 45 billion barrels. In short, there could be much more oil under the sea than previously thought.

The Interior Department has already taken steps for new offshore oil exploration, announcing plans for a lease program that could open up new areas off the coasts of Florida, Georgia, Texas, North Carolina, Virginia, and other coastal states to drilling if Congress lifts the ban. Randall Luthi, director of the department's Minerals Management Service, which handles offshore oilfield regulations and leases, said, "The technology has improved . . . the safety systems we now require have greatly improved . . . and the industry has a good record." According to Luthi, new tools such as high-tech computers that make exploration easier and tougher building materials on platforms are making offshore drilling safer. Seismic technology and directional drilling techniques let oil companies drill 100 exploratory wells from a single offshore platform, reducing the number of derricks and therefore the potential for problems. Automatic shut-off valves underneath the seabed can cut the flow of oil immediately if there's a problem or a storm coming. Blowout prevention equipment can automatically seal off pipes leading to the surface in the case of an unexpected pressure buildup, and undersea pipelines and wellheads can be monitored with special equipment such as unmanned, camera-equipped, and sensor-laden underwater vehicles. [. . .] New drilling technologies and the industry's track record in the Gulf of Mexico show that offshore drilling for oil is safer than it ever has been, proving that you can drill and still be environmentally friendly.

Conclusion

The demand for energy is going up, not down, and for a long time, even as alternative sources of energy are developed, more oil will be needed. The strongest argument against drilling is that it could distract the country from the pursuit of alternative sources of energy. The United States cannot drill its way to energy independence. But with the developing economies of China and India steadily increasing their oil needs in their latter-day industrial revolutions, the United States can no longer afford to turn its back on finding all the sources of fuel necessary to maintain its economy and standard of living. What is required is a long-term, comprehensive plan that includes wind, solar, geothermal, biofuels, and nuclear—and that acknowledges that oil and gas will be instrumental to the United States' well-being for many years to come.

 **NO**

The Environmental Impacts of Offshore Oil Drilling

Stephen L. Baird's article in the November 2008 issue of *The Technology Teacher* describes a contemporary debate about opening more U.S. land and coastal regions to oil and gas exploration and production (E&P). While Baird's thesis—"informed and rational decisions can be reached through the understanding of how complex technological systems can impact the environment, our economy, our politics, and ultimately our culture" [. . .] epitomizes the goal of a technologically literate citizen, his article is a stark contradiction to this call for understanding. His one-sided argument is built upon public opinion driven by consumer demand for cheap energy, economic trade imbalance, and politics. Baird fails to connect the offshore oil and gas E&P to the toxins and greenhouse gases these technological processes release to the marine environment and the atmosphere. Decades of empirical evidence indicates that all stages of offshore oil and gas activity have consequences for the health and survivability of marine plants and animals, humans, and our planet.

In the following, I counter Baird's proposition that "environmental arguments no longer add up" [. . .] by identifying a few of the impacts of offshore oil E&P on the environment. My hope is that this analysis will better prepare teachers to foster the development of *critical-thinking* skills in their students. These skills are prerequisite to assessing the impacts of technology upon the environment and society [. . .] and essential for making environmentally sustainable choices.

Technology Assessment

When we ask our students to assess the impacts of technology, we ask them to engage in a process of inquiry, a cognitive journey of questioning assumptions, hypothesizing, gathering and reviewing evidence and trends, and testing their hypotheses against the body of evidence. This process, known as Technology Assessment (TA), refers to an examination of the potential or existing risks and consequences of developing, adopting, or using a technology. TA begins by bounding the study to identify time horizons, impact zones, stakeholders, and a host of relevant technical and environmental information. Tools, such as cross-impact analysis, mathematical models, and regression analysis, are used to analyze this data and to predict outcomes and risks associated with possible decisions. TA results in a list of "if/then" statements, options, trade-offs, or alternative future scenarios, which decision makers use to inform policies, make investments, and plan for the future.

Nature of Petroleum (Hydrocarbons)

To examine environmental impacts, we should begin by looking at the nature of crude oil (petroleum). Petroleum is a fossil fuel that forms from the remains of prehistoric vegetation and animals as a result of millions of years of heat and pressure. It is a complex mixture of hydrocarbons, several minor constituents (e.g., sulfur), and trace metals (e.g., chromium). The chemical composition of crude oil varies by the age of the geologic formation from which it came. When crude oil is released into the environment, biological, physical, and chemical processes (referred to as weathering) alter the oil's original characteristics. [. . .] Lighter oils tend to be volatile, reactive, and highly flammable, while heavier crudes tend to be tarry and waxy and contain cancer-causing polycyclic aromatic hydrocarbons (PAH) and other toxic substances.

Offshore Oil Exploration and Production

One challenge for offshore (waters beyond three miles from the shoreline) E&P operators is to control the dynamic changes in temperature and pressures when drilling into rock formations located deep beneath the ocean. As depths increase, the pressure of drilling fluids (muds) is used to counter the deep-sea pressures related to depth and the pockets of high-pressure and high-temperature (HPHT) gases. If not contained, these HPHT gases result in dangerous oil-well blowouts that emit a buoyant plume of oil, produced water, and pressurized natural gas (methane).

The 1969 blowout off the Santa Barbara, California coast that spilled 80-100,000 gallons of crude oil and inundated local beaches was an ecological disaster, killing thousands of birds, fish, and marine mammals. [. . .] However, this pales in comparison to the 1979 blowout at the Ixtoc 1 offshore oil rig in the Gulf of Mexico, which spewed an estimated 140 million gallons of crude oil until that well was capped nine months later. [. . .]

Modern technology and the research and monitoring systems of the Department of Interior's Minerals Management Service (MMS), which manages E&P in the outer continental shelf (OCS), have reduced the frequency of these ecological catastrophes. However, weather, tectonic events, equipment failure, transportation accidents, human error, and deliberate unethical choices continue to make oil spills a reality.

Today, there are nearly 4,000 active platforms in the OCS. [. . .] MMS [. . .] indicates that 115 platforms were destroyed and 600 offshore pipelines were damaged by Hurricanes Katrina and Rita in 2005. For these hurricanes, "124 spills were reported with a total volume of roughly 17,700 barrels of total petroleum products, of which about 13,200 barrels were crude oil and condensate from platforms, rigs, and pipelines, and 4,500 barrels were refined products from platforms and rigs." [. . .] In 2008, 60 platforms were destroyed during Hurricanes Gustav and Ike [. . .]; data on oil spills and damage to pipe-lines has not yet been released.

Offshore E&P also requires transportation vessels and terrestrial storage. As a direct result of Hurricane Katrina, an above-ground storage tank in St. Bernard Parish, Louisiana, spilled over 25,110 barrels of mixed crude oil. [. . .] The oil inundated about 1,700 homes and several canals. Testing conducted in 2006 confirmed that contaminants, including PAHs, arsenic, and other toxics were above acceptable risk standards. [. . .]

Wastes from Offshore E&P

The less dramatic, yet more disturbing impacts of offshore oil E&P relate to the volume and type of wastes these processes generate. Wastes include pro-duced water, drilling fluids (muds), cuttings (crushed rock), diesel emissions, and chemicals associated with operating mechanical, hydraulic, and electrical equipment, such as biocides, solvents, and corrosion inhibitors.

Produced Water. By volume, 98% of the waste from E&P is *produced water,* with estimates at 480,000 barrels per day in 1999. [. . .] Produced water is a water mixture consisting of hydrocarbons (e.g., PAH, organic acids, phenols, and volatiles), naturally occurring radioactive materials, dissolved solids, and chemical additives used during drilling. Glickman [. . .] concluded that "hydro-carbons are likely contributors to produced water toxicity, and their toxicities are additive, so that although individually the toxicities may be insignificant, when combined, aquatic toxicity can occur." [. . .] Furthermore, studies docu-ment that sediments become contaminated with these toxins and that the concentration has a direct correlation with produced water discharges. [. . .]

Citing "relief to coastal waters, which support spawning grounds, nurs-eries, and habitats for commercial and recreational fisheries: reducing doc-umented aquatic 'dead zone' impacts; reduction of potential cancer risks to anglers from consuming seafood contaminated by produced water radionu-clides; and reducing potential exposure of endangered species to toxic con-taminants" the EPA [. . .] banned the release of produced waters to inland and coastal waters (extending to three miles from shore). However, offshore E&P operations in U.S. waters may legally discharge treated produced water directly into the ocean or inject it into underground wells.

Drilling Fluids. Drilling muds and cuttings are of environmental concern because of their potential toxicity and the large volume that are discharged during drilling. Three types occur, including oil-based (OBM; diesel or mineral oil serves as base fluid), water-based (WBM), and synthetic-based muds (SBM). The EPA [. . .] requires zero discharge of OBM—to dispose, OBM is shipped to onshore oil field waste sites or injected into disposal wells at sea. WBM and SBM typically contain arsenic, barium, cadmium, chromium, copper, iron, lead, mercury, and zinc. Barium and barium compounds are used in drilling muds because they act as lubricants and increase the density of mud, thereby sealing gases in the well.

Drilling in deep water (>1,000 ft) uses rotary bits that chip through thousands of feet of rock to access oil and gas deposits. Diesel-powered engines provide the power to operate the drilling rig and drive the drill. As paraphrased from Continental Shelf Associates, Inc. [. . .], the general sequence of events entails initial "open hole" drilling where a drill bit positioned within a drill pipe chips away at the ocean floor. As the drill bit spins, mud (WBM) and water are forced at high velocity around the drill bit to force rock chips (cuttings) up and out to the seabed. After a known distance, the drill bit and pipe are removed and a wellhead is installed. WBM is typically discharged and replaced with SBM. Additionally, a marine riser system is connected to the wellhead to return fluids, muds, and cuttings to the drill rig where they are separated. Cuttings are discharged in a plume from the platform, and mud is recycled back to the drill bit. [. . .]

Continental Shelf Associates, Inc. [. . .] examined the impact of synthetic-based drilling fluids (SBF: mixtures of organic isomers) by comparing indicators from near and far distances from four E&P drilling sites in the Gulf of Mexico. [. . .] Cuttings and SBM extended several hundred meters from the well site and up to 45 cm in thickness. Analyses indicated that near-field sediments were toxic to amphipods (crustaceans). Chemicals associated with both WBM and SBM waste solids in near-field sediments contributed to sediment toxicity. Significantly higher mercury and lead concentrations were found in near-field sediments than in far-field sediments for some sites. Red crabs had high concentrations of toxins, such as arsenic, barium, chromium, and mercury.

Ethical Climate

Offshore E&P is conducted by a small number of operators, most of which are multinational oil corporations with single-year income larger than the GDP of entire nations; e.g., ExxonMobil Corporation [. . .] reported $40.610 billion in net income for 2007. The isolated nature of offshore drilling operations fosters a climate conducive to environmentally irresponsible behaviors. For instance, as reported by the EPA [. . .], two large electrical transformers located on Platform Hondo, Exxon's Santa Ynez Unit, leaked nearly 400 gallons of fluids contaminated with PCBs into the Pacific Ocean. In one instance, a transformer leaked for almost two years before repairs were made. Cleanup workers were not provided protective equipment to protect themselves against direct contact with and inhalation of PCBs. Exxon agreed to a settlement of $2.64 M in violation of the federal *Toxic Substances Control Act*.

Environmental Impacts

There are known detrimental impacts upon the marine environment for all phases of offshore E&P. [. . .] While natural seepages contribute more hydrocarbons to the marine environment by volume, the quick influx and concentration of oil during a spill makes them especially harmful to localized marine organisms and communities. Plants and animals that become coated in oil perish from mechanical smothering, birds die from hypothermia as their feathers lose their waterproofing, turtles die after ingesting oil-coated food, and animals become disoriented and exhibit other behavior changes after breathing volatile organic compounds.

When emitted into the marine environment, oil, produced water, and drilling muds may adversely impact an entire population by disrupting its food chain and reproductive cycle. Marine estuaries are especially susceptible, as hydrocarbons and other toxins tend to persist in the sediments where eggs and young often begin life. However, the severity and effects of oil exposure vary by concentration, season, and life stage. The oil spill from the Ixtoc 1 blowout threatened a rare nesting site of the Kemp's Ridley sea turtle, an endangered species. Field and laboratory data on the nests of turtle eggs found a significant decrease in survival of hatchlings, and some hatchlings had developmental deformities. [. . .]

Marine organisms that live near an existing or sealed wellhead or an oil spill area experience persistent exposure to a complex web of hydrocarbons, petroleum-degrading microbes, and toxic substances associated with drilling muds and produced water. Abundance and diversity of marine life, especially those living near or in the seabed, decline. The growth and reproduction rates of entire populations that live in the water column may decline for months after a spill [. . .], natural defense mechanisms necessary to deal with disease (immune suppression) become compromised [. . .], and genetic mutations may occur. Many of these toxins (e.g., arsenic, chromium, mercury, and PAH) move up the food chain and biomagnify, i.e., increase in concentration.

One of the most disturbing trends is the evidence that common hydro-carbon contaminants (e.g., PAH) act as endocrine disrupters. Endocrine disrupters are chemicals that can act as hormones or anti-hormones in aquatic ecosystems, thus disrupting normal reproductive and developmental patterns. [. . .] Evidence also suggests that polychlorinated biphenyls (PCBs), a known carcinogen, also exhibit these endocrine effects. [. . .]

Climate Change and Ocean Acidification.   Greenhouse gases (GHG) are generated directly by offshore oil E&P and indirectly by enabling future emissions of oil as it is refined, distributed, and consumed, primarily by the transportation sector. In the U.S., 2007 emissions of methane from petroleum E&P was estimated at 22 $MMTCO_2e$ (million metric tons of carbon dioxide equivalents) in addition to total contributions of petroleum at 2,579.9 $MMTCO_2e$. [. . .] These GHGs are primary drivers that disrupt the carbon cycle involving the biosphere, atmosphere, sediments (including fossil fuels), and the ocean. The consequences of this disruption include climate change (global warming, melting of ice at the poles, and ocean acidification). The ocean acts as a carbon sink, absorbing CO_2. As CO_2 increases, the acidity of the ocean increases and carbonate becomes less available to marine organisms that need it to build shells and skeletal material. Corals, calcareous phytoplankton, and mussels are especially susceptible to acidosis, which "can lead to lowered immune response, metabolic depression, behavioral depression affecting physical activity and reproduction, and asphyxiation." [. . .]

Human Health Impacts

Workers, victims of oil spills, and rescue workers are exposed to a host of chemical hazards. When people come in dermal contact with drilling fluids, muds, and cuttings, they can experience dermatitis; as exposure increases, impacts can include hypokalemia, renal toxicity, and cardiovascular and neuromuscular effects. [. . .] Exposure to volatile aromatic hydrocarbons (e.g., benzene) results in respiratory distress and unconsciousness. Long-term exposure can cause anemia, leukemia, reproductive problems, and developmental disorders. [. . .] Exposure to fine particulate matter, nitrogen oxides, sulphur, and dozens of hydrocarbons (e.g., PAH) emitted from diesel and gasoline engines, is linked to a variety of health impacts, including asthma attacks, cancer, endocrine disruption, and cardiopulmonary ailments. Because toxins bioaccumulate in fish, people who eat fish and shellfish from affected waters may experience nervous system effects, such as impairment of peripheral vision and seizure. Children and fetuses are especially vulnerable; exposure to toxins impairs physical and cognitive development.

Conclusion

The environmental impacts of offshore oil exploration and production are complex and far-reaching. Petroleum, produced water, and the chemicals used to extract petroleum from under the ocean have mechanical, toxic, carcinogenic, and mutagenic impacts on marine life and the humans who eat its marvelous bounty. But more profoundly, the combustion of petroleum is a major contributor of carbon dioxide to the atmosphere which, in turn, drives global warming and ocean acidification.

Therefore, the decision to open additional U.S. lands and oceanic territories to oil exploration and production should be based on more than public opinion and some unquestioned assumption that the U.S. has a right to consume about 25% of the world's energy and contribute 21% of the worlds' CO_2 emissions. Engaging students in a process of technology assessment is a viable pedagogy to help students develop dispositions and critical-thinking skills. These skills will enable students to not only recognize narrow, one-sided perspectives, but be better prepared to seek and apply valid data and analytical strategies to the critically important decisions that could impact life on this planet. In an age of complexity, these skills are essential for making environmentally sustainable choices. [. . .]

POSTSCRIPT

Should We Drill for Offshore Oil?

According to Steve Stein, "Energy Independence Isn't Very Green," *Policy Review* (April-May 2008), the effort to make the U.S. independent of foreign sources of oil has little to do with combating global warming or with other environmental issues. It is much more a matter of national security, and arguments in favor of energy independence would make more headway if separated from "green" concerns.

That said, the debate over whether to expand drilling for offshore oil is by no means over. In the wake of President Bush's lifting of the executive order banning offshore drilling, the federal Minerals Management Service (MMS) proposed 31 oil and gas lease sales in areas of the nation's Outer Continental Shelf. These areas are estimated to contain at least 86 billion barrels of oil and 420 trillion cubic feet of natural gas, although specific deposits have not yet been discovered. According to the MMS press release, MMS director Randall Luthi said, "We're basically giving the next Administration a two-year head start. This is a multi-step, multi-year process with a full environmental review and several opportunities for input from the states, other government agencies and interested parties, and the general public."

Reactions from states off whose shores these areas lie have been largely negative. Shortly after President Obama took office, the new Interior Department Secretary, Ken Salazar, canceled the Bush Administration's Utah oil and gas leases and announced a new strategy for offshore energy development that would include oil, gas, and renewable resources such as wind power. The press release (http://www.mms.gov/ooc/press/2009/press0210.htm) said that Salazar's strategy calls for extending the public comment period on a proposed 5-year plan for oil and gas development on the U.S. Outer Continental Shelf by 180 days, assembling a detailed report from Interior agencies on conventional and renewable offshore energy resources, holding four regional conferences to review these findings, and expediting renewable energy rulemaking for the Outer Continental Shelf.

"To establish an orderly process that allows us to make wise decisions based on sound information, we need to set aside the Bush Administration's midnight timetable for its OCS drilling plan and create our own timeline," Salazar said.

By mid-February 2009, the House of Representatives' Committee on Natural Resources had already held the first of three scheduled hearings, with the list of witnesses testifying against offshore drilling including such prominent environmentalists as actor Ted Danson. Subsequent hearings featured representatives of the states and the oil and gas industry. See http://resourcescommittee.house.gov/index.php?option=com_content&task=view&id=502&Itemid=27.

In California, the State Lands Commission rejected a plan to permit off-shore drilling. When Governor Schwarzenlegger attempted to overturn the rejection on the grounds that the recession of 2008–2009 had created a budget crisis that demanded solution, controversy erupted. See Bernie Woodall, "California Board Urges No New Oil Drilling Offshore," Reuters (June 1, 2009) (http://www.reuters.com/article/environmentNews/idUSTRE5507UB20090601). We take it for granted that government should control access to resources such as those on the continental shelf and on public lands. However, see John Bratland, "Externalities, Conflict, and Offshore Lands," *Independent Review* (Spring 2004). Bratland argues that fostering private ownership of offshore lands and requiring that external costs of oil drilling (such as pollution) be internalized could help address the problems of offshore drilling. He finds intriguing the notion that environmental organizations might choose to protect offshore lands by becoming their owners.

ISSUE 8

Is Carbon Capture Technology Ready to Limit Carbon Emissions?

YES: David G. Hawkins, from "Carbon Capture and Sequestration," Testimony before the House Committee on Energy and Commerce, Subcommittee on Energy and Air Quality (March 6, 2007)

NO: Charles W. Schmidt, from "Carbon Capture & Storage: Blue-Sky Technology or Just Blowing Smoke?" *Environmental Health Perspectives* (November 2007)

ISSUE SUMMARY

YES: David G. Hawkins, director of the Climate Center of the Natural Resources Defense Council, argues that we know enough to implement large-scale carbon capture and sequestration for new coal plants. The technology is ready to do so safely and effectively.

NO: Charles W. Schmidt argues that the technology is not yet technically and financially feasible, research is stuck in low gear, and the political commitment to reducing carbon emissions is lacking.

It is now well established that burning fossil fuels is a major contributor to global warming, and thus a hazard to the future well-being of human beings and ecosystems around the world. The reason lies in the release of carbon dioxide, a major "greenhouse gas." It follows logically that if we reduce the amount of carbon dioxide we release to the atmosphere, we must prevent or ease global warming. Such a reduction would of course follow if we shifted away from fossil fuels as an energy source. Another option is to capture the carbon dioxide before it reaches the atmosphere and put it somewhere else.

The question is: Where? Lal Rattan, "Carbon Sequestration," *Philosophical Transactions: Biological Sciences* (February 2008), describes a number of techniques and notes that all are expensive and have leakage risks. One proposal is that supplying nutrients to ocean waters could stimulate the growth of algae, which removes carbon dioxide from the air; when the algae die, the carbon should settle to the ocean floor. The few experiments that have been done so far indicate that though fertilization does in fact stimulate algae growth, carbon does not always settle deep enough to keep it out of the air for long.

The experiments also fail to say whether the procedure would damage marine ecosystems. See Eli Kintisch, "Should Oceanographers Pump Iron?" *Science* (November 30, 2007), and Sandra Upson, "Algae Bloom Climate-Change Scheme Doomed," *IEEE Spectrum* (January 2008). Another proposal is that carbon dioxide be concentrated from power plant exhaust streams, liquefied, and pumped to the deep ocean, where it would remain for thousands of years.

Concern that storage time is not long enough helped shift most attention to underground storage. Carbon dioxide, in either gas or liquid form, can be pumped into porous rock layers deep beneath the surface. Such layers are accessible in the form of depleted oil deposits, and, in fact, carbon dioxide injection can be used to force residual oil out of the deposits. See Robert H. Sokolow, "Can We Bury Global Warming?" *Scientific American* (July 2005); Valerie Brown, "A Climate Change Solution?" *High Country News* (September 3, 2007), and S. Bhatia, "Carbon Capture and Storage: Solution or a Challenge," *Hydrocarbon Processing* (November 2008). Jennie C. Stephens and Bob Van Der Zwann, "The Case for Carbon Capture and Storage," *Issues in Science and Technology* (Fall 2005). Note that the technology exists but industry lacks incentives to implement it; such incentives could be supplied if the federal government established limits for carbon emissions. Additional research is also needed to determine the long-term stability of carbon storage. Long-term risk management will also need careful attention; see Melisa Pollak and Elizabeth J. Wilson, "Risk Governance for Geological Storage of CO_2 under the Clean Development Mechanism," *Climate Policy (Earthscan)* (2009).

In the following selections, David G. Hawkins, Director of the Climate Center of the Natural Resources Defense Council, argues that we know enough to implement large-scale carbon capture and sequestration for new coal plants. The technology is ready to do so safely and effectively. Charles W. Schmidt argues that the technology is not yet technically and financially feasible, research is stuck in low gear, and the political commitment to reducing carbon emissions is lacking. In addition, it has not been shown that carbon dioxide stored in underground reservoirs will stay in place indefinitely, and it has not been decided who will monitor such storage and take responsibility if it fails.

YES

David G. Hawkins

Carbon Capture and Sequestration

Today, the U.S. and other developed nations around the world run their economies largely with industrial sources powered by fossil fuel and those sources release billions of tons of carbon dioxide (CO_2) into the atmosphere every year. There is national and global interest today in capturing that CO_2 for disposal or sequestration to prevent its release to the atmosphere. To distinguish this industrial capture system from removal of atmospheric CO_2 by soils and vegetation, I will refer to the industrial system as carbon capture and disposal or CCD.

The interest in CCD stems from a few basic facts. We now recognize that CO_2 emissions from use of fossil fuel result in increased atmospheric concentrations of CO_2, which along with other so-called greenhouse gases, trap heat, leading to an increase in temperatures, regionally and globally. These increased temperatures alter the energy balance of the planet and thus our climate, which is simply nature's way of managing energy flows. Documented changes in climate today along with those forecasted for the next decades, are predicted to inflict large and growing damage to human health, economic well-being, and natural ecosystems.

Coal is the most abundant fossil fuel and is distributed broadly across the world. It has fueled the rise of industrial economies in Europe and the U.S. in the past two centuries and is fueling the rise of Asian economies today. Because of its abundance, coal is cheap and that makes it attractive to use in large quantities if we ignore the harm it causes. However, per unit of energy delivered, coal today is a bigger global warming polluter than any other fuel: double that of natural gas; 50 per cent more than oil; and, of course, enormously more polluting than renewable energy, energy efficiency, and, more controversially, nuclear power. To reduce coal's contribution to global warming, we must deploy and improve systems that will keep the carbon in coal out of the atmosphere, specifically systems that capture carbon dioxide (CO_2) from coal-fired power plants and other industrial sources for safe and effective disposal in geologic formations. . . .

The Need for CCD

Turning to CCD, my organization supports rapid deployment of such capture and disposal systems for sources using coal. Such support is not a statement about how dependent the U.S. or the world should be on coal and for how

From *U.S. House of Representatives Committee on Energy and Commerce* by David G. Hawkins, (March 6, 2007).

long. Any significant additional use of coal that vents its CO_2 to the air is fundamentally in conflict with the need to keep atmospheric concentrations of CO_2 from rising to levels that will produce dangerous disruption of the climate system. Given that an immediate world-wide halt to coal use is not plausible, analysts and advocates with a broad range of views on coal's role should be able to agree that, if it is safe and effective, CCD should be rapidly deployed to minimize CO_2 emissions from the coal that we do use.

Today coal use and climate protection are on a collision course. Without rapid deployment of CCD systems, that collision will occur quickly and with spectacularly bad results. The very attribute of coal that has made it so attractive—its abundance—magnifies the problem we face and requires us to act now, not a decade from now. Until now, coal's abundance has been an economic boon. But today, coal's abundance, absent corrective action, is more bane than boon.

Since the dawn of the industrial age, human use of coal has released about 150 billion metric tons of carbon into the atmosphere—about half the total carbon emissions due to fossil fuel use in human history. But that contribution is the tip of the carbon iceberg. Another 4 *trillion* metric tons of carbon are contained in the remaining global coal resources. That is a carbon pool nearly seven times greater than the amount in our pre-industrial atmosphere. Using that coal without capturing and disposing of its carbon means a climate catastrophe. And the die is being cast for that catastrophe today, not decades from now. Decisions being made today in corporate board rooms, government ministries, and congressional hearing rooms are determining how the next coal-fired power plants will be designed and operated. Power plant investments are enormous in scale, more than $1 billion per plant, and plants built today will operate for 60 years or more. The International Energy Agency (IEA) forecasts that more than $5 trillion will be spent globally on new power plants in the next 25 years. Under IEA's forecasts, over 1800 gigawatts (GW) of new coal plants will be built between now and 2030—capacity equivalent to 3000 large coal plants, or an average of ten new coal plants every month for the next quarter century. This new capacity amounts to 1.5 times the total of all the coal plants operating in the world today.

The astounding fact is that under IEA's forecast, 7 out of every 10 coal plants that will be operating in 2030 don't exist today. That fact presents a huge opportunity—many of these coal plants will not need to be built if we invest more in efficiency; additional numbers of these coal plants can be replaced with clean, renewable alternative power sources; and for the remainder, we can build them to capture their CO_2, instead of building them the way our grandfathers built them.

If we decide to do it, the world could build and operate new coal plants so that their CO_2 is returned to the ground rather than polluting the atmosphere. But we are losing that opportunity with every month of delay—10 coal plants were built the old-fashioned way last month somewhere in the world and 10 more old-style plants will be built this month, and the next and the next. Worse still, with current policies in place, none of the 3000 new plants projected by IEA are likely to capture their CO_2.

Each new coal plant that is built carries with it a huge stream of CO_2 emissions that will likely flow for the life of the plant—60 years or more. Suggestions that such plants might be equipped with CO_2 capture devices later in life might come true but there is little reason to count on it. As I will discuss further in a moment, while commercial technologies exist for pre-combustion capture from gasification-based power plants, most new plants are not using gasification designs and the few that are, are not incorporating capture systems. Installing capture equipment at these new plants after the fact is implausible for traditional coal plant designs and expensive for gasification processes.

If all 3000 of the next wave of coal plants are built with no CO_2 controls, their lifetime emissions will impose an enormous pollution lien on our children and grandchildren. Over a projected 60-year life these plants would likely emit 750 billion tons of CO_2, a total, from just 25 years of investment decisions, that is 30% greater than the total CO_2 emissions from all previous human use of coal. Once emitted, this CO_2 pollution load remains in the atmosphere for centuries. Half of the CO_2 emitted during World War I remains in the atmosphere today. In short, we face an onrushing train of new coal plants with impacts that must be diverted without delay. What can the U.S. do to help? The U.S. is forecasted to build nearly 300 of these coal plants, according to reports and forecasts published by the U.S. EIA. By taking action ourselves, we can speed the deployment of CO_2 capture here at home and set an example of leadership. That leadership will bring us economic rewards in the new business opportunities it creates here and abroad and it will speed engagement by critical countries like China and India.

To date our efforts have been limited to funding research, development, and limited demonstrations. Such funding can help in this effort if it is wisely invested. But government subsidies—which are what we are talking about—cannot substitute for the driver that a real market for low-carbon goods and services provides. That market will be created only when requirements to limit CO_2 emissions are adopted. This year in Congress serious attention is finally being directed to enactment of such measures and we welcome your announcement that you intend to play a leadership role in this effort.

Key Questions about CCD

I started studying CCD in detail ten years ago and the questions I had then are those asked today by people new to the subject. Do reliable systems exist to capture CO_2 from power plants and other industrial sources? Where can we put CO_2 after we have captured it? Will the CO_2 stay where we put it or will it leak? How much disposal capacity is there? Are CCD systems "affordable"? To answer these questions, the Intergovernmental Panel on Climate Change (IPCC) decided four years ago to prepare a special report on the subject. That report was issued in September, 2005 as the IPCC Special Report on Carbon Dioxide Capture and Storage. I was privileged to serve as a review editor for the report's chapter on geologic storage of CO_2.

CO_2 Capture

The IPCC special report groups capture or separation of CO_2 from industrial gases into four categories: post-combustion; pre-combustion; oxyfuel combustion; and industrial separation. I will say a few words about the basics and status of each of these approaches. In a conventional pulverized coal power plant, the coal is combusted using normal air at atmospheric pressures. This combustion process produces a large volume of exhaust gas that contains CO_2 in large amounts but in low concentrations and low pressures. Commercial post-combustion systems exist to capture CO_2 from such exhaust gases using chemical "stripping" compounds and they have been applied to very small portions of flue gases (tens of thousands of tons from plants that emit several million tons of CO_2 annually) from a few coal-fired power plants in the U.S. that sell the captured CO_2 to the food and beverage industry. However, industry analysts state that today's systems, based on publicly available information, involve much higher costs and energy penalties than the principal demonstrated alternative, pre-combustion capture.

New and potentially less expensive post-combustion concepts have been evaluated in laboratory tests and some, like ammonia-based capture systems, are scheduled for small pilot-scale tests in the next few years. Under normal industrial development scenarios, if successful such pilot tests would be followed by larger demonstration tests and then by commercial-scale tests. These and other approaches should continue to be explored. However, unless accelerated by a combination of policies, subsidies, and willingness to take increased technical risks, such a development program could take one or two decades before post-combustion systems would be accepted for broad commercial application.

Pre-combustion capture is applied to coal conversion processes that gasify coal rather than combust it in air. In the oxygen-blown gasification process coal is heated under pressure with a mixture of pure oxygen, producing an energy-rich gas stream consisting mostly of hydrogen and carbon monoxide. Coal gasification is widely used in industrial processes, such as ammonia and fertilizer production around the world. Hundreds of such industrial gasifiers are in operation today. In power generation applications as practiced today this "syngas" stream is cleaned of impurities and then burned in a combustion turbine to make electricity in a process known as Integrated Gasification Combined Cycle or IGCC. In the power generation business, IGCC is a relatively recent development—about two decades old and is still not widely deployed. There are two IGCC power-only plants operating in the U.S. today and about 14 commercial IGCC plants are operating, with most of the capacity in Europe. In early years of operation for power applications a number of IGCC projects encountered availability problems but those issues appear to be resolved today, with Tampa Electric Company reporting that its IGCC plant in Florida is the most dispatched and most economic unit in its generating system.

Commercially demonstrated systems for pre-combustion capture from the coal gasification process involve treating the syngas to form a mixture of hydrogen and CO_2 and then separating the CO_2, primarily through the use of solvents. These same techniques are used in industrial plants to separate CO_2 from natural gas and to make chemicals such as ammonia out of gasified coal. However, because CO_2 can be released to the air in unlimited amounts under today's laws, except in niche applications, even plants that separate CO_2 do not capture it; rather they release it to the atmosphere. Notable exceptions include the Dakota Gasification Company plant in Beulah, North Dakota, which captures and pipelines more than one million tons of CO_2 per year from its lignite gasification plant to an oil field in Saskatchewan, and ExxonMobil's Shute Creek natural gas processing plant in Wyoming, which strips CO_2 from sour gas and pipelines several million tons per year to oil fields in Colorado and Wyoming.

Today's pre-combustion capture approach is not applicable to the installed base of conventional pulverized coal in the U.S. and elsewhere. However, it is ready today for use with IGCC power plants. The oil giant BP has announced an IGCC project with pre-combustion CO_2 capture at its refinery in Carson, California. When operational the project will gasify petroleum coke, a solid fuel that resembles coal more than petroleum to make electricity for sale to the grid. The captured CO_2 will be sold to an oil field operator in California to enhance oil recovery. The principal obstacle for broad application of pre-combustion capture to new power plants is not technical, it is economic: under today's laws it is cheaper to release CO_2 to the air rather than capturing it. Enacting laws to limit CO_2 can change this situation, as I discuss later.

While pre-combustion capture from IGCC plants is the approach that is ready today for commercial application, it is not the only method for CO_2 capture that may emerge if laws creating a market for CO_2 capture are adopted. I have previously mentioned post-combustion techniques now being explored. Another approach, known as oxyfuel combustion, is also in the early stages of research and development. In the oxyfuel process, coal is burned in oxygen rather than air and the exhaust gases are recycled to build up CO_2 concentrations to a point where separation at reasonable cost and energy penalties may be feasible. Small scale pilot studies for oxyfuel processes have been announced. As with post-combustion processes, absent an accelerated effort to leapfrog the normal commercialization process, it could be one or two decades before such systems might begin to be deployed broadly in commercial application.

Given the massive amount of new coal capacity scheduled for construction in the next two decades, we cannot afford to wait until we see if these alternative capture systems prove out, nor do we need to. Coal plants in the design process today can employ proven IGCC and pre-combustion capture systems to reduce their CO_2 emissions by about 90 percent. Adoption of policies that set a CO_2 performance standard now for such new plants will not anoint IGCC as the technological winner since alternative approaches can be employed when they are ready. If the alternatives prove superior to IGCC and pre-combustion capture, the market will reward them accordingly. As I will discuss later, adoption of CO_2 performance standards is a critical step to improve today's capture methods and to stimulate development of competing systems.

I would like to say a few words about so-called "capture-ready" or "capture-capable" coal plants. I will admit that some years ago I was under the impression that some technologies like IGCC, initially built without capture equipment could be properly called "capture-ready." However, the implications of the rapid build-out of new coal plants for global warming and many conversations with engineers since then have educated me to a different view. An IGCC unit built without capture equipment can be equipped later with such equipment and at much lower cost than attempting to retrofit a conventional pulverized coal plant with today's demonstrated post-combustion systems. However, the costs and engineering reconfigurations of such an approach are substantial. More importantly, we need to begin capturing CO_2 from new coal plants without delay in order to keep global warming from becoming a potentially runaway problem. Given the pace of new coal investments in the U.S. and globally, we simply do not have the time to build a coal plant today and think about capturing its CO_2 down the road.

Implementation of the Energy Policy Act of 2005 approach to this topic needs a review in my opinion. The Act provides significant subsidies for coal plants that do not actually capture their CO_2 but rather merely have carbon "capture capability." While the Act limits this term to plants using gasification processes, it is not being implemented in a manner that provides a meaningful substantive difference between an ordinary IGCC unit and one that genuinely has been designed with early integration of CO_2 capture in mind. Further, in its FY2008 budget request, the administration seeks appropriations allowing it to provide $9 billion in loan guarantees under Title XVII of the Act, including as much as $4 billion in loans for "carbon sequestration optimized coal power plants." The administration request does not define a "carbon sequestration optimized" coal power plant and it could mean almost anything, including, according to some industry representatives, a plant that simply leaves physical space for an unidentified black box. If that makes a power plant "capture-ready," Mr. Chairman, then my driveway is "Ferrari-ready." We should not be investing today in coal plants at more than a billion dollars apiece with nothing more than a hope that some kind of capture system will turn up. We would not get on a plane to a destination if the pilot told us there was no landing site but options were being researched.

Geologic Disposal

We have a significant experience base for injecting large amounts of CO_2 into geologic formations. For several decades oil field operators have received high pressure CO_2 for injection into fields to enhance oil recovery, delivered by pipelines spanning as much as several hundred miles. Today in the U.S. a total of more than 35 million tons of CO_2 are injected annually in more than 70 projects. (Unfortunately, due to the lack of any controls on CO_2 emissions, about 80 per cent of that CO_2 is sources from natural CO_2 formations rather than captured from industrial sources. Historians will marvel that we persisted so long in pulling CO_2 out of holes in the ground in order to move it hundreds of miles and stick in back in holes at the same time we were recognizing the harm being caused by emissions of the same molecule from nearby large industrial sources.) In addition to this enhanced oil recovery experience, there are several other large injection projects in operation or announced. The longest running of these, the Sleipner project, began in 1996.

But the largest of these projects injects on the order of one million tons per year of CO, while a single large coal power plant can produce about five million tons per year. And of course, our experience with man-made injection projects does not extend for the thousand year or more period that we would need to keep CO_2 in place underground for it to be effective in helping to avoid dangerous global warming. Accordingly, the public and interested members of the environmental, industry and policy communities rightly ask whether we can carry out a large scale injection program safely and assure that the injected CO_2 will stay where we put it.

. . . In its 2005 report the IPCC concluded the following with respect to the question of whether we can safely carry out carbon injection operations on the required scale:

> With appropriate site selection based on available subsurface information, a monitoring programme to detect problems, a regulatory system and the appropriate use of remediation methods to stop or control CO_2 releases if they arise, the local health, safety and environment risks of geological storage would be comparable to the risks of current activities such as natural gas storage, EOR and deep underground disposal of acid gas.

The knowledge exists to fulfill all of the conditions the IPCC identifies as needed to assure safety. While EPA has authority regulate large scale CO_2 injection projects its current underground injection control regulations are not designed to require the appropriate showings for permitting a facility intended for long-term retention of large amounts of CO_2. With adequate resources applied, EPA should be able to make the necessary revisions to its rules in two to three years. We urge this Committee to act to require EPA to undertake this effort this year.

Do we have a basis today for concluding that injected CO_2 will stay in place for the long periods required to prevent its contributing to global warming? The IPCC report concluded that we do, stating:

> Observations from engineered and natural analogues as well as models suggest that the fraction retained in appropriately selected and managed geological reservoirs is very likely to exceed 99% over 100 years and is likely to exceed 99% over 1,000 years.

Despite this conclusion by recognized experts there is still reason to ask what are the implications of imperfect execution of large scale injection projects, especially in the early years before we have amassed more experience? Is this reason enough to delay application of CO_2 capture systems to new power plants until we gain such experience from an initial round of multi-million ton "demonstration" projects? To sketch an answer to this question, my colleague Stefan Bachu, a geologist with the Alberta Energy and Utilities Board, and I wrote a paper for the Eighth International Conference on Greenhouse Gas Control Technologies in June 2006. The obvious and fundamental point we made is that without CO_2 capture, new coal plants built during any "delay and research" period will put 100 per cent of their CO_2 into the air and may do so for their operating life if they were "grandfathered" from retrofit requirements. Those releases need to be compared to hypothetical leaks from early injection sites.

Our conclusions were that even with extreme, unrealistically high hypothetical leakage rates from early injection sites (10% per year), a long period to leak detection (5 years) and a prolonged period to correct the leak (1 year), a policy that delayed installation of CO_2 capture at new coal plants to await further research would result in cumulative CO_2 releases twenty times greater than from the hypothetical faulty injection sites, if power plants built during the research period were "grandfathered" from retrofit requirements. If this wave of new coal plants were all required to retrofit CO_2 capture by no later than 2030, the cumulative emissions would still be four times greater than under the no delay scenario. I believe that any objective assessment will conclude that allowing new coal plants to be built without CO_2 capture equipment on the ground that we need more large scale injection experience will always result in significantly greater CO_2 releases than starting CO_2 capture without delay for new coal plants now being designed.

The IPCC also made estimates about global storage capacity for CO_2 in geologic formations. It concluded as follows:

> Available evidence suggests that, worldwide, it is likely that there is a technical potential of at least about 2,000 $GtCO_2$ (545 GtC) of storage capacity in geological formations. There could be a much larger potential for geological storage in saline formations, but the upper limit estimates are uncertain due to lack of information and an agreed methodology.

Current CO_2 emissions from the world's power plants are about 10 Gt (billion metric tons) per year, so the IPCC estimate indicates 200 years of capacity if power plant emissions did not increase and 100 years capacity if annual emissions doubled.

Policy Actions to Speed CCD

As I stated earlier, research and development funding is useful but it cannot substitute for the incentive that a genuine commercial market for CO_2 capture and disposal systems will provide to the private sector. The amounts of capital that the private sector can spend to optimize CCD methods will almost certainly always dwarf what Congress will provide with taxpayer dollars. To mobilize those private sector dollars, Congress needs a stimulus more compelling than the offer of modest handouts for research. Congress has a model that works: intelligently designed policies to limit emissions cause firms to spend money finding better and less expensive ways to prevent or capture emissions.

Where a technology is already competitive with other emission control techniques, for example, sulfur dioxide scrubbers, a cap and trade program like that enacted by Congress in 1990, can result in more rapid deployment, improvements in performance, and reductions in costs. Today's scrubbers are much more effective and much less costly than those built in the 1980s. However, a CO_2 cap and trade program by itself may not result in deployment of CCD systems as rapidly as we need. Many new coal plant design decisions are being made literally today. Depending on the pace of required reductions under a global warming bill, a firm may decide to build a conventional coal plant and purchase credits from the cap and trade market rather than applying CCD systems to the plant. While this may appear to be economically rational in the short term, it is likely to lead to higher costs of CO_2 control in the mid and longer term if substantial amounts of new conventional coal construction leads to ballooning demand for CO_2 credits. Recall that in the late 1990's and the first few years of this century, individual firms thought it made economic sense to build large numbers of new gas-fired power plants. The problem is too many of them had the same idea and the resulting increase in demand for natural gas increased both the price and volatility of natural gas to the point where many of these investments are idle today.

Moreover, delaying the start of CCD until a cap and trade system price is high enough to produce these investments delays the broad demonstration of the technology that the U.S. and other countries will need if we continue substantial use of coal as seem likely. The more affordable CCD becomes, the more widespread its use will be throughout the world, including in rapidly growing economies like China and India. But the learning and cost reductions for CCD that are desirable will come only from the experience gained by building and operating the initial commercial plants. The longer we wait to ramp up this experience, the longer we will wait to see CCD deployed here and in countries like China.

Accordingly, we believe the best policy package is a hybrid program that combines the breadth and flexibility of a cap and trade program with well-designed performance measures focused on key technologies like CCD. One such performance measure is a CO_2 emissions standard that applies to new power investments. California enacted such a measure in SB1368 last year. It requires new investments for sale of power in California to meet a performance standard that is achievable by coal with a moderate amount of CO_2 capture.

Another approach is a low-carbon generation obligation for coal-based power. Similar in concept to a renewable performance standard, the low-carbon generation obligation requires an initially small fraction of sales from coal-based power to meet a CO_2 performance standard that is achievable with CCD. The required fraction of sales would increase gradually over time and the obligation would be tradable. Thus, a coal-based generating firm could meet the requirement by building a plant with CCD, by purchasing power generated by another source that meets the standard, or by purchasing credits from those who build such plants. This approach has the advantage of speeding the deployment of CCD while avoiding the "first mover penalty." Instead of causing the first builder of a commercial coal plant with CCD to bear all of the incremental costs, the tradable low-carbon generation obligation would spread those costs over the entire coal-based generation system. The builder of the first unit would achieve far more hours of low-carbon generation than required and would sell the credits to other firms that needed credits to comply. These credit sales would finance the incremental costs of these early units. This approach provides the coal-based power industry with the experience with a technology that it knows is needed to reconcile coal use and climate protection and does it without sticker shock.

A bill introduced in the other body, S. 309, contains such a provision. It begins with a requirement that one-half of one per cent of coal-based power sales must meet the low-carbon performance standard starting in 2015 and the required percentage increases over time according to a statutory minimum schedule that can be increased in specified amounts by additional regulatory action.

A word about costs is in order. With today's off the shelf systems, estimates are that the production cost of electricity at a coal plant with CCD could be as much as 40% higher than at a conventional plant that emits its CO_2. But the impact on average electricity prices of introducing CCD now will be very much smaller due to several factors. First, power production costs represent about 60% of the price you and I pay for electricity; the rest comes from transmission and distribution costs. Second, coal-based power represents just over half of U.S. power consumption. Third, and most important, even if we start now, CCD would be applied to only a small fraction of U.S. coal capacity for some time. Thus, with the trading approach I have outlined, the incremental costs on the units equipped with CCD would be spread over the entire coal-based power sector or possibly across all fossil capacity depending on the choices made by Congress. Based on CCD costs available in 2005 we estimate that a low-carbon generation obligation large enough to cover all forecasted new U.S. coal capacity through 2020 could be implemented for about a two per cent increase in average U.S. retail electricity rates.

Conclusions

To sum up, since we will almost certainly continue using large amounts of coal in the U.S. and globally in the coming decades, it is imperative that we act now to deploy CCD systems. Commercially demonstrated CO_2 capture systems exist today and competing systems are being researched. Improvements in current systems and emergence of new approaches will be accelerated by requirements to limit CO_2 emissions. Geologic disposal of large amounts of CO_2 is viable and we know enough today to conclude that it can be done safely and effectively. EPA must act without delay to revise its regulations to provide the necessary framework for efficient permitting, monitoring and operational practices for large scale permanent CO_2 repositories.

Finally CCD is an important strategy to reduce CO_2 emissions from fossil fuel use but it is not the basis for a climate protection program by itself.

Increased reliance on low-carbon energy resources is the key to protecting the climate. The lowest carbon resource of all is smarter use of energy; energy efficiency investments will be the backbone of any sensible climate protection strategy. Renewable energy will need to assume a much greater role than it does today. With today's use of solar, wind and biomass energy, we tap only a tiny fraction of the energy the sun provides every day. There is enormous potential to expand our reliance on these resources.

We have no time to lose to begin cutting global warming emissions. Fortunately, we have technologies ready for use today that can get us started.

Charles W. Schmidt  **NO**

Carbon Capture & Storage: Blue-Sky Technology or Just Blowing Smoke?

Towering 650 feet over the sea surface and spouting an impressive burning flare, it would be easy to mistake the Sleipner West gas platform for an environmental nightmare. Its eight-story upper deck houses 200 workers and supports drilling equipment weighing 40,000 tons. Located off the Norwegian coast, it ranks among Europe's largest natural gas producers, delivering more than 12 billion cubic feet of the fuel annually to onshore terminals by pipeline. Roughly 9% of the natural gas extracted here is carbon dioxide (CO_2), the main culprit behind global warming. But far from a nightmare, Sleipner West is actually a bellwether for environmental innovation. Since 1996, the plant's operators have stripped CO_2 out of the gas on-site and buried it 3,000 feet below the sea floor, where they anticipate it will remain for at least 10,000 years.

Operated by StatoilHydro, Norway's largest company, Sleipner is among the few commercial-scale facilities in the world today that capture and bury CO_2 underground. Many experts believe this practice, dubbed carbon capture and storage (sometimes known as carbon capture and sequestration, but in either case abbreviated CCS), could be crucial for keeping industrial CO_2 emissions out of the atmosphere. Sleipner injects 1 million tons of CO_2 annually into the Utsira Formation, a saline aquifer big enough to store 600 years' worth of emissions from all European power plants, company representatives say.

With mounting evidence of climate change—and predictions that fossil fuels could supply 80% of global energy needs indefinitely—the spotlight on CCS is shining as brightly as the Sleipner flare. A panel of experts from the Massachusetts Institute of Technology (MIT) recently concluded that CCS is "the critical enabling technology to reduce CO_2 emissions significantly while allowing fossil fuels to meet growing energy needs." The panel's views were presented in *The Future of Coal,* a report issued by MIT on 14 March 2007.

Environmental groups are split on the issue. Speaking for the Natural Resources Defense Council (NRDC), David Hawkins, director of the council's Climate Center and a member of the MIT panel's external advisory committee, says, "We believe [CCS] is a viable way to cut global warming pollution. . . . We

From *Environmental Health Perspectives*, 115(11), November 2007, pp. A538–A545. Copyright © 2007 by National Institute of Environmental Health Sciences. Reprinted by permission.

have the knowledge we need to start moving forward." Other environmental groups, including the World Resources Institute, Environmental Defense, and the Pew Center on Global Climate Change, have also come out in support of CCS. These groups view CCS as one among many alternatives (including renewable energy) for reducing CO_2 emissions.

Greenpeace is perhaps the most vocal critic of CCS. Truls Gulowsen, Greenpeace's Nordic climate campaigner, stresses that CCS deflects attention from renewable energy and efficiency improvements, which, he says, offer the best solutions to the problem of global warming. "Companies are doing a lot of talking about CCS, but they're doing little to actually put it into place," he says. "So, they're talking about a possible solution that they don't really want to implement now, and at the same time, they're trying to push for more coal, oil, and gas development instead of renewables, which we already know can deliver climate benefits."

Coal Use Drives CCS Adoption

The pressure to advance on CCS has been fueled by soaring coal use world-wide. China, which is building coal-fired power plants at the rate of two per week, surpassed the United States as the world's largest producer of greenhouse gases in June 2007, years earlier than predicted. Coal use in India and other developing nations is also on the rise, while the United States sits on the larg-est coal reserves in the world, enough to supply domestic energy needs for 300 years, states the MIT report. Coal already supplies more than 50% of U.S. electricity demand and could supply 70% by 2025, according to the Interna-tional Energy Agency. Meanwhile, coal-fired power plants already account for nearly 40% of CO_2 emissions worldwide, a figure that—barring some dramatic advance in renewable energy technology—seems poised to rise dramatically. During a 6 September 2007 hearing of the House Select Committee on Energy Independence and Global Warming, Chairman Edward Markey (D–MA) noted that more than 150 new coal plants are being planned in the United States alone, with another 3,000 likely to be built worldwide by 2030.

A mature CCS system would capture, transport, and inject those emis-sions underground to depths of at least 1 km, where porous rock formations in geologically favorable locations absorb CO_2 like a sponge. At those depths, high pressures and temperatures compress the gas into a dense, liquid-like "supercritical" state that displaces brine and fills the tiny pores between rock grains. Three types of geological formations appear especially promising for sequestration: saline (and therefore nonpotable) aquifers located beneath freshwater deposits; coal seams that are too deep or thin to be extracted eco-nomically; and oil and gas fields, where CO_2 stripped from fuels on-site can be injected back underground to force dwindling reserves to the surface, a process called "enhanced recovery." Using CO_2 for enhanced recovery has a long his-tory, particularly in southwestern Texas, where oil yields have been declining for decades.

Of these three options, saline aquifers—with their large storage capac-ity and broad global distribution—are considered the most attractive. Thomas

Sarkus, director of the Applied Science and Energy Technology Division of the DOE National Energy Technology Laboratory (NETL), suggests saline aquifers in the central United States could conceivably store 2,000 years' worth of domestic CO_2 emissions.

Apart from Sleipner, only two other industrial-scale CCS projects are in operation today. In Algeria, a joint venture involving three energy companies—StatoilHydro, BP, and Sonatrach—stores more than 1 million tons of CO_2 annually under a natural gas platform near In Salah, an oasis town in the desert. And in Weyburn, Canada, comparable volumes are being used by EnCana Corporation, a Canadian energy company, for enhanced recovery at an aging oil field. The CO_2 sequestered at Weyburn comes by pipeline from a coal gasification plant in Beulah, North Dakota, 200 miles away. Unlike other enhanced recovery projects—wherein the ultimate fate of CO_2 is not the primary concern—Weyburn combines fossil fuel recovery with research to study sequestration on a large scale.

What's needed now, says Jim Katzer, a visiting scholar at MIT's Laboratory for Energy and the Environment, are more large-scale demonstrations of CCS in multiple geologies, integrated with policies that address site selection, licensing, liability, and other issues. Katzer says there are a number of investigations that are investigating storage in the 5,000- to 20,000-ton-capacity range, and they're generating some useful information. "But," he says, "none of them are getting us to the answer we really need: how are we going to manage storage in the millions of tons over long periods of time?"

Paying for Storage

The task of managing carbon storage is nothing if not daunting: in the United States alone, coal plants produce more than 1.5 billion tons of CO_2 every year. Sequestering that amount of gas will require not only a vast new infrastructure of pipelines and storage sites but also that the country's coal plants adopt costly technologies for carbon capture. Most existing U.S. plants—indeed, most of the world's 5,000 coal-fired power plants, including the ones now being built in China—burn pulverized coal (PC) using technologies essentially unchanged since the Industrial Revolution. CO_2 can be extracted from PC plants only after the fuel has been burned, which is inefficient because the combustion emissions are highly diluted with air.

A more efficient approach is to capture highly concentrated streams of CO_2 from coal before it's burned. Precombustion capture is usually applied at integrated gasification combined cycle (IGCC) coal plants, which are extremely rare, numbering just five worldwide, according to Sarkus. IGCC plants cost roughly 20% more to operate because the gasification process requires additional power, which explains why there are so few of them.

Although they don't rule out the possibility, none of the industry sources interviewed for this article welcome the prospect of retrofitting traditional PC plants for carbon capture. That would require major plant modifications and could potentially double the cost of electricity to consumers, they say. But by ignoring existing facilities, industry will set back CCS expansion by decades—most PC plants in use today have been designed for lifetimes of 30 to 40 years.

Whatever path it takes, the transition to CCS will require enormous sums of money. When used for enhanced recovery, CO_2 is a commodity that pays for its own burial. But only a small fraction of the CO_2 generated by coal plants and other industrial processes is used for that purpose. Creating a broad CCS infrastructure will ultimately require a charge on carbon emissions that, according to calculations described in *The Future of Coal*, should total at least $30 per ton—$25 per ton for CO_2 capture and pressurization and $5 per ton for transportation and storage—with this figure rising annually in accordance with inflation.

Sally Benson, a professor of energy resources engineering at Stanford University, points to different ways to pay that charge. One is a tax on CO_2 emissions, an option she concedes has little political support. Funds could also be raised with a "cap-and-trade" system, which sets area-wide limits on CO_2 emissions that industries can meet by trading carbon credits on the open market. A cap-and-trade system for CO_2 has already been established by the European Union, which regulates the greenhouse gas to meet obligations under the Kyoto Protocol. Jeff Chapman, chief executive officer of the Carbon Capture and Storage Association, a trade group based in London, suggests the European cap-and-trade system could ultimately raise €62 billion.

In the United States, a national cap-and-trade system likely won't appear until the federal government regulates CO_2 as a pollutant, says Luke Popovich, vice president of external communications with the National Mining Association, a coal industry trade group in Washington, DC. In the meantime, individual states—for instance, California, which sets its own air quality standards per a waiver under the Clean Air Act—are planning for their own cap-and-trade systems. California regulates CO_2 under a state law called AB32, which directs industries to reduce all greenhouse gas emissions by 25% over the next 13 years. CCS may ultimately emerge on a state-by-state basis in this country, where charges on carbon emissions allow it, Benson suggests.

Technical Questions Remain

Until the early 1990s, most researchers involved in CCS worked in isolation. But in March 1992, more than 250 gathered for the first International Conference on Carbon Dioxide Removal in Amsterdam. Howard Herzog, a principal research engineer at the MIT Laboratory for Energy and the Environment and a leading expert on CCS, says attendees arrived as individuals but left as a research community that now includes funding agencies, industries, and nongovernmental organizations throughout the world. Unfortunately, that community doesn't have nearly the resources it needs to study CCS on a realistic scale, Katzer says. Indeed, *The Future of Coal* states emphatically that "government and private-sector programs to implement on a timely basis the large-scale integrated demonstrations needed to confirm the suitability of carbon sequestration are completely inadequate."

Absent sufficient evidence, most experts simply assume that vast amounts of sequestered CO_2 will stay in place without leaking to the atmosphere. They base that assumption on available monitoring data from the big three industrial projects—none of which have shown any evidence of CO_2 leakage from their underground storage sites, according to *The Future of Coal*—and also on expectations that buried CO_2 will behave in essentially the same way as underground fossil fuel deposits. "We're optimistic it will work," says Jeffrey Logan, a senior associate in the Climate, Energy, and Transport Program at the World Resources Institute. "The general theory is that if oil and gas resources can remain trapped for millions of years, then why not CO_2?"

Franklin Orr, director of the Global Climate and Energy Project at Stanford University, says monitoring data show that CO_2 injected underground for enhanced oil and gas recovery remains trapped there by the same geological structures that trapped the fuels for millions of years; specifically overlying shale deposits through which neither fossil fuels nor CO_2 can pass. Decades of research by the oil and gas industries, in addition to basic research in geology, have revealed the features needed for CO_2 sequestration, he says: "You're looking for deep zones with highly porous rocks—for instance, sandstone—capped by shale seals with low permeability. Sleipner and Weyburn are both good examples; both have thick shale caps that keep the CO_2 from getting out."

But Orr concedes that questions remain about how large amounts of CO_2 might behave underground. A key risk to avoid, he says, is leakage through underlying faults or abandoned wells that provide conduits to the atmosphere. Yousif Kharaka, a research hydrologist with the USGS in Menlo Park, California, says an unknown but possibly large number of orphaned or abandoned wells in the United States could pose a risk of leakage to the atmosphere. And that, he warns, would negate the climate benefits of sequestration.

The likelihood that CO_2 levels could accumulate and cause health or ecological injuries is minimal, Kharaka says, echoing the conclusions reached in *The Future of Coal*. He says CO_2 in air only becomes harmful to humans at concentrations of 3% or above, which is far higher than might be expected from slow leaks out of the ground. Nonetheless, the possibility that CO_2 leaking from underground storage sites might accumulate to harmful levels in basements or other enclosed spaces can't be discounted entirely, cautions Susan Hovorka, a senior research scientist at the Bureau of Economic Geology, a state-sponsored research unit at the University of Texas at Austin. "It's important that we manage this substance correctly," she says. "If you determine that there's a risk to confined places, then you have to provide adequate ventilation. But we have a high level of confidence that CO_2 will be retained at depth."

The greater concern, says Kharaka, is that migrating CO_2 might mix with brine, forming carbonic acid that could leach metals such as iron, zinc, or lead from the underlying rock. In some cases, acidified brine alone could migrate and mix with fresh groundwater, posing health risks through drinking or irrigation water, he says. Results from an investigation conducted near Houston, Texas, led by Hovorka as principal investigator along with Kharaka and other scientists from 21 organizations, indicate that CO_2 injected into saline aquifers produced sharp drops in brine pH, from 6.5 to around 3.5. These results were published in the September 2007 report *Water–Rock Interaction: Proceedings of the 12th International Symposium on Water–Rock Interaction, Kunming, China, 31 July–5 August, 2007.* Chemical analyses showed the brine contained high concentrations of iron and manganese, which suggests toxic metal contamination can't be ruled out, Hovorka says. "I'd describe this as a nonzero concern," she adds. "It's not something we should write off, but it's not a showstopper."

Experts in this area consistently point to the need for more detailed investigations of CO_2 movements at depth and their geochemical consequences. Hovorka's investigation was among the first of this kind, but its scale—just 1,600 tons—paled in comparison to realistic demands for CO_2 mitigation to combat climate change.

Constrained by inadequate funding, the DOE has put much of its CCS investment into a project dubbed "FutureGen." This initiative seeks to build a prototype coal-fired power plant that will integrate all three features of a CCS system, namely, carbon capture (achieved with IGCC technology), CO_2 transportation, and sequestration. Supported by the DOE and an alliance of industry partners, the four-year, $1.5 billion project was announced formally by President Bush in his 2002 State of the Union Address. Once operational, the plant will supply 275 megawatts of power (compared with the 600–1,300 megawatts supplied by typical U.S. coal plants), enough for 275,000 households. Sarkus, who is also the FutureGen director, says four potential sites for the plant and its CO_2 reservoirs—including two in Illinois and two in Texas—are under consideration. Final site selection, he says, will depend on community support, adequate transportation lines, and proximity to underground storage reservoirs.

The Bush administration's stance is that FutureGen will promote CCS advancements throughout the coal and utility industries. But many stakeholders don't think it goes far enough toward meeting existing needs; the project is "too much 'future' and not enough 'generation,'" quips Hawkins. "What we need is legislation that specifies future power plants must be outfitted with CCS, period." To that, Katzer adds, "FutureGen was announced in 2002, and they still haven't settled on site selection, nor have they resolved key design issues. Operations were set to begin in 2012, and now that's slipping back even further. Assuming you start in 2012 and operate for four years, you're looking at 2016 before you complete a single demonstration project. That stretches things out too far, and speaks to the need for several demonstration projects funded now by the U.S. government so we can deal with CO_2 emissions in a timely fashion."

The Developing Country Factor

With U.S. research efforts stuck in low gear, concerns over a comparable lack of progress in the developing world are growing. China already obtains more than 80% of its domestic electricity from coal. And with a relentless push for economic growth, lowering CO_2 emissions from its coal plants is a low priority. It's likely that none of China's coal-fired plants are outfitted for carbon capture, says Richard Lester, a professor of nuclear science and engineering at MIT.

"Given the scale and expansion of China's electric power sector, the eventual introduction of CCS there is going to be absolutely critical to global efforts to abate or reduce the atmospheric carbon burden," he says.

Meanwhile, India lags just a decade or less behind China in terms of its own economic growth, which is increasingly fueled by coal use, Katzer says. The key difference between the two countries, he says, has to do with planning for environmental and energy development. In China, Katzer explains, growth and environment strategies seem to be dictated at regional levels without any central coordination, which is ironic considering the country's socialist political structure. India, on the other hand, seems to have what Katzer calls a "master plan" for growth. "But they have no clue how to move forward in terms of CO_2 reductions," Katzer says. "What officials in India say to me is, 'We'll manage CO_2 if it doesn't cost too much.' That's the downside in all of this."

In the end, CCS seems to be stuck in a catch-22: In the view of the developing world, the United States and other wealthier nations should take the lead with respect to emissions reduction technology. Governments in wealthier nations, meanwhile—particularly the United States—look to industries in the free market for solutions to the problem. But U.S. industries say they can't afford large-scale research; in industry's opinion, the government should pay for additional studies that lay the groundwork for industry research and the technology's future implementation. The government, however, doesn't fund the DOE and other agencies at nearly the amounts required to achieve this. And at the same time, the two mechanisms that could possibly generate sufficient revenues for CCS—carbon taxes and cap-and-trade systems for CO_2 emissions—are trapped by perpetual political gridlock.

Leslie Harroun, a senior program officer at the Oak Foundation, a Geneva-based organization that funds social and environmental research, warns that industry might leverage the promise of CCS as a public relations strategy today while doing little to ensure its broad-based deployment tomorrow. "The coal industry's many proposals to build 'clean' coal plants that are 'capture ready' across the U.S. is a smokescreen," she asserts. "Coal companies are hoping to build new plants before cap-and-trade regulations go into effect—and they will, soon—with the idea that the plants and their greenhouse gas emissions will be grandfathered in until sequestration is technically and financially feasible. This is an enormously risky investment decision on their part, and morally irresponsible, but maybe they think there is power in numbers."

In a sense, the inertia surrounding CCS might reflect a collective wilt in the face of a seemingly overwhelming technical and social challenge. To make a difference for climate change, a CCS infrastructure will have to capture and store many billions of tons of CO_2 throughout the world for hundreds of years. Those buried deposits will have to be monitored by unknown entities far into the future. Many questions remain about who will "own" these deposits and thereby assume responsibility for their long-term storage. Meanwhile, industry and the government are at an impasse, with neither taking a leading role toward making large-scale CCS a reality. How this state of affairs ultimately plays out for health of the planet remains to be seen.

Whatever Happened to Deep Ocean Storage?

One CCS option that appears to have fallen by the wayside is deep ocean storage. Scientists have long speculated that enormous volumes of CO_2 could be stored in the ocean at depths of 3 km or more. High pressure would compress the CO_2, making it denser than seawater and thus enabling it to sink. So-called CO_2 lakes would hover over the sea floor, suggests Ken Caldeira, a Stanford University professor of global ecology.

"A coal-fired power plant produces a little under one kilogram of CO_2 for each kilowatt-hour of electricity produced," says Caldeira. "An individual one-gigawatt coal-fired power plant, . . . if completely captured and the CO_2 stored on the sea floor, would make a lake ten meters deep and nearly one kilometer square—and it [would grow] by that much each year."

But Caldeira and others acknowledge that deep ocean storage doesn't offer a permanent solution. Unless the gas is somehow physically confined, over time—perhaps 500 to 1,000 years—up to half the CO_2 would diffuse through the ocean and be released back into the atmosphere. Moreover, most life within CO_2 lakes would be extinguished. However, Caldeira believes this consequence would be balanced by the benefits of keeping the greenhouse gas out of the atmosphere, where under global warming scenarios it acidifies and endangers sea life at the surface.

No one knows precisely what would happen during deep ocean storage because it's never been tested. A planned experiment off the coast of Hawaii in the late 1990s, with participation of U.S., Norwegian, Canadian, and Australian researchers, was canceled because of opposition of local environmental activists. According to Caldeira, who previously co-directed the DOE's now-defunct Center for Research on Ocean Carbon Sequestration, government program managers who backed the Hawaiian study were laterally transferred, sending a signal that advocating for this type of research was politically dangerous for career bureaucrats. "Today, there's zero money going into it," Caldeira says. "Right now, ocean sequestration is dead in the water."

POSTSCRIPT

Is Carbon Capture Technology Ready to Limit Carbon Emissions?

One of the great concerns about carbon capture and sequestration (CCS) is that once immense amounts of carbon dioxide have been stored underground, it will leak out again, either slowly or—perhaps after an earthquake—suddenly. Study of past eras has suggested that sudden releases of carbon dioxide from volcanoes have led to rapid greenhouse warming, which reduced oxygen levels in the ocean and caused the buildup of toxic hydrogen sulfide, which in turn reached the air and killed plants and animals on land, resulting in mass extinctions such as the one 250 million years ago. See Peter D. Ward, "Impact from the Deep," *Scientific American* (October 2006). A smaller scale threat, exemplified by Cameroon's Lake Nyos, which released so much dissolved carbon dioxide in 1986 that it flowed downhill and suffocated almost 2,000 people, along with their domestic animals, has been cited by environmental justice groups protesting CCS legislation in California. See Valerie J. Brown, "Of Two Minds: Groups Square Off on Carbon Mitigation," *Environmental Health Perspectives* (November 2007).

Such threats should concern us, but so should the threat of global warming itself. In the long run, we must move to non-fossil fuel sources of energy, because even coal, as plentiful as it is, will not last forever. In the short run, we have coal-burning power plants that continue to emit carbon dioxide, and we are planning to build more. As the Royal Society of Chemistry (RSC) notes in "Can We Bury Our Carbon Dioxide Problem?" *Bulletin 3* (Spring 2006), CCS will require the use of energy and will therefore increase the burning of fossil fuels and the price of energy to the consumer. The RSC also notes that researchers are not sure that there is enough underground capacity for all the carbon dioxide that CCS would endeavor to keep out of the atmosphere. More research is needed in this area, as well as in finding better, more efficient methods of capturing carbon dioxide, which can account for three-quarters of the cost of CCS. Fortunately, researchers are developing new materials that may lower that cost significantly. See Rahul Banerjee, et al., "High-Throughput Synthesis of Zeolitic Imidazolate Frameworks and Application to CO_2 Capture," *Science* (February 15, 2008), Kevin Bullis, "A Better Way to Capture Carbon," *Technology Review* online (February 15, 2008) (http://www.technologyreview.com/Energy/20295/?a=f), and Sid Perkins, "Down with Carbon," *Science News* (May 10, 2008).

Even with much improved technology, any CCS program will require commitment from both industry and government. In this connection, it is discouraging to note that the U.S. Department of Energy has backed out of the

FutureGen project due to high costs and an inability to agree on funding with industry partners. FutureGen's objective was a clean coal–fired power plant incorporating the latest CCS technologies; it was scheduled to begin operation in 2012. See Jeff Tollefson, "Carbon Burial Buried," *Nature* (January 24, 2008).

In 2009, however, the U.S. Government Accounting Office urged the Department of Energy to reconsider; see Steve Blankinship, "Back to the FutureGen," *Power Engineering* (April 2009). Similar technology is being tested in Europe, though it is seen as an interim step on the way to alternative energy technologies; see Jocelyn Rice, "Putting Clean Coal to the Test," *Discover* (February 2009).

It is worth stressing that even those who favor CCS also believe, as David G. Hawkins says in his last paragraph, that though CCS "is an important strategy to reduce CO_2 emissions from fossil fuel . . . it is not the basis for a climate protection program by itself. Increased reliance on low-carbon energy resources is the key to protecting the climate." On the other hand, the Obama Administration is already making Charles Schmidt's comments about lack of research funding and political commitment look dated; see Amanda Ruggeri, "A Huge Cash Infusion in Tough Times," *U.S. News & World Report* (April 1, 2009).

ISSUE 9

Is It Time to Put Geothermal Energy Development on the Fast Track?

YES: Susan Petty, from testimony on the National Geothermal Initiative Act of 2007 before the Senate Committee on Energy and Natural Resources (September 26, 2007).

NO: Alexander Karsner, from testimony on the National Geothermal Initiative Act of 2007 before the Senate Committee on Energy and Natural Resources (September 26, 2007).

ISSUE SUMMARY

YES: Susan Petty, president of AltaRock Energy, Inc., argues that the technology already exists to greatly increase the production and use of geothermal energy. Supplying 20 percent of U.S. electricity from geothermal energy by 2030 is a very realistic goal.

NO: Alexander Karsner, Assistant Secretary for Energy Efficiency and Renewable Energy at the U.S. Department of Energy, argues that it is not feasible to supply 20 percent of U.S. electricity from geothermal energy by 2030.

In June 2007, Senator Jeff Bingaman (D-NM) introduced the National Geothermal Initiative Act of 2007. Hearings were held in September of that year. The bill was to establish a national goal of achieving "20 percent of total electrical energy production in the United States from geothermal resources by not later than 2030." To accomplish that goal, the Department of Energy and the Department of the Interior would characterize the complete U.S. geothermal resource base by 2010 and, among other things, develop policies and programs to sustain a 10 percent annual growth rate in the use of geothermal energy. The bill never made it past the hearing stage, but geothermal energy remains an important alternative energy source. Unlike solar and wind power, it is available 24 hours a day, 12 months a year.

Geothermal heat pumps have immense potential for home and commercial heating, according to Joanna R. Turpin, "Commercial Geothermal: Bright Spot in a Gloomy Economy," *Air Conditioning Heating & Refrigeration*

News (May 4, 2009), and the federal government is using the Federal Economic Stimulus Bill to provide investment tax credits to businesses that install them.

Geothermal energy can also be used to meet demand for electricity. "Conventional" geothermal sources are those where the Earth's heat comes fairly close to the surface and the rock of the heated zone has numerous cracks. Groundwater in the cracks may be heated to the point of boiling and erupt as geysers. Surface water may also be pumped into hot rock zones where it is heated. In both cases, steam can be captured and used to generate electricity.

Hot rock may also exist close to the surface but without cracks. In such cases, it may be "enhanced" by such measures as drilling and the use of explosives to create cracks in the rock. Where the hot rock is further from the surface—and there is hot rock everywhere in the world, if one goes deep enough—enhancement must involve drilling more deeply. The more enhancement that is needed, the more expensive the geothermal resource is to tap (which affects feasibility), but even conventional geothermal sources are not cheap. The hot water and steam coursing through the pipes, pumps, and turbines of geothermal plants can be highly corrosive, which makes maintenance expensive. Even so, however, geothermal energy is highly renewable, it does not depend on foreign sources, it uses no fuel, and it does not contribute to global warming.

The first geothermal energy plant was established in Lardarello, Italy, in 1904. In the United States, the first plant started operations in 1922, at the Geysers, California. Today, the Geysers has the largest complex of geothermal power plants in the world and produces over 850 megawatts of electricity— enough to power a million homes.

When Olafur Ragnar Grimsson, president of Iceland, testified before the Senate Committee on Energy and Natural Resources on September 26, 2007, he said that Iceland meets over half of its total energy needs with geothermal power and that:

> For the United States of America, geothermal energy can become a major energy resource, contributing to the security of the country, limiting dependence on the import of fossil fuels, reducing the risks caused by fluctuating oil prices and providing opportunities for new infrastructures. . . .
>
> [G]eothermal energy is a reliable, flexible and green energy resource which can supply significant amounts of power to households and industry. . . .

It sounds enticing, but even though tapping geothermal energy is almost entirely an engineering problem, engineering requires time, money, and political commitment. In the following selections, Susan Petty, president of AltaRock Energy, Inc., a company that hopes to profit from efforts to increase the use of geothermal energy, argues that the technology already exists to greatly increase the production and use of geothermal energy. Supplying 20 percent of U.S. electricity from geothermal energy by 2030 is a very realistic goal. Alexander Karsner, Assistant Secretary for Energy Efficiency and Renewable Energy at the U.S. Department of Energy, argues that it is not feasible to supply 20 percent of U.S. electricity from geothermal energy by 2030.

YES

Susan Petty

Statement of Susan Petty President–AltaRock Energy, Inc. Before the Senate Committee on Energy and Natural Resources Regarding Senate Bill 1543 National Geothermal Initiative Act of 2007 September 26, 2007

One of the goals of S. 1543 is to achieve 20% of electric power generation from geothermal energy by 2030. You may be asking yourself if this a realistic goal? In the fall of 2004, I was included in a 12 member panel led by Dr. Jefferson Tester of the Massachusetts Institute of Technology that looked at the Future of Geothermal Energy. Our group consisted of members from both industry and academia. While some of us started the study convinced that it was possible to engineer or enhance geothermal systems (EGS) with today's technology, many of us, including myself, were skeptical. As we reviewed data, and listened to experts who were actively researching new methods, testing them in the field, and starting commercial enterprises to develop power projects from geothermal energy using this emerging technology, I believe all of us became convinced that a way had been found to tap into the vast geothermal resource under our feet.

Everywhere on Earth, the deeper you go, the hotter it gets. In some places, high temperatures are closer to the surface than others. We have all heard of the "Ring of Fire," characterized by volcanoes, hot springs and fumaroles around the rim of the Pacific Ocean, including the Cascades, the Aleutian Islands, Japan, the Philippines and Indonesia. We know that along the tectonic rifts such as the Mid-Atlantic Ridge including Iceland and the Azores, the East African Rift Valley, the East Pacific Rise, the Rio Grande Rift running up through New Mexico and Colorado and the Juan de Fuca Ridge the earth's heat is right at the surface. But other geologic settings allow high temperatures to occur at shallow depths, such as the faulted mountains and valleys of the Basin and Range, the deep faults in the Rocky Mountains and the Colorado Plateau. In addition, the sedimentary basins that insulate granites heated by radioactive decay along the Gulf Coast, in the Midwest, along the Chesapeake Bay and just west of the Appalachians can not only provide oil and gas, but hot water as well. [. . .]

U.S. Senate, September 26, 2007.

The heat contained in this vast resource is so large that it is really difficult to contemplate. Even with very conservative calculations, the MIT study panel found that the amount of heat that could be realistically recovered in the U.S. from rocks at depths of 3 km to 10 km (about 2 miles to 6 miles) is almost 3,000 times the current energy consumption of the country. [. . .] Listening to the experience of those developing the Soultz project in France, the Rosemanowes project in the UK and the Cooper Basin project in Australia, the panel members began to understand that the technology to recover this heat was here today. We can drill wells into high temperature rocks at depths greater than 3 km. We can fracture large volumes of hot rock. We can target wells into these man-made fractures and intersect them. We can circulate water through these created fractures, picking up heat and produce it at the other side heated to the temperature of reservoir rocks. We can produce what we inject without having to add more water. Long term tests have been conducted at fairly modest flow rates on these created reservoirs without change in temperature over time. No power plants have yet been built, but several are in progress in Europe.

Does this mean that we can build economic geothermal power plants based on EGS technology right now? At the best sites, where high temperatures occur at shallow depths in large rock masses with similar properties, geothermal power production from EGS technology is economic today. But to bring on line the huge resource stretching across the country from coast to coast, we need to do some work.

I'd like to talk about the economics of geothermal power production so you can better understand what needs to happen to enable widespread development of power projects using EGS.

At some places in the Earth's crust, faults and fractures allow water to circulate in contact with hot rock naturally. These are hydrothermal systems where natural fractures and high permeability allow high production rates. Even low temperature systems can be economic if the flow rates produced are high enough. The capital cost for the wells and wellfield-related equipment generally is between 25%–50% of the total capital cost of the power project. The capital cost for hydrothermal projects can range from around $2,500/installed kW to over $5,000/kW, largely depending on the flow rate per well and the depth of the wells. The levelized break-even cost of energy for commercially viable hydrothermal projects currently ranges from $35/MWh to over $80/MWh. Of this, about $15–25/MWh is operating cost. The rest is the cost to amortize the power generation equipment and the wellfield.

Hydrothermal power is a good deal: Clean, small foot print, cost-effective. So why isn't more power from hydrothermal sources on line? The issue for hydrothermal power is risk. Because the risk related to finding the resource and successfully drilling and completing wells into the resource is high, development by utilities is unlikely. In order to accept this risk, independent power producers need a long-term contract at a guaranteed price and a high return on their investment. Utilities are loath to give a long-term contract because the payments to the generator will be treated as debt in determining their debt-to-equity ratio for credit and bond ratings.

Hydrothermal projects also tend to be small in size. While some of the potential future hydrothermal projects might be large, many of these are associated with scenic volcanic features protected as national parks or revered by Native Americans. A large scale project might mitigate the risk by spreading it over a much larger number of MW. In addition, there is a true economy of scale for geothermal power projects. For instance, the same number of people are needed to operate a 10 MW geothermal project as operate a 120 MW, or even a 250 MW, project.

Most of the really good (i.e. economic) hydrothermal systems are in the arid West. Not only is cooling water, which improves project economics by improving plant efficiency, an issue in this part of the country, but also the wide open spaces mean high-potential sites are often far from transmission, operators, supplies and large population centers with a high demand for power. Little potential for producing power from conventional geothermal, i.e., hydrothermal, sources exists in the Midwest, Southeast or East Coast.

Still, hydrothermal power has the potential to supply the country with more than 20,000 MW, or about 2% of our current installed capacity. However, the very high reliability of geothermal power means that this would be about 4% of our current annual generation. And this power is baseload or power that is available night and day.

Over the years, the cost of generating electricity from hydrothermal sources has dropped from around $130/MWh to less than $50/MWh. This was facilitated by incentives provided both by the market during the mid-1980s oil crisis, and by the government in the form of tax subsidies encourage the construction of over 2,000 MW of geothermal power that went on line from 1986–1995. Some of this drop in cost is due to research conducted by the U.S. Department of Energy (DOE). For instance, in 1980 the DOE completed the first demonstration binary power plant at Raft River. This plant enabled the use of fluids at temperatures much lower than had been developed in the past. Industry commercialized this technology, and now most of the new geothermal power plants being built today are binary plants. DOE research, together with industry, developed high-temperature tools that are now essential to the evaluation of geothermal wells. A combination of DOE-supported research and industry effort as improved binary power plant efficiency by almost 50% from the earliest commercial plants in the 1980s, and flash power-plant efficiency by almost 35% over the same time period. This translates directly into reduction in overall project cost and power prices because fewer wells and less equipment is needed to generate the same amount of energy.

The MIT study started with the current state of the geothermal industry. The first task we realized we needed to undertake was a realistic look at the size and potential cost of developing geothermal power across the continent. It has long been realized by scientists that a vast geothermal resource exists everywhere as long as technology allows us to drill deep enough, develop a reservoir by creating fractures or enhancing natural fractures, and connect wells to circulate fluid through that reservoir. The U.S. Geological Survey has been tasked with a detailed evaluation of the U.S. geothermal resource, but this could not be finished in time for our study. The MIT panel, therefore, undertook a preliminary assessment of the geothermal resource in the U.S.

Using data collected over the years with DOE support, maps of the temperature at depth were developed by Dr. David Blackwell's group at SMU. Temperature at the midpoint of 1 km thick slices was projected at 1 km intervals starting at a depth of 3 km and extending down to 10 km, a reasonable limit for drilling using today's technology. The heat resource contained in each cubic kilometer of rock at these temperatures at each depth was then calculated. The amount of energy stored in this volume of rock is so enormous that it is really impossible to comprehend. . . . We then looked at the studies that had estimated what fraction of this heat might be recovered, and at what efficiency this recovered heat might be turned into electric power. Studies showed that for economic systems, 40% or more of the total heat stored in the rock is recoverable. We also considered the more conservative recoverable estimates of 2% and 20%. Even at 2%, the amount of energy that could be realistically recovered, leaving economics and cost considerations aside, is more than 3,000 times the current total energy consumption of the U.S., including transportation uses.

In order to understand the technology needed to recover this energy, we turned to the published literature on the experiments done in the past at Fenton Hill, Rosemanowes, Hijiori, Ogachi and Soultz. We also brought in experts who are currently working on the Soultz project and on commercial engineered and enhanced geothermal projects in Europe and in Australia to tell us about the status of their work and their future efforts and needs. By the end of the study, we had concluded that EGS technology is technically feasible today. We can:

- Drill wells deep enough and successfully using standard geothermal and oil-and-gas drilling technology with existing infrastructure to tap the geothermal resource across the U.S., including areas in the Midwest, East and Southeast
- Consistently fracture large rock volumes of rock
- Monitor and map these created or enhanced fractures
- Drill production wells into the fractured rock
- Circulate cold water into the injection well and produce heated water from the production wells
- Operate the system without having to add significant amounts of water over time
- Operate the circulation system over extended test periods without measurable drop in temperature
- Generate power from the circulating water at Fenton Hill and Ogachi

In addition, EGS power projects are scalable. Once the first demonstration unit has been tested at a site, the potential exists to develop a really large scale project of 250 to 1000 MW. Combined with the fact that good EGS sites where large bodies of hot rock with fairly uniform properties can be found across the U.S., that the sites are so many that they can be selected to avoid places with no transmission capacity or those located near areas of scenic beauty or environmental sensitivity, generating power from EGS technology looks like a winning proposition.

The real question then becomes, not is it realistic to anticipate generating 20% of our nation's electric power from geothermal energy, but can we make it cost effective?

The MIT panel included members from industry and research who are experts in the economics of power generation. The panel developed a list of key technologies that could help reduce the cost of generating power from EGS. They considered the changes in the cost of power generation from hydro-thermal systems over the last 20 years, and the current state of EGS technology. They also considered research currently underway, not only that sponsored by DOE through universities and the national laboratories, but that being done by industry. Using models developed by both DOE and MIT, the cost of power and the impact on that cost of these possible technology improvements was examined. In addition, the panel looked at the impact of "learning by doing" on the cost of power.

We concluded that at the best sites, those with very high temperatures at depths of around 3–4 km in areas with low permeability natural fractures, EGS is economic today. . . . With current technology power from [a 300°C site at a depth of 3 km] this site could be generated for a levelized cost of power of about $74/MWh. This isn't the price that power could be sold for, since it doesn't include profit. It does, however, include financing charges at higher than utility rates, operating costs and the cost of amortizing the capital investment in the welfield and power plant. At deeper depths and lower temperatures, the cost of generating power using EGS technology is much higher, about $192/MWh. . . .

With incremental technology improvement, the cost of power could be cut in half or more, particularly for the deeper high temperature systems. These incremental technology improvements include things like improving conversion cycle efficiency, being able to isolate the part of the wellbore that has been treated so that untreated parts can be fractured, redesigning wells to reduce the number of casing strings and improved understanding of rock/fluid interaction to prevent or repair short circuiting through the reservoir. None of these technology improvements require game changing strategies, just the kind of advancement that comes from persisting in extending our knowledge to the next level. Looking at the high temperature example . . ., the levelized cost of power could be cut to $54/MWh or about 27% with these technology improvements implemented. The moderate temperature site could see a much larger reduction of over 60% to $74/MWh. . . .

We concluded that at the best sites, those with very high temperatures at depths of around 3–4 km in areas with low-permeability natural fractures, EGS is economic today. With incremental technology improvement, the cost of power could be cut in half or more, particularly for the deeper high temperature systems. These incremental technology improvements include things such as improving conversion cycle efficiency, being able to isolate the part of the wellbore that has been treated so that untreated parts can be fractured, redesigning wells to reduce the number of casing strings and improved understanding of rock/fluid interaction to prevent or repair short circuiting through the reservoir. None of these technology improvements require game-changing or revolutionary strategies, just the kind of advancement that comes from persisting in extending our knowledge to the next level.

The cost of this type of technology improvement is not high. The panel felt that an investment of ~$368,000,000 over a period of about 8–10 years combined with industry involvement could result in 100,000 MW on line by 2030. This would be 10% of the current installed capacity and over 20% of the current electric generation of the country. Combined with the hydrothermal resource, it is a very realistic goal to have geothermal energy provide 20% of the nation's electricity by 2030. However, the effort would require federal support, university, laboratory and industry research, and development and a real commitment to renewable energy use.

Currently more than eight companies are developing EGS power projects in Europe and more than 20 companies are working to get power on line using this technology in Australia. AltaRock Energy Inc. is the only company focused on commercializing power generation from EGS technology in the U.S. In Europe, price subsidies and European Union-sponsored research are helping to start more than 50 EGS projects. In Australia, government grants, help with transmission access, research, and legislation requiring generation from renewable energy sources are driving EGS technology to commercialization. Other countries with fewer economic geothermal resources are planning to include geothermal energy in their generation portfolio. The U.S. needs to commit to this clean, baseload, renewable power source for our own energy future. . . .

Statement of Alexander Karsner Assistant Secretary for Energy Efficiency and Renewable Energy U.S. Department of Energy Before the Committee on Energy and Natural Resources United States Senate

S. 1543 establishes a national goal of achieving "20 percent of total electrical energy production in the United States from geothermal resources by not later than 2030." To accomplish that goal, the legislation requires the Department of Energy and the Department of the Interior to characterize the complete U.S. geothermal resource base by 2010; develop policies and programs to sustain an annual growth rate in geothermal power, heat, and heat pump applications of at least 10 percent, and to achieve new power or commercial heat production from geothermal resources in at least 25 states; demonstrate state-of-the-art geothermal energy production; and develop tools and techniques to construct an engineered geothermal system power plant. Additionally, the legislation directs the Secretary to establish a geothermal research, development, demonstration, commercialization, outreach and education program in support of the 20 percent national goal.

The Department has significant concerns with the feasibility of the national goal established in this legislation. Generating 20 percent of our nation's electricity from geothermal resources would require more than 165,000 megawatts of geothermal power plant capacity by 2030, in Energy Information Administration's (EIA) reference case electricity demand forecast.[1] The 1978 USGS National Geothermal Resource Assessment estimated 23,000 megawatts of identified conventional geothermal resources, also called hydrothermal technology, that can be developed for electricity. The difference, more than 142,000 megawatts, would have to come from new discoveries, conventional resources that were not viable at the time of the 1978 assessment, and unconventional means such as Enhanced Geothermal Systems (EGS), coproduced fluid from oil and gas wells, and geopressured-geothermal resources, as well as and avoided electricity use from heat, and heat pump

U.S. Senate, September 26, 2007.

applications. With the exception of one small co-production generator, none of these unconventional resources are being used currently to generate commercial power. A recent report by the Massachusetts Institute of Technology (MIT), The Future of Geothermal Energy, estimates that 100,000 megawatts of electricity could be installed by 2050 using EGS technology. The MIT projection assumes a 15-year technology development program is conducted by the public and private sector prior to wide-scale installations.

While the Department shares the Committee's interest in rapidly accelerating market penetration of all renewable energy technologies, including geothermal, this particular goal may be technically unattainable within the timeframe specified. The Department looks forward to working with the Committee to resolve these and other technical concerns with S. 1543.

Since the founding of the Department of Energy, the agency has supported geothermal research and development. Over that period, a number of key accomplishments have contributed to increased commercial development of hydrothermal resources—to a point where it has reached market maturity. The Department's investment contributed to the identification of those resources, accurate characterization and modeling of hydrothermal reservoirs, improved drilling techniques, and advanced means of converting the energy for productive uses. The Federal government has realized many successes in hydrothermal technology development, as evidenced by winning eight R&D 100 Awards in the past ten years. I would like to share with the Committee the Department's current assessment of the geothermal industry, and discuss briefly the future potential for geothermal development as a part of a diversified, domestic clean energy portfolio.

Geothermal Industry

Geothermal energy is the heat from deep inside the earth, coming in large part from the decay of radioactive elements. Geothermal heat is considered a base load renewable energy source, and can be used for electricity generation and direct use (space heating, district heating, snow melting, aquaculture, etc.). While geothermal energy is available at some depth everywhere, in the U.S., it is most accessible in western states such as California, Nevada, Utah, and Hawaii, where it is found at shallow depths as hydrothermal resources. This is where the bulk of conventional, commercial geothermal development is taking place, but a number of other states, notably Idaho, Oregon, Arizona and New Mexico, could see new power projects coming online in the very near future.

Geothermal resources can be subdivided into four categories: 1. hydrothermal; 2. deep geothermal (Enhanced Geothermal Systems or EGS); 3. geopressured; and 4. fluid co-produced with oil and gas. Of these, hydrothermal resources, which are characterized by ample heat, fluid, and permeability, have been developed commercially around the world. The other resource categories have not reached commercial maturity and are less accessible through conventional geothermal processes. The United States has been and continues to be the world leader in online capacity of hydrothermal resources for electric power generation.

Currently, the U.S. has approximately 2850 MWe of installed capacity and about 2,900 MWe of new geothermal power plants under development in 74 projects in the Western U.S., according to industry estimates. In 2006, EIA estimates that geothermal energy generated approximately 14,842 gigawatt-hours (GWh) of electricity. The geothermal industry presently accounts for approximately 5% of renewable energy-based electricity consumption in the U.S. Most of the balance is split between hydropower and biomass, with wind and solar contributing a small portion.

In general, conventional hydrothermal technology is sufficiently mature, based on the following:

- The Western Governors Association geothermal task force recently identified over 140 sites with an estimated 13,000 MWe of power with near-term development potential.
- Hydrothermal reservoirs discovered at shallow depths using existing drilling technology, based upon similar available oil and gas practices used in the industry, are cost-effective.
- Power plant technology is based on standard cycles and can be bought off-the-shelf. Major development of binary-cycle power plant technology has enabled the development of increasingly lower temperature hydrothermal resources.
- Hydrothermal-generated electricity is cost competitive in certain regions of the country, where the resource can be maximized.

Favorable provisions of the Energy Policy Act of 2005 (EPACT 2005) and other federal and local incentives encourage industry to develop hydrothermal resources. EPACT 2005 contains significant provisions to promote the installation of geothermal power plants and geothermal heat pumps. These include:

- Resource Assessment—USGS has been directed to update its 1978 assessment of geothermal resources (Circular 790). EPACT 2005 mandates that USGS complete the Resource Assessment report by September 2008. To date, the Department of Energy has contributed over $1 million in financial support as well as technical support through its national laboratories and the Department's Geothermal Resources Exploration and Definitions activity.
- Programmatic Environmental Impact Statement (PEIS)—A PEIS is being developed for the major geothermal areas in the Western U.S. by the Bureau of Land Management (BLM), in partnership with the U.S. Forest Service. DOE is a cooperating agency for the PEIS and the Department anticipates that completion of the PEIS will encourage geothermal production.
- Streamlined Permitting and Royalty Structure—EPACT changed the royalty structure for leasing on Federal land from a 50/50 State/Federal split to a 50/25/25 split for State/Federal/local, providing an incentive for local governments to attract geothermal resource developers. EPAct also streamlined leasing requirements, which lowers costs for potential developers.

- Federal Purchases of Renewable Energy—EPAct 2005 requires that the Secretary of Energy seek to ensure that federal consumption of electric energy during any fiscal year should include the following amounts of renewable energy; 1) not less than 3 percent in fiscal years 2007 through 2009, 2) not less than 5% in fiscal years 2010 through 2012 and 3) not less than 7.5% in fiscal year 2013 and each fiscal year thereafter.
- Loan Guarantees—EPACT 2005 authorizes the Department to issue loan guarantees to eligible projects that "avoid, reduce, or sequester air pollutants or anthropogenic emissions of greenhouse gases" and "employ new or significantly improved technologies as compared to technologies in service in the United States at the time the guarantee is issued." On May 16, 2007, the Department issued a Notice of Proposed Rulemaking to establish the loan guarantee program. The comment period for that rulemaking has closed, and the Department anticipates finalizing the rule shortly. In addition, on August 3, 2007, the Department named David G. Frantz as the Director of the Loan Guarantee Office, reporting directly to the Department's Chief Financial Officer. By providing the full faith and credit of the United States government, loan guarantees will enable the Department to share some of the financial risks of projects that employ new or significantly improved technologies. DOE is currently authorized to provide $4 billion in loan guarantees, and the 2008 President's Budget requested $9 billion in loan volume limitation.

In addition, the Tax Relief and Health Care Act of 2006 extended the production tax credit for geothermal and other renewables that are put into service through December 31, 2008. This provision has had a significant impact on encouraging new installations of conventional geothermal power facilities; as I mentioned previously, over 2,900 MWe are now under development in the U.S. An investment tax credit of 10 percent is also available to the industry, but cannot be combined with the production tax credit. Because conventional geothermal is a mature technology and favorable policy changes have clearly resulted in the growth of the industry, the FY 2008 Budget Request terminates the current Geothermal Technology program.

Enhanced Geothermal Systems (EGS)

Enhanced Geothermal Systems (EGS) involves technology that enables geothermal resources that lack sufficient water or permeability (compared to conventional hydrothermal resources) to be developed. The ultimate intent is to tap energy from hot impermeable rocks that are at a depth of between 3 and 10 kilometers in the earth's crust. Such rock formations require engineered enhancements to enable productive reservoirs.

DOE funded MIT to conduct a study of EGS potential in the U.S. MIT made the following key findings:

- EGS has the potential to produce up to approximately 100,000 MW of new electric power by 2050 based in part on an abundance of available geothermal resources.

- Elements of the technology to capture EGS are in place.
- Multiple reservoir experiments are required.
- Successful R&D could provide performance verification at a commercial scale within a 15-year period nationwide.

The Department is currently considering the findings of the MIT study. DOE is holding discussions with industry and academic experts, further defining technical barriers and gaps, and determining the technical and commercial actions that can help industry overcome the barriers and to bridge the gaps. Input has come from oil and gas companies, service companies, academia, the geothermal industry, international experts, government agencies, and the national laboratories. We expect to release this evaluation by the end of 2007.

Conclusion

In conclusion, Mr. Chairman, the Department anticipates that geothermal resources will continue to play an important and potentially growing role in our nation's energy portfolio, as we look to rapidly expand the availability of clean, secure, reliable domestic energy. The industry currently benefits from tax incentives and regulatory streamlining in EPACT 2005, and future industry investments in enhanced geothermal have the potential to significantly expand domestic geothermal energy production. The Department looks forward to working with this Committee to resolve concerns related to S. 1543, and to continue our national commitment to clean, renewable energy production. . . .

Note

1. The Energy Information Administration projects Total Electric Power Sector Capacity in 2030 to be 1159 GW. This projection is based on an assumption that geothermal power plant has a capacity factor of 80–85 percent.

POSTSCRIPT

Is It Time to Put Geothermal Energy Development on the Fast Track?

MIT's 2006 report, "The Future of Geothermal Energy: Impact of Enhanced Geothermal Systems (EGS) on the United States in the 21st Century"

> Geothermal energy from EGS represents a large, indigenous resource that can provide baseload electric power and heat at a level that can have a major impact on the United States, while incurring minimal environmental impacts. With a reasonable investment in R&D, EGS could provide 100 GW$_e$ or more of cost competitive generating capacity in the next 50 years. Further, EGS provides a secure source of power for the long term that would help protect America against economic instabilities resulting from fuel price fluctuations or supply disruptions. Most of the key technical requirements to make EGS work economically over a wide area of the country are in effect, with remaining goals easily within reach . . . within a 10 to 15 year period nationwide.
>
> In spite of its enormous potential, the geothermal option for the United States has been largely ignored. . . .

The National Geothermal Initiative Act was proposed in 2007 and hearings were promptly held. A similar bill in the House of Representatives—the Advanced Geothermal Energy Research and Development Act of 2007—had a hearing on May 17, 2007, before the Subcommittee on Energy and Environment of the House Committee on Science and Technology. Both bills filled to leave committee. The reason may lie in the fact that, as Mark D. Myers, Director of the U.S. Geological Survey, said in his Senate testimony, although increasing the use of geothermal energy is a laudable goal, funds may not be available to achieve the goals of the National Geothermal Initiative Act of 2007 in the specified time frame. The shortage of funds is due to competing demands, a record federal deficit, and a weakened national economy.

However, the Geothermal Technologies Program of the U.S. Department of Energy's Energy Efficiency and Renewable Energy office announced in July 2008 a new $30.5 billion loan guarantee program for renewable energy (including geothermal energy) projects. On June 2, 2009, the Department of Energy announced that $50 million of Recovery Act funding would go to helping the deployment of geothermal heat pumps. Enhanced Geothermal Systems (EGS) would receive $80 million. How quickly and how extensively geothermal energy will be developed remains to be seen.

Despite the great potential benefits—large amounts available, continuous availability, zero use of fossil fuels and hence no impact on global warming—some researchers and activists do worry about unforeseen consequences. For instance, Michael N. Bates, et al., "Cancer Incidence, Morbidity and Geothermal Air Pollution in Rotorua, New Zealand," *International Journal of Epidemiology* (vol. 27, No. 1, 1998), reported elevated incidences of some cancers and diseases of the nervous system and eye, consistent with exposure to hydrogen sulfide and mercury, both of which can be found in the vapors given off by geothermal systems. Water pumped from below ground may be contaminated with arsenic and other minerals, requiring that the water be reinjected into the ground or otherwise kept out of waterways; see Alper Baba and Halldor Armannsson, "Environmental Impact of the Utilization of Geothermal Areas," *Energy Sources Part B: Economics, Planning & Policy* (July/September 2006). Ernest L. Majer, et al., "Induced Seismicity Associated with Enhanced Geothermal Systems," *Geothermics* (June 2007), note that concerns that the drilling and fluid injection associated with EGS may cause earthquakes "has been the cause of delays and threatened cancellation of at least two EGS projects worldwide." There is a need for careful site selection and attention to public acceptance.

ISSUE 10

Should Cars Be More Efficient?

YES: David Friedman, from "CAFE Standards," Testimony before Committee on Senate Commerce, Science and Transportation, March 6, 2007

NO: Charli E. Coon, from "Why the Government's CAFE Standards for Fuel Efficiency Should Be Repealed, Not Increased," The Heritage Foundation Backgrounder #1458, July 11, 2001

ISSUE SUMMARY

YES: David Friedman, Research Director at the Union of Concerned Scientists, argues that the technology exists to improve the fuel efficiency standards for new cars and trucks and requiring improved efficiency can cut oil imports, save money, create jobs, and help with global warming.

NO: Charli E. Coon, Senior Policy Analyst with the Heritage Foundation, argues that the 1975 Corporate Average Fuel Economy (CAFE) program failed to meet its goals of reducing oil imports and gasoline consumption and has endangered human lives. It needs to be abolished and replaced with market-based solutions.

Automobiles have been a much-beloved feature of modern technology for over a century. Their advent increased mobility, made it possible for city workers to live outside the city, enabled goods to reach stores near almost everyone, and solved a growing environmental problem. Few realize how dirty the streets can be, how smelly cities can be, or how many flies can fill the air when transportation relies on horses! Ralph Turvey, "Horse Traction in Victorian London," *Journal of Transport History* (September 2005), says that these problems drew little mention at the time, but neither did most people mention the noisiness and smokiness (and poor gas mileage) of cars and trucks through the first half of the twentieth century. Still, most people thought cars and trucks were a vast improvement over horses. Those who had made millions in the oil business—beginning in Texas, Oklahoma, California, and some other states, and later in the Middle East and elsewhere—were quite happy with the change. See Leonardo Maugeri, *The Age of Oil: The Mythology, History, and Future of the World's Most Controversial Resource* (Praeger, 2006), and

Lisa Margonelli, *Oil on the Brain*: *Adventures from the Pump to the Pipeline* (Nan A. Talese, 2007).

By the 1970s, there were a great many cars and trucks on the road. Local railways were extinct, and horses were no longer bred in the large numbers of the past. There really was no substitute for gasoline-powered transportation, even though it consumed large amounts of oil imported from distant parts of the world. American dependence on these imports was highlighted when the Organization of Petroleum-Exporting Countries (OPEC) cut supplies and raised prices in the Oil Crisis of 1973. In the wake of the crisis, foreign cars with better gas mileage increased their sales at the expense of American-made cars, highway speed limits were reduced, and the U.S. Congress passed the Energy Policy and Conservation Act of 1975, which included the Corporate Average Fuel Economy (CAFE) program. The goal was to double average fuel efficiency by 1985. Fuel efficiency standards for passenger cars started at 18 mpg in 1978 and rose to 27.5 mpg for 1985; these standards were lowered in the late 1980s, but rose again to 27.5 mpg in 1990, where they remained. Light trucks had lower standards; in 2007, the standard was 22.2 mpg for light trucks.

Prompted by these requirements, as well as by foreign competition for the U.S. market, manufacturers successfully improved the performance of drive trains and engines and developed lighter materials for bodies. But as oil supplies became ample, even though the price of gasoline continued to rise, they also converted light truck designs to passenger versions, now known as Sports Utility Vehicles or SUVs, and sold them as roomier, more powerful, and even safer (largely because of size and weight) automobiles. SUVs are infamous for poor gasoline mileage, compared to passenger cars, but they are still classified as light trucks and held only to that standard. Improving their performance offers "the greatest potential to reduce fuel consumption on a total-gallons-saved basis" (*Effectiveness and Impact of Corporate Average Fuel Economy [CAFE] Standards*, National Academy Press, 2002).

In the following selections, David Friedman, Research Director at the Union of Concerned Scientists, argues that the technology exists to improve the fuel efficiency standards for new cars and trucks, and improved efficiency can cut oil imports, save money, create jobs, and reduce global warming. In addition, more fuel-efficient vehicles are actually safer to drive. Charli E. Coon, Senior Policy Analyst with the Heritage Foundation, argues that the 1975 Corporate Average Fuel Economy (CAFE) program failed to meet its goals of reducing oil imports and gasoline consumption and has endangered human lives. He contends that it needs to be abolished and replaced with market-based solutions.

YES

David Friedman

CAFE Standards

. . . I think we have reached an important milestone on fuel economy. It would appear that some leaders in Congress, including members of this committee, and the president are basically in agreement on how far we should increase fuel economy standards in about a ten year period.

In the president's state of the union speech, he set a goal for America to conserve up to 8.5 billion gallons of gasoline by 2017. To do so, we would need to increase fuel economy standards for cars and trucks to about 34 miles per gallon by 2017, or about 4 percent per year. At the same time, the bill recently introduced by the chairman and many members of this committee establishes a fuel economy target of 35 mpg by 2019. . . . The oil savings benefits of S. 357 are almost the same as the president's goal. Other members of the Senate and House have put forth bills with similar requirements in this and recent years.

In addition, Senator Stevens has introduced a bill to raise fuel economy standards for passenger cars to 40 mpg by 2017, or about a 39 percent increase compared to the average fuel economy of cars today. If Senator Stevens applied the same improvement to the rest of the fleet, it would average just over 34 mpg by 2017. As it stands, the oil savings from S. 183 are half of the others since only half the fleet is included.

I consider this a milestone because this significant agreement on fuel economy goals means that we can focus now on how best to reach them. By reforming and strengthening fuel economy standards for cars and trucks, this committee has a significant opportunity to help cut our oil addiction, save consumers money, create new jobs, and tackle the largest long term environmental threat facing the country and the world today, global warming.

Global Warming

Carbon dioxide, the main heat trapping gas blanketing our planet and warming the earth, has reached a concentration of about 380 parts per million. That is higher than the globe has experienced in the past 650,000 years. We are already seeing the impacts of these elevated concentrations as eleven of the last 12 years rank among the 12 hottest on record.

The worldwide costs of global warming could reach at least five percent of global GDP each year if we fail to take steps to cut emissions. These costs would come in lives and resources as tropical diseases and agricultural pests

From FDCH *Congressional Testimony* by David Friedman (Union of Concerned Scientists), (March 6, 2007).

move north due to our warming continent. These costs could also come from losing 60–80 percent of the snow cover in the Sierras by the end of the century and the resulting impacts on agriculture in California and similar states that rely on snow melt for water. We will also see increased asthma and lung disease because higher temperatures will make urban smog worse than it is today.

Global warming is a worldwide problem and our cars and trucks have impacts that are worldwide in scale. Only the entire economies of the United States, China, and Russia exceed the global warming pollution resulting from our cars and trucks alone. It is clear that the scope of pollution from our cars and trucks requires special attention as we begin to address climate change.

Oil Addiction

In addition to the costs created by the pollution from our cars and trucks, our vehicles also contribute to 40 percent of our oil addiction. Overall, data from the Energy Information Administration indicates that we imported about sixty percent of our oil and other petroleum products in 2006. Last year alone, our net imports were more than 12 million barrels per day. When oil is at $60 per barrel, every minute that passes means over $500,000 that could have been spent creating U.S. jobs and strengthening our economy instead leaves this country. At the end of the day, high oil and gasoline prices and continued increases in our oil addiction represent one of the single biggest threats to U.S. auto jobs today.

Fuel Economy Background

One of the main reasons our vehicles contribute so much to U.S. oil dependence and global warming is that the average fuel economy of the fleet of new cars and trucks sold in the U.S. in 2006 was lower than it was in 1986. And while automakers note the number of models on the market that get more than 30 miles per gallon on the highway, a look at EPA's 2007 fuel economy guide shows that there are more than 300 car and truck configurations that get 15 mpg or less in the city. Even if you exclude pickups and work vans, automakers still flood the market with nearly 200 car, minivan, and SUV configurations that get 15 mpg or less in the city. Consumers simply do not have enough high fuel economy choices when it comes to cars, minivans, SUVs and pickups.

Fuel economy standards were created to solve this exact problem. Just as we see today, automakers were not ready for the problems created in part by our gas guzzling in the early 1970s. As a result consumers jumped on the only option they had at the time, relatively poorly designed smaller cars. However, as fuel economy standards were fully phased in automakers switched from giving consumers poor choices to putting technology in all cars and trucks so car buyers could have options in the showroom with 70% higher fuel economy than they had in 1975 (*2006 EPA Fuel Economy Trends Report*). If the fuel economy of today's cars and trucks was at the level the fleet experienced in 1975 instead of today's 25 miles per gallon, we would be using an additional 80 billion gallons of gasoline on top of the 140 billion gallons we will use this year. That would represent an increase in oil demand by 5.2 million barrels of oil per day, or a 25 percent increase in our oil addiction. At today's average price for regular gasoline, about $2.50 per gallon, that represents $200 billion dollars saved. That number could have been much better, however, if fuel economy standards had not remained essentially unchanged for the past two decades.

Technology to Create Consumer Choice

Driving in America has become a necessity. Because of this and a lack of options, even the spikes in gasoline prices over the past five years have not been enough to push consumers to significantly reduce their gasoline consumption. Better fuels and more alternatives to driving are important to helping consumers and cutting pollution, but the quickest route to reduced gasoline consumption and saving consumers money is put to more high fuel economy choices in the showroom.

The automobile industry has been developing technologies that can safely and economically allow consumers to get more miles to the gallon in cars, minivans, pickups and SUVs of all shapes and sizes. . . . These technologies [can] dramatically increase the fuel economy of an SUV with the size and acceleration of a Ford Explorer. This could be achieved using direct injection gasoline engines, high efficiency automatic manual transmissions, engines that shut off instead of wasting fuel while idling, improved aerodynamics, better tires and other existing efficiency technologies. These technologies have no influence on the safety of the vehicle. Others, such as high-strength steel and aluminum and unibody construction could actually help make highways safer.

For just over $2,500 a consumer could have the choice of an SUV that gets more than 35 mpg. This is an SUV that alone could meet the fuel economy targets laid out by members of this committee and the president. At $2.50 per gallon, this SUV would save consumers over $7,800 on fuel costs during the vehicle's lifetime. The technologies needed for this better SUV would even pay for themselves in about three years. Automakers do already have vehicles on the road that can match this fuel economy, but most are compact cars. That leaves a mother with three children in car-seats or a farmer who needs a work truck with few vehicle choices until these technologies are packaged into higher fuel economy minivans, SUVs, pickups and other vehicles.

The technologies in this better SUV could be used across the fleet to reach more than 40 miles per gallon over the next ten years. The 2002 study by the National Academies on CAFE showed similar results. Data in the report indicate that the technology exists to reach 37 mpg in a fleet of the same make-up as the NAS analyzed, even ignoring hybrids and cleaner diesels.

The question now is whether automakers will use these tools to increase fuel economy. Automakers have spent the past twenty years using similar technologies to nearly double power and increase weight by twenty-five percent instead of increasing fuel economy (*EPA Fuel Economy Trends Report, 2006*). As a result, consumers today have cars and trucks with race-car like acceleration and plenty of room for children, pets and weekend projects. What consumers need now is to keep the size and performance they have today, while getting higher fuel economy. Without increased fuel economy standards, however, this future is unlikely. We are already seeing automakers market muscle hybrids, vehicles that use hybrid technology for increased power instead of increased fuel economy. And technologies such as cylinder cut-off, which increases fuel economy by shutting off engine cylinders when drivers need less power, are being used to offset increased engine power rather than increased fleetwide fuel economy.

This committee is in a position to ensure that consumers can keep the power, size and safety they have in their vehicles today, and save thousands of dollars while cutting both global warming pollution and our oil addiction through deployment of technology aimed at better fuel economy across the vehicle fleet.

Economic and Employment Impacts of Setting Fuel Economy Targets

Contrary to claims by the auto industry, investments in fuel economy technology, just like other investments in the economy, will lead to prosperity. No automaker would simply shut down a plant if it was making gas guzzlers that don't meet national fuel economy targets. Instead, they would make investments to upgrade their tooling to build more fuel efficient vehicles. A 2006 study from Walter McManus at the University of Michigan shows that automakers that invest in fuel economy, even as early as 2010, will improve their competitive position (*Can Proactive Fuel Economy Strategies Help Automakers Mitigate Fuel-Price Risks?*). According to the study, Detroit's Big Three could increase profits by $1.3 billion if they invest in fuel economy, even if gasoline costs only $2 per gallon. However, if they follow a business-as-usual approach their lost profits could be as large as $3.6 billion if gasoline costs $3.10 a gallon.

UCS has also sought to quantify the benefits of increased fuel economy (Friedman, 2004, *Creating Jobs, Saving Energy and Protecting the Environment*). We estimated the effect of moving existing technologies into cars and trucks over 10 years to reach an average of 40 miles per gallon (mpg). We found that:

In 10 years, the benefits resulting from investments in fuel economy would lead to 161,000 more jobs throughout the country, with California, Michigan, New York, Florida, Ohio, and Illinois topping the list.

In the automotive sector, projected jobs would grow by 40,800 in 10 years. A similar analysis done by the economic-research firm Management Information Services (MIS) evaluated the potential job impacts of increasing fuel economy to about 35–36 mpg by 2015 and found even greater growth at more than 350,000 new jobs in 2015 (Bezdek, 2005, *Fuel Efficiency and the Economy*). This job growth included all of the major auto industry states.

In both the UCS and the MIS studies these new jobs would be created both because of investments in new technologies by the automakers and because consumers would shift spending away from gasoline to more productive products and services.

Requiring all automakers to improve fuel economy will increase the health of the industry. Companies like Ford, General Motors and the Chrysler division are currently in bad financial condition due to poor management decisions and elevated gas prices, not fuel economy standards, which have been stagnant for the past two decades. Those poor decisions have put them in a place where, just as in the 1970s, they do not have the products consumers need at a time of increased gasoline prices, and they are continuing the slide in market share that began the first time they made this mistake.

By requiring Ford, GM, Chrysler and all automakers [to] give consumers the choices they need, Congress can ensure automaker jobs stay in the U.S. and models like the Ford Explorer and Chevrolet Tahoe are still on the market ten years from now though they will go farther on a gallon of gas.

Safety Impacts of Setting Fuel Economy Targets

While the NAS study clearly states that fuel economy can be increased with no impact on the safety of our cars and trucks, critics of fuel economy standards often point to the chapter, which takes a retrospective look at safety. Despite the fact that this chapter did not represent a consensus of the committee (a dissenting opinion from two panel members was included in the appendices) and the fact that three major analyses have since shown that fuel economy and safety are not inherently linked, claims are still made to the contrary.

First, David Greene (one of the NAS panel members) produced a report with Sanjana Ahmad in 2004 (*The Effect of Fuel Economy on Automobile Safety: A Reexamination*), which demonstrates that fuel economy is not linked with increased fatalities. In fact, the report notes that, "higher mpg is significantly correlated with fewer fatalities." In other words, a thorough analysis of data from 1966 to 2002 indicates that Congress can likely increase fuel economy without harming safety if the past is precept.

Second, Marc Ross and Tom Wenzel produced a report in 2002 (*An Analysis of Traffic Deaths by Vehicle Type and Model*), which demonstrates that large vehicles do not have lower fatality rates when compared to smaller vehicles. Ross and Wenzel analyzed federal accident data between 1995 and 1999 and showed that, for example, the Honda Civic and VW Jetta both had lower fatality rates for the driver than the Ford Explorer, the Dodge Ram, or the Toyota 4Runner. Even the largest vehicles, the Chevrolet Tahoe and Suburban had fatality rates that were no better than the VW Jetta or the Nissan Maxima. In other words, a well-designed compact car can be safer than an SUV or a pickup. Design, rather than weight, is the key to safe vehicles.

Finally, a study by Van Auken and Zellner in 2003 (*A Further Assessment of the Effects of Vehicle Weight and Size Parameters on Fatality Risk in Model Year 1985–98 Passenger Cars and 1985–97 Light Trucks*) indicates that increased weight is associated with increased fatalities, while increased size is associated with decreased fatalities. While this study was not able to bring in the impacts of design as well as size, it helped inform NHTSA as they rejected weight-based standards in favor of size-based standards based on the vehicle footprint.

These studies further back up Congress' ability to set fuel economy targets as high as 40 mpg for the fleet in the next ten years without impacting highway safety.

Getting Fuel Economy Policy Right

Given broader agreement on how far fuel economy must increase, we now need policies to lay out how to get there. Congress should follow four key steps to ensure that the country gets the benefits of existing fuel economy technology:

> Establish a concrete fuel economy goal
> Provide NHTSA with additional flexibility to establish size based standards
> Institute a backstop to ensure that the fuel savings benefits are realized
> Provide consumers and/or automakers with economic incentives to invest in technology for increased fuel economy, Set a target of 34–35 mpg

Congress can set a standard either meeting the president's goals of 34 mpg by 2017 or 35 mpg by 2019 as in S. 357. Both of these fuel economy levels are supported by the guidance requested and received from the NAS and UCS analysis. By adopting S. 357, Congress would cut global warming pollution by more than 230 million metric tons by 2020, the equivalent of taking more than 30 million of today's automobiles off the road. The bill would also cut oil dependency by 2.3 million barrels of oil per day in 2027, as much oil as we currently import from the Persian Gulf.

The key to reaching these goals, however, is that Congress must set these targets and not leave it up to NHTSA. NHTSA has proven to have a poor track record when setting fuel economy standards so far. Their recent rulemaking on light trucks will save less than two weeks of gasoline each year for the next two decades. This happened in part because they did not value the important benefits of cutting oil dependence and reducing global warming pollution from cars and trucks. By setting specific standards based on where technology can take us, Congress can make clear the importance of tackling these important problems which are hard to quantify analytically, but easy to qualify based on consumer discontent with gasoline prices last summer, political instability from dependence on oil from the Persian Gulf, and the surge in concern over global warming.

Congress should not defer its regulatory authority to the Administration and it need not as it can base such standards on the scientific research it requested. Congress can be confident that the goals are technically feasible, cost effective, and safe.

Provide NHTSA Authority to Establish Size-Based Standards

The bills in the Senate and the president's plans include the ability for NHTSA to set car and light truck standards based on vehicle attributes such as vehicle size. These size-based standards give manufacturers who make everything from compact cars to minivans to large pickups the flexibility they have been asking for and eliminate any arguments automakers have made about CAFE standards treating them inequitably.

Size-based standards designed to increase fleet fuel economy to 35 mpg might require a family car to reach 40 mpg, but a pickup would only have to reach about 28 mpg because it is larger. This is good news for farmers and contractors who rely on these vehicles. With existing technology, pickups could readily reach 28 mpg and would save their owners over $6,000 on gasoline during the life of the vehicle. The pickup would have the same power, performance, size and safety it has today, and would cost an additional $1,500. However, the added fuel economy technology would pay for itself in less than two years with gasoline at $2.50 per gallon. Higher fuel economy standards will help farmers and small businesses who rely on trucks as much or even more than the average consumer.

Ensure No Backsliding

The one challenge with size-based standards is that automakers can game the system and drive down fuel economy. Much as automakers switched to marketing SUVs because of the lower standards required of light trucks to date, automakers may also upsize their vehicles to classes with lower fuel economy targets when they redesign their vehicles every four to seven years. Our analysis of NHTSA's most recent light truck rule shows that we could lose as much as half of the promised fuel economy gains, as small as they are, if the fleet of light trucks increased in size by just 10 percent over ten years. Congress must require a backstop to ensure that fuel savings that would be generated from a 10 mpg fuel economy increase are not lost due to automakers who game the system.

Provide Incentives

Because increased fuel economy will provide a wide variety of benefits for the nation, it is in the nation's interest to help automakers and suppliers who make cars and trucks in the U.S. that go farther on a gallon of gasoline. One way to help the auto industry is to provide tax credits, loan guarantees, or grants to companies that guarantee fuel economy improvements by investing in the equipment and people who will be needed to make these more efficient vehicles. This policy could be further supported by a set of charges and rebates applied to vehicles based on their fuel economy. These "feebates" will send market signals to producers and consumers in support of higher fuel economy standards and can even be made revenue neutral.

Conclusions

Climate change represents the largest long term environmental threat facing our country and the world today and the costs of our oil addiction continue to grow. Setting a fleet-wide target sufficient to meet the president's goal and guarantee fuel economy improvements of at least 10 mpg over the next decade while giving the president the authority to reach that target through size-based standards will save consumers money, stimulate the economy, create and protect jobs and preserve the safety of our vehicles. All of these benefits will come in addition to cutting our oil dependence and emissions of global warming pollutants from our cars and trucks.

Consumers are clearly happy with the size and acceleration of their vehicles today. We don't have to change that. But consumers are clearly unhappy with the growing impacts of global warming and the high cost of gasoline and the pumps and on our economy and security.

Congress has the opportunity to ensure that automakers spend the next decade or more using technology to curb our oil addiction. This is not a surprising role for Congress, the Federal government has helped drive every major transportation revolution this country has seen, whether it was trains, planes, or automobiles. The next transition will be no different.

Charli E. Coon  **NO**

Why the Government's CAFE Standards for Fuel Efficiency Should Be Repealed, Not Increased

Congress may soon decide to increase the standards for fuel economy imposed on manufacturers of vehicles sold in the United States. This would be a mistake.

In 1975, Congress reacted to the 1973 oil embargo imposed by the Organization of Petroleum Exporting Countries (OPEC) by establishing the Corporate Average Fuel Economy (CAFE) Program as part of the Energy Policy and Conservation Act. The goal of the program was to reduce U.S. dependence on imported oil and consumption of gasoline. Advocates also hoped it would improve air quality. But the evidence shows that it has failed to meet its goals; worse, it has had unintended consequences that increase the risk of injury to Americans. Instead of perpetuating such a program, Congress should consider repealing the CAFE standards and finding new market-based solutions to reduce high gasoline consumption and rising prices.

There is significant pressure on Members of Congress, however, not only to continue this failed program, but also to raise fuel efficiency standards even higher. The current CAFE standards require auto manufacturers selling in the United States to meet certain fuel economy levels for their fleets of new cars and light trucks (pickups, minivans, and sport utility vehicles, or SUVs). The standard for passenger cars is currently 27.5 miles per gallon; for light trucks, it is 20.7 mpg.

Manufacturers face stiff fines for failing to meet these standards based on the total number of vehicles in each class sold, but compliance is taken out of their hands. The government measures compliance by calculating a sales-weighted mean of the fuel economies for the fleets of new cars and light trucks a manufacturer sells each year, and it measures domestically produced and imported vehicles separately.

Clearly, the CAFE program has failed to accomplish its purposes. Oil imports have not decreased. In fact, they have increased from about 35 percent of supply in the mid-1970s to 52 percent today. Likewise, consumption has not decreased. As fuel efficiency improves, consumers have generally increased their driving, offsetting nearly all the gains in fuel efficiency. Not only has the

CAFE program failed to meet its goals; it has had tragic even if unintended consequences. As vehicles were being made lighter to achieve more miles per gallon and meet the standards, the number of fatalities from crashes rose.

Politicians should stop distorting the marketplace with unwise policies and convoluted regulations and allow the market to respond to consumer demand for passenger vehicles. In addition to free-market considerations, there are other compelling reasons to reject the CAFE standards. For example:

- CAFE standards endanger human lives;
- CAFE standards fail to reduce consumption; and
- CAFE standards do not improve the environment.

How Cafe Increases Risks to Motorists

The evidence is overwhelming that CAFE standards result in more highway deaths. A 1999 USA TODAY analysis of crash data and estimates from the National Highway Traffic Safety Administration and the Insurance Institute for Highway Safety found that, in the years since CAFE standards were mandated under the Energy Policy and Conservation Act of 1975, about 46,000 people have died in crashes that they would have survived if they had been traveling in bigger, heavier cars. This translates into 7,700 deaths for every mile per gallon gained by the standards.

While CAFE standards do not mandate that manufacturers make small cars, they have had a significant effect on the designs manufacturers adopt—generally, the weights of passenger vehicles have been falling. Producing smaller, lightweight vehicles that can perform satisfactorily using low-power, fuel-efficient engines is the most affordable way for automakers to meet the CAFE standards.

More than 25 years ago, research established that drivers of larger, heavier cars have lower risks in crashes than do drivers of smaller, lighter cars. A 2000 study by Leonard Evans, now the president of the Science Serving Society in Michigan, found that adding a passenger to one of two identical cars involved in a two-car frontal crash reduces the driver fatality risk by 7.5 percent. If the cars differ in mass by more than a passenger's weight, adding a passenger to the lighter car will reduce total risk.

The Evans findings reinforce a 1989 study by economists Robert Crandall of the Brookings Institution and John Graham of the Harvard School of Public Health, who found that the weight of the average American automobile has been reduced 23 percent since 1974, much of this reduction a result of CAFE regulations. Crandall and Graham stated that "the negative relationship between weight and occupant fatality risk is one of the most secure findings in the safety literature."

Harvard University's John Graham reiterated the safety risks of weight reduction in correspondence with then-U.S. Senator John Ashcroft (R-MO) in June 2000. Graham was responding to a May 2000 letter distributed to Members of the House from the American Council for an Energy-Efficient Economy (ACEEE) and the Center for Auto Safety. Graham sought to correct its misleading statements, such as its discussion of weight reduction as a compliance strategy without reference to the safety risks associated with the use of lighter steel. For example, an SUV may be more likely to roll over if it is constructed with lighter materials, and drivers of vehicles that crash into guardrails are generally safer when their vehicle contains more mass rather than less. Further, according to Graham, government studies have found that making small cars heavier has seven times the safety benefit than making light trucks lighter.

The evidence clearly shows that smaller cars have significant disadvantages in crashes. They have less space to absorb crash forces. The less the car absorbs, the more the people inside the vehicle must absorb. Consequently, the weight and size reductions resulting from the CAFE standards are linked with the 46,000 deaths through 1998 mentioned above, as well as thousands of injuries. It is time that policymakers stop defending the failed CAFE program and start valuing human lives by repealing the standards.

Why Cafe Fails to Reduce Consumption

Advocates of higher CAFE standards argue that increasing miles per gallon will reduce gas consumption. What they fail to mention is the well-known "rebound effect"—greater energy efficiency leads to greater energy consumption. A recent article in *The Wall Street Journal* noted that in the 19th century, British economist Stanley Jevons found that coal consumption initially decreased by one-third after James Watt's new, efficient steam engine began replacing older, more energy-hungry engines. But in the ensuing years (1830 to 1863), consumption increased tenfold—the engines were cheaper to run and thus were used more often than the older, less efficient models. In short, greater efficiency produced more energy use, not less.

The same principle applies to CAFE standards. A more fuel-efficient vehicle costs less to drive per mile, so vehicle mileage increases. As the author of *The Wall Street Journal* article notes, "[s]ince 1970, the United States has made cars almost 50% more efficient; in that period of time, the average number of miles a person drives has doubled." This increase certainly offsets a portion of the gains made in fuel efficiency from government mandated standards.

Why Cafe Standards Do Not Improve the Environment

Proponents of higher CAFE standards contend that increasing fuel economy requirements for new cars and trucks will improve the environment by causing less pollution. This is incorrect.

Federal regulations impose emissions standards for cars and light trucks, respectively. These standards are identical for every car or light truck in those two classes regardless of their fuel economy. These limits are stated in grams per mile of acceptable pollution, not in grams per gallon of fuel burned. Accordingly, a Lincoln Town Car with a V-8 engine may not by law emit more emissions in a mile, or 10 miles, or 1,000 miles, than a Chevrolet Metro with a three-cylinder engine.

As noted by the National Research Council (NRC) in a 1992 report on automobile fuel economy, "Fuel economy improvements will not directly affect vehicle emissions." In fact, the NRC found that higher fuel economy standards could actually have a negative effect on the environment:

Improvements in vehicle fuel economy will have indirect environmental impacts. For example, replacing the cast iron and steel components of vehicles with lighter weight materials (e.g., aluminum, plastics, or composites) may reduce fuel consumption but would generate a different set of environmental impacts, as well as result in different kinds of indirect energy consumption.

Nor will increasing CAFE standards halt the alleged problem of "global warming." Cars and light trucks subject to fuel economy standards make up only 1.5 percent of all global man-made greenhouse gas emissions. According to data published in 1991 by the Office of Technology Assessment, a 40 percent increase in fuel economy standards would reduce greenhouse emissions by only about 0.5 percent, even under the most optimistic assumptions.

The NRC additionally noted that "greenhouse gas emissions from the production of substitute materials, such as aluminum, could substantially offset decreases of those emissions achieved through improved fuel economy."

Conclusion

The CAFE program has failed to achieve its goals. Since its inception, both oil imports and vehicle miles driven have increased while the standards have led to reduced consumer choice and lives lost that could have survived car crashes in heavier vehicles.

The CAFE standards should not be increased. They should be repealed and replaced with free market strategies. Consumers respond to market signals. As past experience shows, competition can lead to a market that makes gas guzzlers less attractive than safer and more fuel-efficient vehicles. That is the right way to foster energy conservation.

POSTSCRIPT

Should Cars Be More Efficient?

In June 2007, the U.S. Senate passed the Energy Act with an amendment that increased the average fuel economy standard for cars, trucks, and SUVs by 10 miles per gallon over ten years, reaching 35 mpg for passenger cars by 2020. After that date, further increases in the standard will be at the discretion of the Secretary of Transportation. A second amendment requires the Secretary to establish a plan to make half of all automobiles sold in 2015 run on alternative fuels such as biodiesel, alcohol, or hydrogen. The title of Mark Clayton's and Mark Trumbull's "Fuel Economy Back on U.S. Agenda," *Christian Science Monitor* (May 10, 2007), was clearly apt. Whether the Act's goals can be achieved depends on a number of factors. *Effectiveness and Impact of Corporate Average Fuel Economy (CAFE) Standards* (National Academy Press, 2002) says that while the technology exists to improve gasoline mileage by 20 to 40 percent or more, consumers may have to pay more, and if gasoline prices drop they may refuse. In addition, CAFE standards apply to vehicles sold in a particular year, not to all the vehicles on the road in that year. It can take a decade or more (depending on the economy) to replace all the vehicles on the road. Thus, efficiency of fuel use cannot possibly rise as rapidly as the standards for new cars, although Christopher Evans, Lynette Cheah, Anup Bandivadekar, and John Heywood, "Getting More Miles per Gallon," *Issues in Science and Technology* (Winter 2009), note that increased fuel taxes can motivate consumers to replace older cars with newer, more efficient ones more rapidly.

At the end of 2008, the U.S. Department of Transportation announced standards even tougher that those called for by the Act. Bruce Geiselman, "Government Hikes CAFE Standards," *Waste News* (December 22, 2008), reports that fuel efficiency standards would have to increase by 4.5 percent per year from 2011 through 2015. "Passenger vehicles currently averaging 27.5 miles per gallon would need to achieve an average of 35.7 mpg by 2015." However, in May 2009, the Obama Administration raised the target to 39 mpg for cars and 30 mpg for light trucks by 2016. Not surprisingly, critics continue to claim that more efficient cars will be less safe (see, e.g., Gateway Pundit, "Obama's CAFE Standards Will Kill More Americans Than Iraq War" (http://gatewaypundit .blogspot.com/2009/05/obamas-cafe-standards-will-kill-more.html) (May 20, 2009).

Some of the technologies that can improve fuel efficiency are still in the research and development stage. Some, however, are well established. The Smart Car (developed by Daimler-Benz engineers; see http://www.smartusa.com/) is very common on European streets; using a diesel engine, it gets about 70 mpg. A gasoline-powered version now available in the U.S. gets over 40 mpg. The secret is in part size: The Smart Car is a small car designed for commuters and others who do not have to haul much cargo. It carries two, plus a bag or two of groceries. Mark Clayton, "Safe Cars versus Fuel Efficiency? Not So Fast," *Christian Science Monitor* (June 12, 2007), notes that some automakers are insisting that raising fuel efficiency requires reducing vehicle size, which then reduces safety (see also Moira Herbst, "Fighting for the Right to Make Big Cars," *Business Week Online*, May 30, 2007). But good design can change that. The Smart Car, as just one example, has an egg-shaped frame that holds up very well in crashes despite its size.

ISSUE 11

Are Biofuels Responsible for Rising Food Prices?

YES: Donald Mitchell, from "A Note on Rising Food Prices," The World Bank Development Prospects Group (July 2008)

NO: Keith Kline, Virginia H. Dale, Russell Lee, and Paul Leiby, from "In Defense of Biofuels, Done Right," *Issues in Science and Technology* (Spring 2009)

ISSUE SUMMARY

YES: Donald Mitchell argues that although many factors contributed to the increase in internationally traded food prices from January 2002 to June 2008, the most important single factor was the large increase in biofuels production from grains and oilseeds in the U.S. and EU.

NO: Keith Kline, Virginia H. Dale, Russell Lee, and Paul Leiby argue that the impact of biofuels production on food prices is much less than alarmists claim. There would be greater impact if biofuels development focused on converting biowastes and fast-growing trees and grasses into fuels.

T he threat of global warming has spurred a great deal of interest in finding new sources of energy that do not add to the amount of carbon dioxide in the air. Among other things, this has meant a search for alternatives to fossil fuels, which modern civilization uses to generate electricity, heat homes, and power transportation. Finding alternatives for electricity generation or home heating is easier than finding alternatives for transportation (which relies on oil, refined into gasoline and diesel oil). In addition, the transportation infrastructure, consisting of refineries, pipelines, tank trucks, gas stations, and an immense number of cars and trucks that will be on the road for many years, is well designed for handling liquid fuels. It is not surprising that industry and government would like to find non-fossil liquid fuels for cars and trucks (as well as ships and airplanes).

There are many suitably flammable liquids. Among them are the so-called biofuels or renewable fuels, plant oils and alcohols that can be distilled from plant sugars. According to Daniel M. Kammen, "The Rise of Renewable Energy," *Scientific American* (September 2006), the chief biofuel in the U.S. so far is ethanol, distilled from corn and blended with gasoline. Production is subsidized with $2 billion of federal funds, and "when all the inputs and outputs were correctly factored in, we found that ethanol" contains about 25 percent more energy (to be used when it is burned as fuel) than was used to produce it. At least one study says the "net energy" is actually less than the energy used to produce ethanol from corn; see Dan Charles, "Corn-Based Ethanol Flunks Key Test," *Science* (May 1, 2009). If other sources, such as cellulose-rich switchgrass or cornstalks, can be used, the "net energy" is much better. However, generating ethanol requires first converting cellulose to fermentable sugars, which is so far an expensive process (although many people are working on making the process cheaper; see e.g., Jennifer Chu, "Reinventing Cellulosic Ethanol Production," *Technology Review* online, June 10, 2009, http://www.technologyreview.com/energy/22774/?nlid=2091), and George W. Huber and Bruce E. Dale, "Grassoline at the Pump," *Scientific American* (July 2009). A significant additional concern is the amount of land needed for growing crops to be turned into biofuels; in a world where hunger is widespread, this means land is taken out of food production. If additional land is cleared to grow biofuel crops, this must mean loss of forests and wildlife habitat, increased erosion, and other environmental problems. See "Ethanol: Energy Well Spent, A Survey of Studies Published Since 1990," Natural Resources Defense Council and Climate Solutions (February 2006) (http://www.nrdc.org/air/transportation/ethanol/ethanol.pdf).

Under the Energy Policy Act of 2005, the U.S. Environmental Protection Agency (EPA) requires that gasoline sold in the United States contain a minimum volume of renewable fuel. Under the Renewable Fuel Program (also known as the Renewable Fuel Standard Program, or RFS Program), that volume will increase over the years, reaching 36 billion gallons by 2022. According to the EPA (http://www.epa.gov/otaq/renewablefuels/), "the RFS program was developed in collaboration with refiners, renewable fuel producers, and many other stakeholders." However, some think it is premature and even dangerous to put so much emphasis on biofuels. William Tucker, "Food Riots Made in the USA," *The Weekly Standard* (April 28, 2008), blames the food riots seen in many countries in 2008 on price increases due in large part to the shift in agricultural production from food grains to biofuels crops.

In the following selections, Donald Mitchell argues that although many factors contributed to the increase in internationally traded food prices from January 2002 to June 2008, the most important single factor—accounting for as much as 70 percent of the rise in food prices—was the large increase in biofuels production from grains and oilseeds in the U.S. and EU. Without these increases, global wheat and maize stocks would not have declined appreciably, and price increases due to other factors would have been moderate. Keith Kline, Virginia H. Dale, Russell Lee, and Paul Leiby argue that the impact of biofuels production on food prices is much less than alarmists claim. If biofuels development focused on converting biowastes and fast-growing trees and grasses into fuels, the overall impact would be even better, with a host of benefits in reduced fossil fuel use and greenhouse gas emissions, increased employment, enhanced wildlife habitat, improved soil and water quality, and more stable land use.

YES

Donald Mitchell

A Note on Rising Food Prices

I. Introduction

Internationally traded food commodities prices have increased sharply since 2002 and especially since late-2006, and prices of major staples, such as grains and oilseeds, have doubled in just the past two years. Rising prices have caused food riots in several countries and led to policy actions such as the banning of grain and other food exports by a number of countries and tariff reductions on imported foods in others. The policy actions reflect the concern of governments about the impact of food price increases on the poor in developing countries who, on average, spend half of their household incomes on food. This paper examines how internationally traded food commodities prices (maize, wheat, rice, soybeans, etc.) have changed, and analyzes the factors contributing to these increases. In particular, it looks at the contribution of biofuels production to food price increases. In this paper biofuels refer to ethanol and biodiesel.

II. The Rise in Global Food Prices

The IMF's index of internationally traded food commodities prices increased 130 percent from January 2002 to June 2008 and 56 percent from January 2007 to June 2008. Prior to that, food commodities prices had been relatively stable after reaching lows in 2000 and 2001 following the Asia financial crisis. The low levels of global grain stocks had been identified as a cause for concern in a number of fora and the risk of higher food prices was highlighted in a recent World Bank publication and online.

The increase in food commodities prices was led by grains which began sustained price increases in 2005 despite a record global crop in the 2004/05 crop year that was 10.2 percent larger than the average of the three previous years and a near record crop in 2005/06 that was still 8.9 percent larger. Global stocks of grain increased in 2004/05 but declined in 2005/06 as demand increased more than production. From January 2005 until June 2008, maize prices almost tripled, wheat prices increased 127 percent and rice prices increased 170 percent. The increase in grain prices was followed by increases in fats & oils prices in mid-2006, and that also followed a record 2004/05 global oilseed crop that was 13 percent larger than in the previous year and an even larger crop in 2005/06. Fats & oils prices have shown similar increases to grains, with palm oil prices up 200 percent from January 2005 until June 2008, soybean oil prices up 192 percent, and other vegetable oils prices increasing by similar amounts. Other foods prices (sugar, citrus, bananas, shrimp and meats) increased 48 percent from January 2005 to June 2008.

III. Recent Estimates of the Contribution of Biofuels Production to Food Prices

Estimates of the contribution of biofuels production to food price increases are difficult, if not impossible to compare. Estimates can differ widely due to different time periods considered, different prices (export, import, wholesale, retail) considered, and different coverage of food products. . . .

[But] [d]espite all the differences in approach, many studies recognize biofuels production as a major driver of food prices. The USDA's chief economist in testimony before the Joint Economic Committee of Congress on May 1, attributed much of the increase in farm prices of maize and soybeans to biofuels production. The IMF estimated that the increased demand for biofuels accounted for 70 percent of the increase in maize prices and 40 percent of the increase in soybean. Collins used a mathematical simulation to estimate that about 60 percent of the increase in maize prices from 2006 to 2008 may have been due to the increase in maize used in ethanol. Rosegrant, et al., using a general equilibrium model, calculated the long-term impact on weighted cereal prices of the acceleration in biofuel production from 2000 to 2007 to be 30 percent in real terms. Maize prices were estimated to have increased 39 percent in real terms, wheat prices increased 22 percent and rice prices increased 21 percent. During this period, the U.S. CPI increased by 20.4 percent, which would imply nominal prices increases of 47, 26, and 25, respectively, for maize, wheat and rice prices. This is the same order of magnitude as was calculated with the World Bank's linkages model. Differences in the estimates of the impact of biofuels on the price index of all food depend largely on how broadly the food basket is defined and what is assumed about the interaction between prices of maize and vegetable oils (directly influenced by demand for biofuels) to prices of other crops such as rice through substitution on the supply or demand side. For example, the Council of Economic Advisors estimated that retail food prices increased only about 3 percent over the past 12 months due to ethanol production, in part because they only considered the impact of maize prices, directly and indirectly, on retail prices.

Many other potential drivers of the escalating food prices are mentioned in discussions, but there are few quantitative estimates of their impacts. For example, a recent USDA report attributed the increase in world market prices for major food commodities such as grains and vegetable oils to many factors including biofuels as well as other factors including the declining dollar, rising energy prices, increasing agricultural costs of production, growing foreign exchange holdings by major food-importing countries, and recent policies by some exporting countries to mitigate their own food-price inflation.

IV. Estimates of Factors Contributing to the Rise in Food Commodities Prices

There are a number of factors that have contributed to the rise in food prices. Among these are the increase in energy prices and the related increases in prices of fertilizer and chemicals, which are either produced from energy or are heavy users of energy in their production process. This has increased the cost of production, which ultimately gets reflected in higher food prices. Higher energy prices have also increased the cost of transportation, and increased the incentive to produce biofuels and encouraged policy support for biofuels production. The increase in biofuels production has not only increased demand for food commodities, but also led to large land use changes which reduced supplies of wheat and crops that compete with food commodities used for biofuels. Drought in Australia in 2006 and 2007 and poor crops in Europe in 2007 added to the grain and oilseed price increases, and rapid import demand increases for oilseeds by China to feed its growing livestock and poultry industry contributed to oilseed price increases. Other factors, including the decline of the dollar, and the increased investment in commodities by institutional investors to hedge against inflation and diversify portfolios may have also contributed to the price increases. The remainder of this section will examine these factors.

High energy prices have contributed about 15–20 percent to higher U.S. food commodities production and transport costs. Production costs per acre for U.S. corn, soybeans and wheat increased 32.3, 25.6 and 31.4 percent, respectively, from 2002 to 2007, according to the USDA's cost-of-production surveys and forecasts. However, yield increases during this period reduced the per bushel cost increases to 17.0, 24.1 and 6.7 percent, respectively. The contribution of the energy-intensive components of production costs—fertilizer, chemicals, fuel, lubricants and electricity—were 13.4 percent for corn, 6.7 percent for soybeans and 9.4 percent for wheat per bushel. The production-weighted average increase in the cost of production due to these energy-intensive inputs for these crops was 11.5 percent between 2002 and 2007. In addition to the increase in production costs, transport costs also increased due to higher fuel costs and the margin between domestic and export prices reflect this cost. However, these margins also include handling and other charges, such as insurance, which increase with crop prices. The margin for corn between central Illinois cash and the Gulf Ports barge increased from $0.36 to $0.72 per bushel for an increase of 15.5 percent, while the margin between Kansas City and the Gulf Ports wheat increased only $1 per metric ton. An export weighted average of these prices suggests that transport costs could have added as much as 10.2 percent to the export prices of corn and wheat. Comparable data was not available for soybeans. Thus, the combined increase in production costs and transport costs for the major U.S. food commodities—corn, soybeans and wheat—was at most 21.7 percent, and this amount likely overstates the increase, because transport costs are not estimated separately. It therefore seems reasonable to conclude that higher energy and related costs increased export prices of major U.S. food commodities by about 15–20 percent between 2002 and 2007.

Increased biofuel production has increased the demand for food commodities. The use of maize for ethanol grew especially rapidly from 2004 to 2007 and used 70 percent of the increase in global maize production. In contrast, feed use of maize, which accounts for 65 percent of global maize use, grew by only 1.5 percent per year from 2004 to 2007 while ethanol use grew by 36 percent per year. The share of global feed use of total use declined in response to maize price rises from 69 to 64 percent from 2004 to 2007, and from 70 to 67 percent when the feed by-products from biofuel production are included in feed use.

The United States is the largest producer of ethanol from maize and is expected to use about 81 million tons for ethanol in the 2007/08 crop year. Canada, China, and the European Union used roughly an additional 5 million tons of maize for ethanol in 2007, bringing the total use of maize for ethanol to 86 million tons, which was about 11 percent of global maize production. The large use of maize for ethanol in the U.S. has important global implications, because the U.S. accounts for about one-third of global maize production and two-thirds of global exports and used 25 percent of its production for ethanol in 2007/08.

About 7 percent of global vegetable oil supplies were used for biodiesel production in 2007 and about one-third of the increase in consumption from 2004 to 2007 was due to biodiesel. The largest biodiesel producers were the European Union, the United States, Argentina, Australia, and Brazil, with a combined use of vegetable oils for biodiesel of about 8.6 million tons in 2007 compared with global vegetable oils production of 132 million tons according to the USDA (2008f). From 2004 to 2007, global consumption of vegetable oils for all uses increased by 20.8 million tons, with food use accounting for 80 percent of total use and 60 percent of the increase. Industrial uses of vegetable oils (which include biodiesel) grew by 15 percent per annum from 2004 to 2007, compared with 4.2 percent per annum for food use. The share of industrial use of total use rose from 14.4 percent in 2004 to 18.7 percent in 2007.

Imports of vegetable oils by the EU and U.S. have increased substantially, with the EU-27 increasing imports from 4.4 to 6.9 million tons from 2000 to 2007 and the U.S. increasing imports from 1.7 to 2.9 million tons. The large imports coincided with the increase in biodiesel production in the EU-27 from .45 billion gallons in 2004 to 1.9 billion gallons in 2007 and from .03 billion gallons in the U.S. in 2004 to an estimated .44 billion gallons in 2007.

Brazilian ethanol production from sugar cane has not contributed appreciably to the recent increase in food commodities prices, because Brazilian sugar cane production has increased rapidly and sugar exports have nearly tripled since 2000. Brazil uses approximately half of its sugar cane to produce ethanol for domestic consumption and exports and the other half to produce sugar. The increase in cane production has been large enough to allow sugar production to increase from 17.1 million tons in 2000 to 32.1 million tons in 2007 and exports to increase from 7.7 million tons to 20.6 million tons. Brazil's share of global sugar exports increased from 20 percent in 2000 to 40 percent in 2007, and that was sufficient to keep sugar price increases small except for 2005 and early 2006 when Brazil and Thailand had poor crops due to drought.

The increases in biofuels production in the EU, U.S., and most other bio-fuel-producing countries have been driven by subsidies and mandates. The U.S. has a tax credit available to blenders of ethanol of $0.51 per gallon and an import tariff of $0.54 per gallon, as well as a biodiesel blenders tax credit $1.00 per gallon. The U.S. mandated 7.5 billion gallons of renewable fuels by 2012 in its 2005 legislation and raised the mandate to 15 billion gallons of ethanol from conventional sources (maize) by 2022 and 1.0 billion gallons of biodiesel by 2012 in energy legislation passed in late-2007. The new U.S. mandates will require ethanol production to more than double and biodiesel production to triple if they are met from domestic production. The EU has a specific tariff of €0.192/liter of ethanol (€0.727 or about $1.10 per gallon) and an ad valorem duty of 6.5 percent on biodiesel. EU member states are permit-ted to exempt or reduce excise taxes on biofuels, and several EU member states have introduced mandatory blending requirements. Individual member states have also provided generous excise tax concessions without limit, and Ger-many for example, has provided tax exemptions of €0.4704/ ($0.64) per liter of biodiesel and €0.6545 ($0.88) per liter of ethanol prior to new legislation in 2006. These strong incentives and mandates encouraged the rapid expansion of biofuels in both the EU and U.S.

The EU began to rapidly expand biodiesel production after the EU directive on biofuels entered into effect in October 2001 stipulating that national measures must be taken by EU countries aimed at replacing 5.75 percent of all transport fossil fuels with biofuels by 2010. This led to an increase in biodiesel production from 0.28 billion gallons in 2001 to 1.78 billion gallons in 2007 (FAPRI 2008). Rapeseed was the primary feedstock, followed by soybean oil and sunflower oil. The combined use of vegetable oils for biodiesel was 6.1 million tons in 2007 compared with about 1.0 million tons in 2001.

The U.S. expanded its biodiesel production following legislation passed in 2004 which took effect in January 2005, providing an excise tax credit of US$1.00 per gallon of biodiesel made from agricultural products. This contrib-uted to an increase in biodiesel production in the U.S. from 0.03 billion gallons in 2005 to .44 billion gallons in 2007 and used 3.0 million tons of soybean oil and 0.3 million tons of other fats and oils. These two policies encouraged the rapid expansion of oilseeds production for biodiesel and contributed to the surge in vegetable oils prices, with annual average soybean oil prices rising from $354/ton in 2001 to $881 per ton in 2007. Monthly soybean oil prices rose to $1,522/ton in June 2008. Since oilseeds are close substitutes and prices highly correlated, this led to similar increases in other oilseeds prices.

Land use changes due to expanded biofuel's feedstock production have been large and have led to reduced production of other crops. The U.S. expanded maize area 23 percent in 2007 in response to high maize prices and rapid demand growth for maize for ethanol production. This expansion resulted in a 16 percent decline in soybean area, which reduced soybean pro-duction and contributed to a 75 percent rise in soybean prices between April 2007 and April 2008.

While maize displaced soybeans in the U.S., other oilseeds displaced wheat in the EU and other wheat exporting countries. The expansion of biodiesel production in the EU diverted land from wheat and slowed the increase in wheat production, which would have otherwise kept wheat stocks higher. In response to the increased demand and rising prices for oilseeds, land planted to oilseeds increased, especially rapeseed and to a lesser extent sunflower. The increase was primarily in the countries that are also major wheat exporters such as Argentina, Canada, the EU, Russia, and Ukraine. Oilseeds and wheat are grown under similar climatic conditions and in similar areas and most of the expansion of rapeseed and sunflower displaced wheat or was on land that could have grown wheat. The 8 largest wheat exporting countries expanded area in rapeseed and sunflower by 36 percent (8.4 million hectares) between 2001 and 2007 while wheat area fell by 1.0 percent. The wheat production potential of this land was 26 million tons in 2007 based on average wheat yields in each country, and the cumulative wheat production potential of that land totaled 92 million tons from 2002 to 2007. To illustrate the impact of this land shift on wheat stocks, . . . [simulations show that] if the land planted to rapeseed and sunflower had been planted to wheat and if wheat stocks had increased by the same amounts . . . wheat stocks would have been almost as large in 2007 as in 2001 rather than lower by almost half.

Export bans and restrictions fueled the price increases by restricting access to supplies. A number of countries have imposed export restrictions or bans on grain exports to contain domestic price increases. These include Argentina, India, Kazakhstan, Pakistan, Ukraine, Russia, and Vietnam. . . . According to the USDA and the International Grains Council, there were no other important market developments at that time that could account for the subsequent rice price increases. The USDA had projected India to export 4.1 million tons in the month prior to the ban and that was revised to 3.4 million tons in the month following the ban. The ban on exports led to a steady increase in prices over the following weeks. While it is probably not correct to say that all of the price increases were due to the ban, it likely focused attention on the market fundamentals and the rise in wheat prices and caused market participants to reconsider their imports and exports.

Rice is not used for biofuels, but the increase in prices of other commodities contributed to the rapid rise in rice prices. Rice prices almost tripled from January to April 2008 despite little change in production or stocks. This increase was mostly in response to the surge in wheat prices in 2007 (up 88 percent from January to December) which raised concerns about the adequacy of global grain supplies and encouraged several countries to ban rice exports to protect consumers from international price increases, and caused others to increase imports.

Weather-related production shortfalls have been identified as a major factor underpinning world cereals prices, especially in Australia, U.S., EU, Canada, Russia, and Ukraine. The back-to-back droughts in Australia in 2006 and 2007 reduced grain exports by an average of 9.2 million tons per year compared with 2005, and poor crops in the EU and Ukraine reduced their exports by an additional 10 million tons in 2007. However, these declines were more than offset by large crops in Argentina, Kazakhstan, Russia, and the U.S. Total grain exports from these countries in 2007 increased by about 22 million tons compared with 2006. Global grain production did decline by 1.3 percent in 2006, but it then increased 4.7 percent in 2007. Thus, the production shortfall in grains would not, by itself, have been a major contributor to the increase in grain prices. But when combined with large increases in biofuels production, land use changes, and stock declines it undoubtedly contributed to higher prices. The production shortfall was most significant in wheat, where global production declined 4.5 percent in 2006 and then increased only 2 percent in 2007. Global oilseed production rose 5.4 percent in 2006/07 and declined 3.4 percent in 2007/08.

Rapid income growth in developing countries has not led to large increases in global grain consumption and was not a major factor responsible for the large grain price increases. However, it has contributed to increased oilseed demand and higher oilseed prices as China increased soybean imports for its livestock and poultry industry. Both China and India have been net grain exporters since 2000, although exports have declined as consumption has increased. Global consumption of wheat and rice grew by only 0.8 and 1.0 percent per annum, respectively, from 2000 to 2007 while maize consumption grew by 2.1 percent (excluding the demand for biofuels in the U.S.). This was slower than demand growth during 1995–2000 when wheat, rice, and maize consumption increased by 1.4, 1.4, and 2.6 percent per annum, respectively.

Other factors, such as the decline of the dollar, contributed to food commodity price increases. . . .

Speculative and investor activity has also increased and could have contributed to food price increases. A reflection of this increased activity was the quadrupling of the number of wheat futures contacts traded on the Chicago Board of Trade from 2002 to 2006. However, the increase in futures contracts does not coincide closely with the increase in wheat prices, which raises doubts about the impact on prices. The impact on prices is hard to quantify and most studies do not find that such activity changes prices from the levels which would have prevailed without such activity, however, they may change the rate of adjustment to a new equilibrium when fundamental factors change.

Summary and Conclusions

The increase in internationally traded food prices from January 2002 to June 2008 was caused by a confluence of factors, but the most important was the large increase in biofuels production from grains and oilseeds in the U.S. and EU. Without these increases, global wheat and maize stocks would not have declined appreciably, and price increases due to other factors would have been moderate. Land use changes in wheat exporting countries in response to increased plantings of oilseeds for biodiesel production limited expansion of wheat production that could have otherwise prevented the large declines in global wheat stocks and the resulting rise in wheat prices. The rapid rise in oilseed prices was caused mostly by demand for biodiesel production in response to incentives provided by policy changes in the EU beginning in 2001 and in the

U.S. beginning in 2004. The large increase in rice prices was largely a response to the increase in wheat prices rather than to changes in rice production or stocks, and was thus indirectly related to the increase in biofuels. Recent export bans on grains and speculative activity would probably not have occurred without the large price increases due to biofuels production because they were largely responses to rising prices. Higher energy and fertilizer prices would have still increased crop production costs by about 15–20 percentage points in the U.S. and lesser amounts in countries with less intensive production practices. The back-to-back droughts in Australia would not have had a large impact because they only reduced global grain exports by about 4 percent and other exporters would normally have been able to offset this loss. The decline of the dollar has contributed about 20 percentage points to the rise in dollar food prices.

Thus, the combination of higher energy prices and related increases in fertilizer prices and transport costs, and dollar weakness caused food prices to rise by about 35–40 percentage points from January 2002 until June 2008. These factors explain 25–30 percent of the total price increase, and most of the remaining 70–75 percent increase in food commodities prices was due to biofuels and the related consequences of low grain stocks, large land use shifts, speculative activity and export bans. It is difficult, if not impossible, to compare these estimates with estimates from other studies because of different methodologies, widely different time periods considered, different prices compared, and different food products examined, however, most other studies have also recognized biofuels production as a major factor driving food prices. The increase in grain consumption in developing countries has been moderate and did not lead to large price increases. Growth in global grain consumption (excluding biofuels) was only 1.7 percent per annum from 2000 to 2007, while yields grew by 1.3 percent and area grew by 0.4 percent, which would have kept global demand and supply roughly in balance. This was slower than growth during 1995–2000 when wheat, rice and maize consumption increased by 1.4, 1.4, and 2.6 percent per annum, respectively.

The large increases in biofuels production in the U.S. and EU were supported by subsidies, mandates, and tariffs on imports. Without these policies, biofuels production would have been lower, and food commodity price increases would have been smaller. Biofuels production from sugar cane in Brazil is lower-cost than biofuels production in the U.S. or EU and has not raised sugar prices significantly because sugar cane production has grown fast enough to meet both the demand for sugar and ethanol. Removing tariffs on ethanol imports in the U.S. and EU would allow more efficient producers such as Brazil and other developing countries, including many African countries, to produce ethanol profitably for export to meet the mandates in the U.S. and EU. Biofuels policies which subsidize production need to be reconsidered in light of their impact on food prices.

Keith Kline, Virginia H. Dale,
Russell Lee, and Paul Leiby

 **NO**

In Defense of Biofuels, Done Right

Biofuels have been getting bad press, not always for good reasons. Certainly important concerns have been raised, but preliminary studies have been misinterpreted as a definitive condemnation of biofuels. One recent magazine article, for example, illustrated what it called "Ethanol USA" with a photo of a car wreck in a corn field. In particular, many criticisms converge around grain-based biofuel, traditional farming practices, and claims of a causal link between U.S. land use and land-use changes elsewhere, including tropical deforestation.

Focusing only on such issues, however, distracts attention from a promising opportunity to invest in domestic energy production using biowastes, fast-growing trees, and grasses. When biofuel crops are grown in appropriate places and under sustainable conditions, they offer a host of benefits: reduced fossil fuel use; diversified fuel supplies; increased employment; decreased greenhouse gas emissions; enhanced habitat for wildlife; improved soil and water quality; and more stable global land use, thereby reducing pressure to clear new land.

Not only have many criticisms of biofuels been alarmist, many have been simply inaccurate. In 2007 and early 2008, for example, a bumper crop of media articles blamed sharply higher food prices worldwide on the production of biofuels, particularly ethanol from corn, in the United States. Subsequent studies, however, have shown that the increases in food prices were primarily due to many other interacting factors: increased demand in emerging economies, soaring energy prices, drought in food-exporting countries, cut-offs in grain exports by major suppliers, market-distorting subsidies, a tumbling U.S. dollar, and speculation in commodities markets.

Although ethanol production indeed contributes to higher corn prices, it is not a major factor in world food costs. The U.S. Department of Agriculture (USDA) calculated that biofuel production contributed only 5% of the 45% increase in global food costs that occurred between April 2007 and April 2008. A Texas A&M University study concluded that energy prices were the primary cause of food price increases, noting that between January 2006 and January 2008, the prices of fuel and fertilizer, both major inputs to agricultural production, increased by 37% and 45%, respectively. And the International Monetary Fund has documented that since their peak in July 2008, oil prices declined by 69% as of December 2008, and global food prices declined by

33% during the same period, while U.S. corn production has remained at 12 billion bushels a month, one-third of which is still used for ethanol production.

In another line of critique, some argue that the potential benefits of biofuel might be offset by indirect effects. But large uncertainties and postulations underlie the debate about the indirect land-use effects of biofuels on tropical deforestation, the critical implication being that use of U.S. farmland for energy crops necessarily causes new land-clearing elsewhere. Concerns are particularly strong about the loss of tropical forests and natural grasslands. The basic argument is that biofuel production in the United States sets in motion a necessary scenario of deforestation.

According to this argument, if U.S. farm production is used for fuel instead of food, food prices rise and farmers in developing countries respond by growing more food. This response requires clearing new land and burning native vegetation and, hence, releasing carbon. This "induced deforestation" hypothesis is based on questionable data and modeling assumptions about available land and yields, rather than on empirical evidence. The argument assumes that the supply of previously cleared land is inelastic (that is, agricultural land for expansion is unavailable without new deforestation). It also assumes that agricultural commodity prices are a major driving force behind deforestation and that yields decline with expansion. The calculations for carbon emissions assume that land in a stable, natural state is suddenly converted to agriculture as a result of biofuels. Finally, the assertions assume that it is possible to measure with some precision the areas that will be cleared in response to these price signals.

A review of the issues reveals, however, that these assumptions about the availability of land, the role of biofuels in causing deforestation, and the ability to relate crop prices to areas of land clearance are unsound. Among our findings:

First, sufficient suitably productive land is available for multiple uses, including the production of biofuels. Assertions that U.S. biofuel production will cause large indirect land-use changes rely on limited data sets and unverified assumptions about global land cover and land use. Calculations of land-use change begin by assuming that global land falls into discrete classes suitable for agriculture—cropland, pastures and grasslands, and forests—and results depend on estimates of the extent, use, and productivity of these lands, as well as presumed future interactions among land-use classes. But several major organizations, including the Food and Agriculture Organization (FAO), a primary data clearinghouse, have documented significant inconsistencies surrounding global land-cover estimates. For example, the three most recent FAO Forest Resource Assessments, for periods ending in 1990, 2000, and 2005, provide estimates of the world's total forest cover in 1990 that vary by as much as 470 million acres, or 21% of the original estimate.

Cropland data face similar discrepancies, and even more challenging issues arise when pasture areas are considered. Estimates for land used for crop production range from 3.8 billion acres (calculated by the FAO) to 9 billion acres (calculated by the Millennium Ecosystem Assessment, an international effort spearheaded by the United Nations). In a recent study attempting to reconcile cropland use circa 2000, scientists at the University of Wisconsin-Madison and McGill University estimated that there were 3.7 billion acres of cropland, of which 3.2 billion were actively cropped or harvested. Land-use studies consistently acknowledge serious data limitations and uncertainties, noting that a majority of global crop lands are constantly shifting the location of cultivation, leaving at any time large areas fallow or idle that may not be captured in statistics. Estimates of idle croplands, prone to confusion with pasture and grassland, range from 520 million acres to 4.9 billion acres globally. The differences illustrate one of many uncertainties that hamper global land-use change calculations. To put these numbers in perspective, USDA has estimated that in 2007, about 21 million acres were used worldwide to produce biofuel feedstocks, an area that would occupy somewhere between 0.4% and 4% of the world's estimated idle cropland.

Diverse studies of global land cover and potential productivity suggest that anywhere from 600 million to more than 7 billion additional acres of underutilized rural lands are available for expanding rain-fed crop production around the world, after excluding the 4 billion acres of cropland currently in use, as well as the world's supply of closed forests, nature reserves, and urban lands. Hence, on a global scale, land per se is not an immediate limitation for agriculture and biofuels.

In the United States, the federal government, through the multiagency Biomass Research and Development Initiative (BRDI), has examined the land and market implications of reaching the nation's biofuel target, which calls for producing 36 billion gallons by 2022. BRDI estimated that a slight net reduction in total U.S. active cropland area would result by 2022 in most scenarios, when compared with a scenario developed from USDA's so-called "baseline" projections. BRDI also found that growing biofuel crops efficiently in the United States would require shifts in the intensity of use of about 5% of pasture lands to more intensive hay, forage, and bioenergy crops (25 million out of 456 million acres) in order to accommodate dedicated energy crops, along with using a combination of wastes, forest residues, and crop residues. BRDI's estimate assumes that the total area allocated to USDA's Conservation Reserve Program (CRP) remains constant at about 33 million acres but allows about 3 million acres of the CRP land on high-quality soils in the Midwest to be offset by new CRP additions in other regions. In practice, additional areas of former cropland that are now in the CRP could be managed for biofuel feedstock production in a way that maintains positive impacts on wildlife, water, and land conservation goals, but this option was not included among the scenarios considered.

Yields are important. They vary widely from place to place within the United States and around the world. USDA projects that corn yields will rise by 20 bushels per acre by 2017; this represents an increase in corn output equivalent to adding 12.5 million acres as compared with 2006, and over triple that area as compared with average yields in many less-developed nations. And there is the possibility that yields will increase more quickly than projected in the USDA baseline, as seed companies aim to exceed 200 bushels per acre by 2020. The potential to increase yields in developing countries offers tremendous opportunities to improve welfare and expand production while reducing or maintaining the area harvested. These improvements are consistent with U.S. trends during the past half century showing agricultural output growth averaging 2% per year while cropland use fell by an average of 0.7% per year. Even without large yield increases, cropland requirements to meet biofuel production targets may not be nearly as great as assumed.

Concerns over induced deforestation are based on a theory of land displacement that is not supported by data. U.S. ethanol production shot up by more than 3 billion gallons (150%) between 2001 and 2006, and corn production increased 11%, while total U.S. harvested cropland fell by about 2% in the same period. Indeed, the harvested area for "coarse grains" fell by 4% as corn, with an average yield of 150 bushels per acre, replaced other feed grains such as sorghum (averaging 60 bushels per acre). Such statistics defy modeling projections by demonstrating an ability to supply feedstock to a burgeoning ethanol industry while simultaneously maintaining exports and using substantially less land. So although models may assume that increased use of U.S. land for biofuels will lead to more land being cleared for agriculture in other parts of the world, evidence is lacking to support those claims.

Second, there is little evidence that biofuels cause deforestation, and much evidence for alternative causes. Recent scientific papers that blame biofuels for deforestation are based on models that presume that new land conversion can be simulated as a predominantly market-driven choice. The models assume that land is a privately owned asset managed in response to global price signals within a stable rule-based economy—perhaps a reasonable assumption for developed nations.

However, this scenario is far from the reality in the smoke-filled frontier zones of deforestation in less-developed countries, where the models assume biofuel-induced land conversion takes place. The regions of the world that are experiencing first-time land conversion are characterized by market isolation, lawlessness, insecurity, instability, and lack of land tenure. And nearly all of the forests are publicly owned. Indeed, land-clearing is a key step in a long process of trying to stake a claim for eventual tenure. A cycle involving incremental degradation, repeated and extensive fires, and shifting small plots for subsistence tends to occur long before any consideration of crop choices influenced by global market prices.

The causes of deforestation have been extensively studied, and it is clear from the empirical evidence that forces other than biofuel use are responsible for the trends of increasing forest loss in the tropics. Numerous case studies document that the factors driving deforestation are a complex expression of cultural, technological, biophysical, political, economic, and demographic interactions. Solutions and measures to slow deforestation have also been analyzed and tested, and the results show that it is critical to improve governance, land tenure, incomes, and security to slow the pace of new land conversion in these frontier regions.

Selected studies based on interpretations of satellite imagery have been used to support the claims that U.S. biofuels induce deforestation in the Amazon, but satellite images cannot be used to determine causes of land-use change. In practice, deforestation is a site-specific process. How it is perceived will vary greatly by site and also by the temporal and spatial lens through which it is observed. Cause-and-effect relationships are complex, and the many small changes that enable larger future conversion cannot be captured by satellite imagery. Although it is possible to classify an image to show that forest in one period changed to cropland in another, cataloguing changes in discrete classes over time does not explain why these changes occur. Most studies asserting that the production and use of biofuels cause tropical deforestation point to land cover at some point after large-scale forest degradation and clearing have taken place. But the key events leading to the primary conversion of forests often proceed for decades before they can be detected by satellite imagery. The imagery does not show how the forest was used to sustain livelihoods before conversion, nor the degrees of continual degradation that occurred over time before the classification changed. When remote sensing is supported by a ground-truth process, it typically attempts to narrow the uncertainties of land-cover classifications rather than research the history of occupation, prior and current use, and the forces behind the land-use decisions that led to the current land cover.

First-time conversion is enabled by political, as well as physical, access. Southeast Asia provides one example where forest conversion has been facilitated by political access, which can include such diverse things as government-sponsored development and colonization programs in previously undisturbed areas and the distribution of large timber and mineral concessions and land allotments to friends, families, and sponsors of people in power. Critics have raised valid concerns about high rates of deforestation in the region, and they often point an accusing finger at palm oil and biofuels.

Palm oil has been produced in the region since 1911, and plantation expansion boomed in the 1970s with growth rates of more than 20% per year. Biodiesel represents a tiny fraction of palm oil consumption. In 2008, less than 2% of crude palm oil output was processed for biofuel in Indonesia and Malaysia, the world's largest producers and exporters. Based on land-cover statistics alone, it is impossible to determine the degree of attribution that oil palm may share with other causes of forest conversion in Southeast Asia. What is clear is that oil palm is not the only factor and that palm plantations are established after a process of degradation and deforestation has transpired. Deforestation data may offer a tool for estimating the ceiling for attribution, however. In Indonesia, for example, 28.1 million hectares were deforested between 1990 and 2005, and oil palm expansion in those areas was estimated to be between 1.7 million and 3 million hectares, or between 6% and 10% of the forest loss, during the same period.

Initial clearing in the tropics is often driven more by waves of illegitimate land speculation than agricultural production. In many Latin American frontier zones, if there is native forest on the land, it is up for grabs, as there is no legal tenure of the land. The majority of land-clearing in the Amazon has been blamed on livestock because, in part, there is no alternative for classifying the recent clearings and, in part, because land holders must keep it "in production" to maintain claims and avoid invasions. The result has been the frequent burning and the creation of extensive cattle ranches. For centuries, disenfranchised groups have been pushed into the forests and marginal lands where they do what they can to survive. This settlement process often includes serving as low-cost labor to clear land for the next wave of better-connected colonists. Unless significant structural changes occur to remove or modify enabling factors, the forest-clearing that was occurring before this decade is expected to continue along predictable paths.

Testing the hypothesis that U.S. biofuel policy causes deforestation else-where depends on models that can incorporate the processes underlying initial land-use change. Current models attempt to predict future land-use change based on changes in commodity prices. As conceived thus far, the computa-tional general equilibrium models designed for economic trade do not ade-quately incorporate the processes of land-use change. Although crop prices may influence short-term land-use decisions, they are not a dominant factor in global patterns of first-time conversion, the land-clearing of chief concern in relating biofuels to deforestation. The highest deforestation rates observed and estimated globally occurred in the 1990s. During that period, there was a surplus of commodities on world markets and consistently depressed prices.

Third, many studies omit the larger problem of widespread global mis-management of land. The recent arguments focusing on the possible defor-estation attributable to biofuels use idealized representations of crop and land markets, omitting what may be larger issues of concern. Clearly, the causes of global deforestation are complex and are not driven merely by a single crop market. Additionally, land mismanagement, involving both initial clearing and maintaining previously cleared land, is widespread and leads to a process of soil degradation and environmental damage that is especially prevalent in the frontier zones. Reports by the FAO and the Millennium Ecosystem Assess-ment describe the environmental consequences of repeated fires in these areas. Estimates of global burning vary annually, ranging from 490 million to 980 million acres per year between 2000 and 2004. The vast majority of fires in the tropics occur in Africa and the Amazon in what were previously cleared, nonforest lands. In a detailed study, the Amazon Institute of Environmental Research and Woods Hole Research Center found that 73% of burned area in the Amazon was on previously cleared land, and that was during the 1990s, when overall deforestation rates were high.

Fire is the cheapest and easiest tool supporting shifting subsistence cul-tivation. Repeated and extensive burning is a manifestation of the lack of ten-ure, lack of access to markets, and severe poverty in these areas. When people or communities have few or no assets to protect from fire and no incentive to invest in more sustainable production, they also have no reason to limit the extent of burning. The repeated fires modify ecosystem structure, penetrate ever deeper into forest margins, affect large areas of understory vegetation (which is not detected by remote sensing), and take an ever greater cumula-tive toll on soil quality and its ability to sequester carbon. Profitable biofuel markets, by contributing to improved incentives to grow cash crops, could reduce the use of fire and the pressures on the agricultural frontier. Biofuels done right, with attention to best practices for sustained production, can make significant contributions to social and economic development as well as environmental protection.

Furthermore, current literature calculates the impacts from an assumed agricultural expansion by attributing the carbon emissions from clearing intact ecosystems to biofuels. If emission analyses consider empirical data reflecting the progressive degradation that occurs (often over decades) before and independently of agriculture market signals for land use, as well as changes in the frequency and extent of fire in areas that biofuels help bring into more stable market economies, then the resulting carbon emission estimates would be worlds apart.

Brazil provides a good case in point, because it holds the globe's largest remaining area of tropical forests, is the world's second-largest producer of biofuel (after the United States), and is the world's leading supplier of biofuel for global trade. Brazil also has relatively low production costs and a growing focus on environmental stewardship. As a matter of policy, the Brazilian government has supported the development of biofuels since launching a National Ethanol Program called Proálcool in 1975. Brazil's ethanol industry began its current phase of growth after Proálcool was phased out in 1999 and the government's role shifted from subsidies and regulations toward increased collaboration with the private sector in R&D. The government helps stabilize markets by supporting variable rates of blending ethanol with gasoline and planning for industry expansion, pipelines, ports, and logistics. The government also facilitates access to global markets; develops improved varieties of sugarcane, harvest equipment, and conversion; and supports improvements in environmental performance.

New sugarcane fields in Brazil nearly always replace pasture land or less valuable crops and are concentrated around production facilities in the developed southeastern region, far from the Amazon. Nearly all production is rainfed and relies on low input rates of fertilizers and agrochemicals, as compared with other major crops. New projects are reviewed under the Brazilian legal framework of Environmental Impact Assessment and Environmental Licensing. Together, these policies have contributed to the restoration or protection of reserves and riparian areas and increased forest cover, in tandem with an expansion of sugarcane production in the most important producing state, Sao Paulo.

Yet natural forest in Brazil is being lost, with nearly 37 million acres lost between May 2000 and August 2006, and a total of 150 million acres lost since 1970. Some observers have suggested that the increase in U.S. corn production for biofuel led to reduced soybean output and higher soybean prices, and that these changes led, in turn, to new deforestation in Brazil. However, total deforestation rates in Brazil appear to fall in tandem with rising soybean prices. This co-occurrence illustrates a lack of connection between commodity prices and initial land clearing. This phenomenon has been observed around the globe and suggests an alternate hypothesis: Higher global commodity prices focus production and investment where it can be used most efficiently, in the plentiful previously cleared and underutilized lands around the world. In times of falling prices and incomes, people return to forest frontiers, with all of their characteristic tribulations, for lack of better options.

Biofuels Done Right

With the right policy framework, cellulosic biofuel crops could offer an alternative that diversifies and boosts rural incomes based on perennials. Such a scenario would create incentives to reduce intentional burning that currently affects millions of acres worldwide each year. Perennial biofuel crops can help stabilize land cover, enhance soil carbon sequestration, provide habitat to support biodiversity, and improve soil and water quality. Furthermore, they can reduce pressure to clear new land via improved incomes and yields. Developing countries have huge opportunities to increase crop yield and thereby grow more food on less land, given that cereal yields in less developed nations are 30% of those in North America. Hence, policies supporting biofuel production may actually help stop the extensive slash-and-burn agricultural cycle that contributes to greenhouse gas emissions, deforestation, land degradation, and a lifestyle that fails to support farmers and their families.

Biofuels alone are not the solution, however. Governments in the United States and elsewhere will have to develop and support a number of programs designed to support sustainable development. The operation and rules of such programs must be transparent, so that everyone can understand them and see that fair play is ensured. Among other attributes, the programs must offer economic incentives for sustainable production, and they must provide for secure land tenure and participatory land-use planning. In this regard, pilot biofuel projects in Africa and Brazil are showing promise in addressing the vexing and difficult challenges of sustainable land use and development. Biofuels also are uniting diverse stakeholders in a global movement to develop sustainability metrics and certification methods applicable to the broader agricultural sector.

Given a priority to protect biodiversity and ecosystem services, it is important to further explore the drivers for the conversion of land at the frontier and to consider the effects, positive and negative, that U.S. biofuel policies could have in these areas. This means it is critical to distinguish between valid concerns calling for caution and alarmist criticisms that attribute complex problems solely to biofuels.

Still, based on the analyses that we and others have done, we believe that biofuels, developed in an economically and environmentally sensible way, can contribute significantly to the nation's—indeed, the world's—energy security while providing a host of benefits for many people in many regions.

POSTSCRIPT

Are Biofuels Responsible for Rising Food Prices?

In March 2007, President Bush visited Brazil, which meets much of its need for vehicle fuel with ethanol from sugarcane, and agreed to work with Brazil in developing and promoting biofuels. According to the U.S. State Department, the agreement "reassures small countries in Central America and the Caribbean that they can reduce their dependence on foreign oil." In both Europe and the U.S., governments are rushing to encourage the production and use of biofuels. L. Pelmans, et al., "European Biofuels Strategy," *International Journal of Environmental Studies* (June 2007), attempts to classify nations according to their strategies so that "the formulation of a strategy to support the advancement of biofuels and alternative motor fuels in general should become more manageable." Corporations and investors see huge potential for profit, and many environmentalists see benefits for the environment.

However, there *are* problems with biofuels, and those problems are getting a great deal of attention. Robin Maynard, "Against the Grain," *The Ecologist* (March 2007), stresses that when food and fuel compete for farmland, food prices will rise, perhaps drastically. The poor will suffer, as will rainforests. Renton Righelato, "Forests or Fuel," *The Ecologist* (March 2007), reminds us that when forests are cleared, they no longer serve as "carbon sinks"; deforestation thus adds to the global warming problem, and it may take a century for the benefit of biofuels to show itself. David Pimentel, "Biofuel Food Disasters and Cellulosic Ethanol Problems," *Bulletin of Science, Technology & Society* (June 2009), says that because using 20 percent of the U.S. corn crop displaces a mere 1 percent of oil consumption, corn ethanol is a disaster, while using crop wastes and other biological materials poses its own problems. Heather Augustyn, "A Burning Issue," *World Watch* (July/August 2007), describes the impact of forest fires to clear land for oil palm plantations in Indonesia. Palm oil holds great promise as a biofuel, but the plantations displace natural ecosystems and destroy habitat for numerous species, as well as for indigenous peoples. Man Kee Lam, et al., "Malaysian Palm Oil: Surviving the Food versus Fuel Dispute for a Sustainable Future," *Renewable & Sustainable Energy Reviews* (August 2009), are more optimistic, saying that with care palm oil may be able to satisfy needs for both edible oils and fuel oils. Gernot Stoeglehner and Michael Narodoslawsky, "How Sustainable Are Biofuels? Answers and Further Questions Arising from an Ecological Footprint Perspective," *Bioresource Technology* (August 2009), note that the sustainability of biofuels production depends very much on regional context.

Laura Venderkam, "Biofuels or Bio-Fools?" *American: A Magazine of Ideas* (May/June 2007), describes the huge amounts of money being invested in companies planning to bring biofuels to market. A great deal of research is also going on, including efforts to use genetic engineering to produce enzymes that can cheaply and efficiently break cellulose into its component sugars (see Matthew L. Wald, "Is Ethanol for the Long Haul?" *Scientific American*, January 2007, and Michael E. Himmel, et al., "Biomass Recalcitrance: Engineering Plants and Enzymes for Biofuels Production," *Science*, February 9, 2007), make bacteria or yeast that can turn a greater proportion of sugar into alcohol (see Francois Torney, et al., "Genetic Engineering Approaches to Improve Bioethanol Production from Maize," *Current Opinion in Biotechnology*, June 2007), and even bacteria that can convert sugar or cellulose into hydrocarbons that can easily be turned into gasoline or diesel fuel (see Neil Savage, "Building Better Biofuels," *Technology Review*, July/August 2007). If these efforts succeed, the price of biofuels may drop drastically, leading investors to abandon the field. Such a price drop would, of course, benefit the consumer and lead to wider use of biofuels. It would also, say C. Ford Runge and Benjamin Senauer, "How Biofuels Could Starve the Poor," *Foreign Affairs* (May/June 2007) ease the impact on food supply.

Peter Rosset, "Agrofuels, Food Sovereignty, and the Contemporary Food Crisis," *Bulletin of Science, Technology & Society* (June 2009), says that biofuels are not a prime cause of the 2008 food crisis but they are "clearly contraindicated." On the other hand, Jose C. Escobar, et al., "Biofuels: Environment, Technology and Food Security," *Renewable & Sustainable Energy Reviews* (August 2009), considers that increased reliance on biofuels is inevitable due to "the imminent decline of the world's oil production, its high market prices and environmental impacts."

Suzanne Hunt, "Biofuels, Neither Saviour nor Scam: The Case for a Selective Strategy," *World Policy Journal* (Spring 2008), argues that all the concerns are real, but the larger problem is that "our current agricultural, energy, and transport systems are failing." John Ohlrogge, et al., "Driving on Biomass," *Science* (May 22, 2009), point to a potential fix for the failure of the transportation system when they show that converting biomass to biofuels is much less efficient than burning it to produce electricity and using the electricity to power electric cars, which are in fact projected to "gain substantial market share in the coming years."

ISSUE 12

Is It Time to Revive Nuclear Power?

YES: Iain Murray, from "Nuclear Power? Yes, Please," *National Review* (June 16, 2008)

NO: Kristin Shrader-Frechette, from "Five Myths About Nuclear Energy," *America* (June 23–30, 2008)

ISSUE SUMMARY

YES: Iain Murray argues that the world's experience with nuclear power has shown it to be both safe and reliable. Costs can be contained, and if one is concerned about global warming, the case for nuclear power is unassailable.

NO: Professor Kristin Shrader-Frechette argues that nuclear power is one of the most impractical and risky of energy sources. Renewable energy sources such as wind and solar are a sounder choice.

The technology of releasing for human use the energy that holds the atom together did not get off to an auspicious start. Its first significant application was military, and the deaths associated with the Hiroshima and Nagasaki explosions have ever since tainted the technology with negative associations. It did not help that for the ensuing half-century, millions of people grew up under the threat of nuclear Armageddon. But almost from the beginning, nuclear physicists and engineers wanted to put nuclear energy to more peaceful uses, largely in the form of power plants. Touted in the 1950s as an astoundingly cheap source of electricity, nuclear power soon proved to be more expensive than conventional sources, largely because safety concerns caused delays in the approval process and prompted elaborate built-in precautions. Safety measures have worked well when needed—Three Mile Island, often cited as a horrific example of what can go wrong, released very little radioactive material to the environment. The Chernobyl disaster occurred when safety measures were ignored. In both cases, human error was more to blame than the technology itself. The related issue of nuclear waste has also raised fears and proved to add expense to the technology.

It is clear that two factors—fear and expense—impede the wide adoption of nuclear power. If both could somehow be alleviated, it might become possible to gain the benefits of the technology. Among those benefits are that nuclear power does not burn oil, coal, or any other fuel, does not emit air pollution and thus contribute to smog and haze, does not depend on foreign sources of fuel and thus weaken national independence, and does not emit carbon dioxide. Avoiding the use of fossil fuels is an important benefit; see Robert L. Hirsch, Roger H. Bezdek, and Robert M. Wendling, "Peaking Oil Production: Sooner Rather Than Later?" *Issues in Science and Technology* (Spring 2005). But avoiding carbon dioxide emissions may be more important at a time when society is concerned about global warming, and this is the benefit that prompted James Lovelock, creator of the Gaia Hypothesis and hero to environmentalists everywhere, to say, "If we had nuclear power we wouldn't be in this mess now, and whose fault was it? It was [the anti-nuclear environmentalists']." See his autobiography, *Homage to Gaia: The Life of an Independent Scientist* (Oxford University Press, 2001). Others have also seen this point. The OECD's Nuclear Energy Agency ("Nuclear Power and Climate Change" [Paris, France, 1998] [http://www.nea.fr/html/ndd/climate/climate.pdf]) found that a greatly expanded deployment of nuclear power to combat global warming was both technically and economically feasible. Robert C. Morris published *The Environmental Case for Nuclear Power: Economic, Medical, and Political Considerations* (Paragon House) in 2000. "The time seems right to reconsider the future of nuclear power," say James A. Lake, Ralph G. Bennett, and John F. Kotek in "Next-Generation Nuclear Power," *Scientific American* (January 2002). Stewart Brand, long a leading environmentalist, predicts in "Environmental Heresies," *Technology Review* (May 2005), that nuclear power will soon be seen as the "green" energy technology. David Talbot, "Nuclear Powers Up," *Technology Review* (September 2005), notes that "While the waste problem remains unsolved, current trends favor a nuclear renaissance. Energy needs are growing. Conventional energy sources will eventually dry up. The atmosphere is getting dirtier." Peter Schwartz and Spencer Reiss, "Nuclear Now!" *Wired* (February 2005), argue that nuclear power is the one practical answer to global warming and forthcoming shortages of fossil fuels.

In the following selections, Iain Murray of the Competitive Enterprise Institute argues that the world's experience with nuclear power has shown it to be both safe and reliable. Costs can be contained, and if one is concerned about global warming, the case for nuclear power is unassailable. Professor Kristin Shrader-Frechette argues that nuclear power is one of the most impractical and risky of energy sources. Renewable energy sources such as wind and solar are a sounder choice.

YES

Iain Murray

Nuclear Power? Yes, Please

My grandfather was a coal miner. My father was an electrical engineer. Energy appears to be in my blood. In the late 1960s and early 1970s, when I was growing up in the north of England, we seemed to be heading for a new age of energy production. Just to the south of us was a community called Seaton Carew, in the county of Durham. It needed more electricity, and a new coal-fired power station was proposed. The people objected. As Ian Fells, emeritus professor of energy conversion at the nearby University of Newcastle-upon-Tyne puts it, they said, "Why can't we have one of those new, clean nuclear power stations?" My neighbors got their wish. In 1969, work began on the Hartlepool Nuclear Plant. Since 1983, it has provided 3 percent of the United Kingdom's energy.

Next year, decommissioning work will begin on the plant. The local people are opposed to a replacement nuclear facility. What happened? The answer is strangely universal. Around the Western world, environmentalists did what they do best: They exploited fear and massaged science to make nuclear power morally unacceptable, impeding progress and innovation. Yet today they find their case weakened—perhaps fatally—by another concern born from another campaign of fear: global warming. The hysteria environmentalists have built up around carbon emissions may lead to a new dawn for nuclear power. Given that, the case for new nuclear is almost unassailable.

Throughout the late 1960s and early 1970s, there was considerable technical discussion about the safety of nuclear power throughout the world, and local court cases against new construction had been common. The Calvert Cliffs Coordinating Committee, for example, delayed the building of a reactor in Maryland after a court victory over the Atomic Energy Commission in 1971. Construction on subsequent projects slowed as a result. Mass popular opposition to the "new, clean nuclear power stations" worldwide, however, seems to have begun in Germany. Shortly after work started in Hartlepool, a new plant was proposed for the tiny southwest German village of Wyhl. Opposition grew and, by the time work began in 1975, a mass movement was ready. Thirty thousand protesters gathered, seized control of the site, and occupied it. The plant was never built.

The German protesters did not rest on their laurels after this victory. In 1977, 20,000 people protested the use of salt mines at Gorleben for nuclear-waste storage. Growing bolder, they turned to open violence. At Brokdorf in

From *The National Review*, June 16, 2008, pp. 32, 34–36. Copyright © 2008 by National Review, Inc, 215 Lexington Avenue, New York, NY 10016. Reprinted by permission.

1981, 100,000 demonstrators surrounded the site of a proposed nuclear plant, confronting 10,000 police. According to the *New York Times* report, "groups of hundreds of demonstrators armed with gasoline bombs, sticks, stones and high-powered slingshots" attacked the police, injuring 21 of them.

It was about this time that I noticed a cultural invasion of England by Germany. Dozens of students at my high school started sporting large yellow buttons depicting a smiley-faced sun surrounded by the German words "ATOMKRAFT? NEIN DANKE!" ("Nuclear power? No thanks!") In 1979, the movement had reached such levels of power that it was able to establish its own sustainable political party. Die Grünen (the Greens) had among their early leaders a roll call of radical leftists. Among them was the tiny and articulate Petra Kelly, a former European Commission bureaucrat who had been educated in the United States and had worked on the Robert F. Kennedy and Hubert Humphrey campaigns in 1968. Others included former student radical Rudi Dutschke, who followed Gramsci in advocating a "long march through the institutions," the novelist Heinrich Böll, and the artist Joseph Beuys. Anti-nuclear activism was the core of their appeal and remains a central plank today. They have indeed marched through the institutions, even claiming the foreign ministry from 1998 to 2005 and entering into a state-level coalition with Germany's nominal conservatives, the Christian Democrats. The Green virus spread dramatically out from Germany, infecting the politics of most developed nations.

I have spent a little time outlining this early history because it is important to recognize that the anti-nuclear movement was already ascendant before the Three Mile Island incident in 1979. On May 2, 1977, no doubt inspired by the German protesters, 1,414 people were arrested protesting a nuclear plant in Seabrook, N.H. Three Mile Island merely galvanized an already-active movement. Nuclear-plant orders had in fact peaked in 1973, and fell off sharply after 1974 as energy-conservation concerns reigned supreme in the wake of the Arab oil embargo. . . . The anti-nuclear movement's achievement was not to stop nuclear power, but to take it off the table as an option after 1979.

Fear and Loathing

How did they succeed? By creating a global zeitgeist—an appropriately German word—holding as an article of faith that nuclear power is a severe danger in all sorts of ways. Their arguments revolved around three main propositions: that nuclear plants are dangerous because they can blow up or melt down; that nuclear waste is extremely and persistently dangerous; and that nuclear power and nuclear weapons are intrinsically linked. All these arguments are overstated.

As to the safety of nuclear power stations, there is now a significant history to demonstrate that these concerns are no longer justified, even if they may have had some precautionary legitimacy in the 1970s. It has long been recognized that the Chernobyl accident was caused by features unique to the Soviet-style RBMK (*reaktor bolshoy moshchnosti kanalniy*—high-power channel reactor). When reactors of that sort get too hot, the rate of the nuclear reaction increases—the reverse of what happens in most Western reactors. Moreover, RBMK reactors do not have containment shells that prevent radioactive material from getting out. The worst incident in the history of nuclear power, Chernobyl killed just 56 people and made 20 square miles of land uninhabitable. (The exclusion zone has now become a haven for wildlife, which is thriving.) There are suggestions that hundreds or thousands more may die because of long-term effects, but these estimates are based on the controversial Linear Non-Threshold (LNT) theory about the effects of radiation.

Official EPA doctrine, based on the LNT theory, holds that no level of radiation is safe, and that the maximum allowable exposure to radiation is an extremely stringent 15 millirems (mrem) per year. After Hiroshima and Nagasaki, researchers discovered that 600,000 mrem was a sufficient dose of radiation to kill anyone exposed to it, and 400,000 mrem killed half the people exposed. Symptoms of radiation sickness develop at 75,000–100,000 mrem. By extrapolating linearly, the model holds that there is no level of radiation at which someone is not adversely affected (hence "nonthreshold"). Therefore, if a million people are exposed to a very low dose of radiation—say 500 mrem—then 6,250 of them will die of cancer brought on by the exposure. At least according to the theory.

But this is mere assumption, with no epidemiological evidence to back it up. As Prof. Donald W. Miller Jr. of the University of Washington School of Medicine wrote in 2004, "Known and documented health-damaging effects of radiation—radiation sickness, leukemia, and death—are only seen with doses greater than 100 rem [which is to say, 100,000 mrem]. The risk of doses less than 100 rem is a black box into which regulators extend 'extrapolated data.' There are no valid epidemiologic or experimental data to support linearly extrapolated predictions of cancer resulting from low doses of radiation."

In fact, Americans are naturally exposed to around 200 mrem a year of background radiation. In some places around the world that background level is much higher. In Ramsar, Iran, thanks to the presence of natural radium in the vicinity, residents get 26,000 mrem a year, but there is no increased incidence of cancer or shortened lifespan. This is a real problem for the LNT theory. The predicted deaths and cancer cases haven't materialized.

In Britain, much hay was made by Greenpeace and other organizations of the emergence of greater incidences of leukemia in children living near the nuclear-reprocessing plant at Sellafield in the early 1990s. But such "cancer clusters" appear all over the place, and are just as likely to appear next to an organic farm—to borrow the formulation of British environment writer Rob Johnston—as next to a nuclear facility. There does not appear to be any greater incidence of leukemia in the children of those who work in the nuclear industry. In fact, there is so little evidence of significant safety risks related to nuclear power that the British government "continues to believe that new nuclear power stations would pose very small risks to safety, security, health and proliferation," according to a recent analysis it undertook. It also believes that "these risks are minimized and sensibly managed by industry."

Nuclear waste is a stickier problem, but one that can be safely managed. In most American reactors, fuel rods need to be replaced every 18 months or so. When they are taken out, they contain large amounts of radioactive fission products and produce enough heat that they need to be cooled in water. Radioactivity declines as the isotopes decay and the rods produce less heat. It is the very nature of radioactivity that, as materials decay, they become less dangerous and easier to handle. The question is what to do with the waste when space runs out.

In most of the rest of the world, fuel reprocessing extracts usable uranium and plutonium. The highly radioactive waste that remains is not a large amount. By 2040, Britain will have just 70,000 cubic feet of such waste. This volume could be contained in a cube measuring 42 feet on each side. Moreover, most of Britain's waste is left over from its nuclear-weapons program. The British government has determined that "geological disposal"—burial deep underground—provides the best available approach to dealing with existing and also with new nuclear waste, arguing that "the balance of ethical considerations does not rule out the option of new nuclear power."

In the United States, unfortunately, reprocessing was stopped during the Carter administration, in the naïve belief that other countries would follow suit and thereby reduce the amount of plutonium available for weapons proliferation. For that reason, the United States has rather more nuclear waste than any other nation, about 144 million pounds of it. Since 1987, the United States has focused its own efforts at geological disposal at a remote Nevada site called Yucca Mountain, located within a former nuclear-test facility.

The story of Yucca Mountain is well-known—it has become a political football as pro- and anti-nuclear forces try to accelerate or delay (or even stop) the facility's commissioning. Legal challenges have focused on the question of how much radiation will escape to the public from the facility—over a timeline of a million years. The Department of Energy has calculated that exposure will be no more than 0.98 mrem per year, up to a million years into the future. Even those who hold to the stringent LNT view of radiation should be satisfied.

With that settled, the Department of Energy announced in 2006 that Yucca Mountain would open for business in 2017. But later that year, when Harry Reid was chosen as Senate majority leader, he announced, "Yucca Mountain is dead. It'll never happen." The project's budget has been slashed. As a result, the question of storing America's nuclear waste remains open, even as more and more of that waste piles up around the country.

On the issue of persistence, bear in mind that reprocessing the fuel means that, after ten years, the fission products are only one-thousandth as radioactive as they were initially. After 500 years, they will be less radioactive than the uranium ore they originally came from. The waste question is therefore simply one of storage. It is a political, not a scientific, dispute.

As for the problem of nuclear proliferation, the unpleasant fact is that every country that has been willing to invest the time and effort required to make a nuclear weapon has succeeded. The existence of nuclear power plants in Western countries has nothing to do with this. In fact, in order to keep plants economical, fuel rods are kept in the reactor long enough that the weapons-grade plutonium, Pu-239, absorbs another neutron and becomes the much less dangerous Pu-240. To be effective in a weapon, a given volume of plutonium must contain no more than 7 percent Pu-240. Spent fuel from civilian nuclear plants is typically composed of about 26 percent Pu-240. This makes it extremely difficult even for experts to use in the manufacture of nuclear weapons—and well nigh impossible for amateurs.

There is some concern that nuclear power plants present an attractive target for terrorists. After the attacks of Osama bin Laden's impromptu air force in 2001, the Department of Energy commissioned a study into the effects of a fully fueled jetliner's hitting a reactor containment vessel at maximum speed. In none of the simulations was containment breached. Given the massive investment that would be needed to compromise a nuclear power station, it is highly unlikely that terrorists would seek to attack such a hard target— especially when their revealed preference has been for soft targets offering the maximum possible loss of civilian life.

The world's experience with nuclear power, therefore, has confounded the arguments of the environmentalists. It has proven safe and reliable—and if you still need convincing of this, remember that the second-worst nuclear incident, Three Mile Island, saw a destroyed reactor confined with no casualties.

Environmental Economics

With safety concerns off the table, the real question remaining as to whether we should move forward with nuclear power is one of economics. Up until now, the high costs of construction and decommissioning have made nuclear more expensive than coal and natural gas. That, however, is likely to change, and it is the environmentalists who will have brought about the change.

The Congressional Budget Office recently released a report finding that nuclear power costs around $72 per megawatt-hour of energy, compared with $55 for coal and $57 for natural gas. But those estimates reflect very high construction costs. When it comes to operating costs, nuclear power is much less expensive. Therefore, the economics of nuclear power depend to a large extent on reducing those construction costs—both the direct costs of building (which is highly dependent on construction time) and the costs of regulation. It should be feasible to reduce significantly the costs of regulation by various means. Under its Nuclear Power 2010 program, for example, the U.S. Department of Energy has offered interested parties the opportunity to operate under a surprisingly non-bureaucratic model for licensing based on the French system. (Less responsibly, for the first six plants, the program also offers to subsidize 25 percent to 50 percent of any construction-cost overruns due to delays.) Nuclear plants have never been built anywhere in an environment of low regulatory costs, so it is hard to estimate to what degree deregulation could reduce expenses without jeopardizing safety.

Those high construction costs have led many to argue that nuclear power is intrinsically uneconomic without subsidy. This is contradicted by the British government's recent analysis, which found that nuclear power is viable without subsidies. The British government received confirmation from potential nuclear operators that subsidy was neither needed nor desired. In the United States, there have been 30 announcements of plans for new nuclear plants, totaling 40,000 megawatts of capacity, which would power 32 million households.

In all the hysteria about global warming, environmentalists have, for the most part, agreed on one thing above all—that the use of fossil fuels must be made more expensive. Every proposal currently under consideration for the reduction of greenhouse-gas emissions seeks to raise prices as a brake on emissions, through either a cap-and-trade system or a carbon tax. Once this expense is included in the calculations, nuclear power becomes extremely competitive, and remains considerably cheaper than wind power. The Congressional Budget Office found that nuclear power is the most attractive source of electricity once the price of carbon emissions reaches $45 a ton. If natural-gas prices increase as rapidly as they have done recently, then that figure will come down even further. The British-government review found that nuclear provides "economic benefit regardless of the carbon price." Moreover, it provides carbon reductions much more cheaply than wind power does. Using nuclear power, it costs 60 cents to eliminate a ton of CO_2 emissions, as opposed to a staggering $100 per ton for onshore wind power. It is true that a carbon tax amounts to a subsidy for nuclear power. But if carbon emissions are to be taxed, then that is the only subsidy that nuclear power will ever need.

Keep in mind that many of the current arguments used against nuclear power by environmentalists are economic in tone—that uranium is running out (not true even in the medium term); that decommissioning is expensive and/or will be a burden on the taxpayer (it is expensive, but the cost could be met by requiring the operator to pay into a fund during the reactor's life); or that building reactors takes too long (true, but most of that is the fault of red tape). The Canadian company ACEL has managed to shorten the time for building a reactor—from groundbreaking to coming online—to four years. Such a schedule should significantly reduce construction costs which, as we have seen, are the main impediment to nuclear cost-effectiveness.

Nuclear Is Greener

Beyond economics, if you take seriously the issue of greenhouse gases—whether as a climate alarmist or as someone with an open mind who believes in a degree of prudence—then the case for nuclear power is unassailable. James Lovelock, famous as the father of the "Gaia Hypothesis," has said nuclear power represents humanity's only hope to escape runaway global warming. Yet most environmental groups refuse to recognize the impracticality of opposing both greenhouse-gas emissions and our most effective way of reducing them. For those groups, the problem isn't "dirty" energy, but energy itself. They presumably agree with one of their sages, Amory Lovins, who told *Playboy* in 1977, "It'd be a little short of disastrous for us to discover a source of clean, cheap, abundant energy because of what we might do with it."

For the rest of us, what we might do with it is the whole point—we might increase human prosperity and welfare. If we're determined to price coal out of the energy market, then nuclear is it. If we're determined to cure our "addiction to oil," then we will need nuclear facilities to power our plug-in hybrid electric cars or to make the hydrogen for our fuel cells. This is not a green pipe dream. In fact, given the way automotive technology is developing, it is plausible that a majority of vehicles sold in the U.S. by 2020 will use electric power trains, increasing our need for electricity. We might not even need to close our coal mines, since we can get more energy from the uranium found in coal than from burning the coal itself.

Those denizens of Seaton Carew had it right. Nuclear power is clean. In a sense it is still new. Thirty years after their radical predecessors took nuclear energy away from the people of the world, environmentalists might have inadvertently given it back. *Atomkraft? Ja, bitte!*

NO

Five Myths About Nuclear Energy

Atomic energy is among the most impractical and risky of available fuel sources. Private financiers are reluctant to invest in it, and both experts and the public have questions about the likelihood of safely storing lethal radioactive wastes for the required million years. Reactors also provide irresistible targets for terrorists seeking to inflict deep and lasting damage on the United States. The government's own data show that U.S. nuclear reactors have more than a one-in-five lifetime probability of core melt, and a nuclear accident could kill 140,000 people, contaminate an area the size of Pennsylvania, and destroy our homes and health.

In addition to being risky, nuclear power is unable to meet our current or future energy needs. Because of safety requirements and the length of time it takes to construct a nuclear-power facility, the government says that by the year 2050 atomic energy could supply, at best, 20 percent of U.S. electricity needs; yet by 2020, wind and solar panels could supply at least 32 percent of U.S. electricity, at about half the cost of nuclear power. Nevertheless, in the last two years, the current U.S. administration has given the bulk of taxpayer energy subsidies—a total of $20 billion—to atomic power. Why? Some officials say nuclear energy is clean, inexpensive, needed to address global climate change, unlikely to increase the risk of nuclear proliferation and safe.

On all five counts they are wrong. Renewable energy sources are cleaner, cheaper, better able to address climate change and proliferation risks, and safer. The government's own data show that wind energy now costs less than half of nuclear power; that wind can supply far more energy, more quickly, than nuclear power; and that by 2015, solar panels will be economically competitive with all other conventional energy technologies. The administration's case for nuclear power rests on at least five myths. Debunking these myths is necessary if the United States is to abandon its current dangerous energy course.

Myth 1. Nuclear Energy Is Clean

The myth of clean atomic power arises partly because some sources, like a pro-nuclear energy analysis published in 2003 by several professors at the Massachusetts Institute of Technology, call atomic power a "carbon-free source" of energy. On its Web site, the U.S. Department of Energy, which is

From *America*, June 23–30, 2008, pp. 12–16. Copyright © 2008 by America Magazine. All rights reserved. Reprinted by permission of America Press. For subscription information, visit http://www.americamagazine.org.

also a proponent of nuclear energy, calls atomic power "emissions free." At best, these claims are half-truths because they "trim the data" on emissions.

While nuclear reactors themselves do not release greenhouse gases, reactors are only part of the nine-stage nuclear fuel cycle. This cycle includes mining uranium ore, milling it to extract uranium, converting the uranium to gas, enriching it, fabricating fuel pellets, generating power, reprocessing spent fuel, storing spent fuel at the reactor and transporting the waste to a permanent storage facility. Because most of these nine stages are heavily dependent on fossil fuels, nuclear power thus generates at least 33 grams of carbon-equivalent emissions for each kilowatt-hour of electricity that is produced. (To provide uniform calculations of greenhouse emissions, the various effects of the different greenhouse gases typically are converted to carbon-equivalent emissions.) Per kilowatt-hour, atomic energy produces only one-seventh the greenhouse emissions of coal, but twice as much as wind and slightly more than solar panels.

Nuclear power is even less clean when compared with energy-efficiency measures, such as using compact-fluorescent bulbs and increasing home insulation. Whether in medicine or energy policy, preventing a problem is usually cheaper than curing or solving it, and energy efficiency is the most cost-effective way to solve the problem of reducing greenhouse gases. Department of Energy data show that one dollar invested in energy-efficiency programs displaces about six times more carbon emissions than the same amount invested in nuclear power. Government figures also show that energy-efficiency programs save $40 for every dollar invested in them. This is why the government says it could immediately and cost-effectively cut U.S. electricity consumption by 20 percent to 45 percent, using only existing strategies, like time-of-use electricity pricing. (Higher prices for electricity used during daily peak-consumption times—roughly between 8 a.m. and 8 p.m.—encourage consumers to shift their time of energy use. New power plants are typically needed to handle only peak electricity demand.)

Myth 2. Nuclear Energy Is Inexpensive

Achieving greater energy efficiency, however, also requires ending the lopsided system of taxpayer nuclear subsidies that encourage the myth of inexpensive electricity from atomic power. Since 1949, the U.S. government has provided about $165 billion in subsidies to nuclear energy, about $5 billion to solar and wind together, and even less to energy-efficiency programs. All government efficiency programs—to encourage use of fuel-efficient cars, for example, or to provide financial assistance so that low-income citizens can insulate their homes—currently receive only a small percentage of federal energy monies.

After energy-efficiency programs, wind is the most cost-effective way both to generate electricity and to reduce greenhouse emissions. It costs about half as much as atomic power. The only nearly finished nuclear plant in the West, now being built in Finland by the French company Areva, will generate electricity costing 11 cents per kilowatt-hour. Yet the U.S. government's Lawrence Berkeley National Laboratory calculated actual costs of new wind

WHAT DOES THE CHURCH SAY?

Though neither the Vatican nor the U.S. bishops have made a statement on nuclear power, the church has outlined the ethical case for renewable energy. In *Centesimus Annus* Pope John Paul II wrote that just as Pope Leo XIII in 1891 had to confront "primitive capitalism" in order to defend workers' rights, he himself had to confront the "new capitalism" in order to defend collective goods like the environment. Pope Benedict XVI warned that pollutants "make the lives of the poor especially unbearable." In their 2001 statement *Global Climate Change,* the U.S. Catholic bishops repeated his point: climate change will "disproportionately affect the poor, the vulnerable, and generations yet unborn."

The bishops also warn that "misguided responses to climate change will likely place even greater burdens on already desperately poor peoples." Instead they urge "energy conservation and the development of alternate renewable and clean-energy resources." They argue that renewable energy promotes care for creation and the common good, lessens pollution that disproportionately harms the poor and vulnerable, avoids threats to future generations and reduces nuclear-proliferation risks.

---

plants, over the last seven years, at 3.4 cents per kilowatt-hour. Although some groups say nuclear energy is inexpensive, their misleading claims rely on trimming the data on cost. The 2003 M.I.T. study, for instance, included neither the costs of reprocessing nuclear material, nor the full interest costs on nuclear-facility construction capital, nor the total costs of waste storage. Once these omissions—from the entire nine-stage nuclear fuel cycle—are included, nuclear costs are about 11 cents per kilowatt-hour.

The cost-effectiveness of wind power explains why in 2006 utility companies worldwide added 10 times more wind-generated, than nuclear, electricity capacity. It also explains why small-scale sources of renewable energy, like wind and solar, received $56 billion in global private investments in 2006, while nuclear energy received nothing. It explains why wind supplies 20 percent of Denmark's electricity. It explains why, each year for the last several years, Germany, Spain and India have each, alone, added more wind capacity than all countries in the world, taken together, have added in nuclear capacity.

In the United States, wind supplies up to 8 percent of electricity in some Midwestern states. The case of Louis Brooks is instructive. Utilities pay him $500 a month for allowing 78 wind turbines on his Texas ranch, and he can still use virtually all the land for farming and grazing. Wind's cost-effectiveness also explains why in 2007 wind received $9 billion in U.S. private investments, while nuclear energy received zero. U.S. wind energy has been growing by nearly 3,000 megawatts each year, annually producing new electricity equivalent to what three new nuclear reactors could generate. Meanwhile, no new U.S. atomic-power reactors have been ordered since 1974.

Should the United States continue to heavily subsidize nuclear technology? Or, as the distinguished physicist Amory Lovins put it, is the nuclear industry dying of an "incurable attack of market forces"? Standard and Poor's, the credit- and investment-rating company, downgrades the rating of any utility that wants a nuclear plant. It claims that even subsidies are unlikely to make nuclear investment wise. *Forbes* magazine recently called nuclear investment "the largest managerial disaster in business history," something pursued only by the "blind" or the "biased."

Myth 3. Nuclear Energy Is Necessary to Address Climate Change

Government, industry and university studies, like those recently from Princeton, agree that wind turbines and solar panels already exist at an industrial scale and could supply one-third of U.S. electricity needs by 2020, and the vast majority of U.S. electricity by 2050—not just the 20 percent of electricity possible from nuclear energy by 2050. The D.O.E. says wind from only three states (Kansas, North Dakota and Texas) could supply all U.S. electricity needs, and 20 states could supply nearly triple those needs. By 2015, according to the D.O.E., solar panels will be competitive with all conventional energy technologies and will cost 5 to 10 cents per kilowatt hour. Shell Oil and other fossil-fuel companies agree. They are investing heavily in wind and solar.

From an economic perspective, atomic power is inefficient at addressing climate change because dollars used for more expensive, higher-emissions nuclear energy cannot be used for cheaper, lower-emissions renewable energy. Atomic power is also not sustainable. Because of dwindling uranium supplies, by the year 2050 reactors would be forced to use low-grade uranium ore whose greenhouse emissions would roughly equal those of natural gas. Besides, because the United States imports nearly all its uranium, pursuing nuclear power continues the dangerous pattern of dependency on foreign sources to meet domestic energy needs.

Myth 4. Nuclear Energy Will Not Increase Weapons Proliferation

Pursuing nuclear power also perpetuates the myth that increasing atomic energy, and thus increasing uranium enrichment and spent-fuel reprocessing, will increase neither terrorism nor proliferation of nuclear weapons. This myth has been rejected by both the International Atomic Energy Agency and the U.S. Office of Technology Assessment. More nuclear plants means more weapons materials, which means more targets, which means a higher risk of terrorism and proliferation. The government admits that Al Qaeda already has targeted U.S. reactors, none of which can withstand attack by a large airplane. Such an attack, warns the U.S. National Academy of Sciences, could cause fatalities as far away as 500 miles and destruction 10 times worse than that caused by the nuclear accident at Chernobyl in 1986.

Nuclear energy actually increases the risks of weapons proliferation because the same technology used for civilian atomic power can be used for weapons, as the cases of India, Iran, Iraq, North Korea and Pakistan illustrate. As the Swedish Nobel Prize winner Hannes Alven put it, "The military atom and the civilian atom are Siamese twins." Yet if the world stopped building nuclear-power plants, bomb ingredients would be harder to acquire, more conspicuous and more costly politically, if nations were caught trying to obtain them. Their motives for seeking nuclear materials would be unmasked as military, not civilian.

Myth 5. Nuclear Energy Is Safe

Proponents of nuclear energy, like Patrick Moore, cofounder of Greenpeace, and the former Argonne National Laboratory adviser Steve Berry, say that new reactors will be safer than current ones—"meltdown proof." Such safety claims also are myths. Even the 2003 M.I.T. energy study predicted that tripling civilian nuclear reactors would lead to about four core-melt accidents. The government's Sandia National Laboratory calculates that a nuclear accident could cause casualties similar to those at Hiroshima or Nagasaki: 140,000 deaths. If nuclear plants are as safe as their proponents claim, why do utilities need the U.S. Price-Anderson Act, which guarantees utilities protection against 98 percent of nuclear-accident liability and transfers these risks to the public? All U.S. utilities refused to generate atomic power until the government established this liability limit. Why do utilities, but not taxpayers, need this nuclear-liability protection?

Another problem is that high-level radioactive waste must be secured "in perpetuity," as the U.S. National Academy of Sciences puts it. Yet the D.O.E. has already admitted that if nuclear waste is stored at Nevada's Yucca Mountain, as has been proposed, future generations could not meet existing radiation standards. As a result, the current U.S. administration's proposal is to allow future releases of radioactive wastes, stored at Yucca Mountain, provided they annually cause no more than one person—out of every 70 persons exposed to them—to contract fatal cancer. These cancer risks are high partly because Yucca Mountain is so geologically unstable. Nuclear waste facilities could be breached by volcanic or seismic activity. Within 50 miles of Yucca Mountain, more than 600 seismic events, of magnitude greater than two on the Richter scale, have occurred since 1976. In 1992, only 12 miles from the site, an earthquake (5.6 on the Richter scale) damaged D.O.E. buildings. Within 31 miles of the site, eight volcanic eruptions have occurred in the last million years. These facts suggest that Alvin Weinberg was right. Four decades ago, the then-director of the government's Oak Ridge National Laboratory warned that nuclear waste required society to make a Faustian bargain with the devil. In exchange for current military and energy benefits from atomic power, this generation must sell the safety of future generations.

Yet the D.O.E. predicts harm even in this generation. The department says that if 70,000 tons of the existing U.S. waste were shipped to Yucca Mountain, the transfer would require 24 years of dozens of daily rail or truck shipments. Assuming low accident rates and discounting the possibility of terrorist attacks on these lethal shipments, the D.O.E. says this radioactive-waste transport likely would lead to 50 to 310 shipment accidents. According to the D.O.E., each of these accidents could contaminate 42 square miles, and each could require a 462-day cleanup that would cost $620 million, not counting medical expenses. Can hundreds of thousands of mostly unguarded shipments of lethal materials be kept safe? The states do not think so, and they have banned Yucca Mountain transport within their borders. A better alternative is onsite storage at reactors, where the material can be secured from terrorist attack in "hardened" bunkers.

Where Do We Go From Here?

If atomic energy is really so risky and expensive, why did the United States begin it and heavily subsidize it? As U.S. Atomic Energy Agency documents reveal, the United States began to develop nuclear power for the same reason many other nations have done so. It wanted weapons-grade nuclear materials for its military program. But the United States now has more than enough weapons materials. What explains the continuing subsidies? Certainly not the market. *The Economist* (7/7/05) recently noted that for decades, bankers in New York and London have refused loans to nuclear industries. Warning that nuclear costs, dangers and waste storage make atomic power "extremely risky," *The Economist* claimed that the industry is now asking taxpayers to do what the market will not do: invest in nuclear energy. How did *The Economist* explain the uneconomical $20 billion U.S. nuclear subsidies for 2005–7? It pointed to campaign contributions from the nuclear industry.

Despite the problems with atomic power, society needs around-the-clock electricity. Can we rely on intermittent wind until solar power is cost-effective in 2015? Even the Department of Energy says yes. Wind now can supply up to 20 percent of electricity, using the current electricity grid as backup, just as nuclear plants do when they are shut down for refueling, maintenance and leaks. Wind can supply up to 100 percent of electricity needs by using "distributed" turbines spread over a wide geographic region—because the wind always blows somewhere, especially offshore.

Many renewable energy sources are safe and inexpensive, and they inflict almost no damage on people or the environment. Why is the current U.S. administration instead giving virtually all of its support to a riskier, more costly nuclear alternative?

POSTSCRIPT

Is It Time to Revive Nuclear Power?

Robert Evans, "Nuclear Power: Back in the Game," *Power Engineering* (October 2005), reports that a number of power companies are considering new nuclear power plants. See also Eliot Marshall, "Is the Friendly Atom Poised for a Comeback?" and Daniel Clery, "Nuclear Industry Dares to Dream of a New Dawn," *Science* (August 19, 2005). Nuclear momentum is growing, says Charles Petit, "Nuclear Power: Risking a Comeback," *National Geographic* (April 2006), thanks in part to new technologies. One motive is stability of energy supply; see "Why Go Nuclear? Perspectives from the United Arab Emirates, Jordan, Egypt, & Thailand," *Bulletin of the Atomic Scientists* (September/October 2008). Diane Farsetta, "The Campaign to Sell Nuclear," *Bulletin of the Atomic Scientists* (September/October 2008), calls the current effort to revive nuclear power just the latest of the industry's public relations campaigns. Karen Charman, "Brave Nuclear World?" (Part I) *World Watch* (May/June 2006), objects that producing nuclear fuel uses huge amounts of electricity derived from fossil fuels, so going nuclear can hardly prevent all releases of carbon dioxide (although using electricity derived from nuclear power would reduce the problem). She also notes that "Although no comprehensive and integrated study comparing the collateral and external costs of energy sources globally has been done, all currently available energy sources have them. . . . Burning coal—the single largest source of air pollution in the U.S.—causes global warming, acid rain, soot, smog, and other toxic air emissions and generates waste ash, sludge, and toxic chemicals. Landscapes and ecosystems are completely destroyed by mountaintop removal mining, while underground mining imposes high fatality, injury, and sickness rates. Even wind energy kills birds, can be noisy, and, some people complain, blights landscapes."

Michael J. Wallace tells us that there are 103 nuclear reactors operating in the United States today. Stephen Ansolabehere, et al., "The Future of Nuclear Power," *An Interdisciplinary MIT Study* (MIT, 2003), note that in 2000 there were 352 in the developed world as a whole, and a mere 15 in developing nations. Even a very large increase in the number of nuclear power plants—to 1,000 to 1,500—will not stop all releases of carbon dioxide. In fact, if carbon emissions double by 2050 as expected, from 6,500 to 13,000 million metric tons per year, the 1,800 million metric tons not emitted because of nuclear power will seem relatively insignificant. Nevertheless, say John M. Deutch and Ernest J. Moniz, "The Nuclear Option," *Scientific American* (September 2006), such a cut in carbon emissions would be "significant." Christine Laurent, in "Beating Global Warming with Nuclear Power?" *UNESCO Courier* (February 2001), notes, "For several years, the nuclear energy industry has attempted to cloak itself in different ecological robes. Its credo: nuclear energy is a formidable asset in battle against global warming because it emits very small amounts of greenhouse gases. This stance, first presented in the late 1980s when the extent of the phenomenon was still the subject of controversy, is now at the heart of policy debates over how to avoid droughts, downpours and floods." Laurent adds that it makes more sense to focus on reducing carbon emissions by reducing energy consumption.

The debate over the future of nuclear power is likely to remain vigorous for some time to come. But as Richard A. Meserve says in a *Science* editorial ("Global Warming and Nuclear Power," *Science* [January 23, 2004]), "For those who are serious about confronting global warming, nuclear power should be seen as part of the solution. Although it is unlikely that many environmental groups will become enthusiastic proponents of nuclear power, the harsh reality is that any serious program to address global warming cannot afford to jettison any technology prematurely. . . . The stakes are large, and the scientific and educational community should seek to ensure that the public understands the critical link between nuclear power and climate change." Paul Lorenzini, "A Second Look at Nuclear Power," *Issues in Science and Technology* (Spring 2005), argues that the goal must be energy "sufficiency for the foreseeable future with minimal environmental impact." Nuclear power can be part of the answer, but making it happen requires that we shed ideological biases. "It means ceasing to deceive ourselves about what might be possible."

Alvin M. Weinberg, former director of the Oak Ridge National Laboratory, notes in "New Life for Nuclear Power," *Issues in Science and Technology* (Summer 2003), that to make a serious dent in carbon emissions would require perhaps four times as many reactors as suggested in the MIT study. The accompanying safety and security problems would be challenging. If the challenges can be met, says John J. Taylor, retired vice president for nuclear power at the Electric Power Research Institute, in "The Nuclear Power Bargain," *Issues in Science and Technology* (Spring 2004), there are a great many potential benefits. Are new reactor technologies needed? Richard K. Lester, "New Nukes," *Issues in Science and Technology* (Summer 2006), says that better centralized waste storage is what is needed, at least in the short term. See also Elizabeth Svoboca, "Back to the Atom," *Discover* (June 2009).

Environmental groups such as Friends of the Earth are adamantly opposed, but there are signs that some environmentalists do not agree; see William M. Welch, "Some Rethinking Nuke Opposition," *USA Today* (March 23, 2007). Judith Lewis, "The Nuclear Option," *Mother Jones* (May/June 2008), concludes that "[w]hen rising seas flood our coasts, the idea of producing electricity from the most terrifying force ever harnessed may not seem so frightening—or expensive—at all." However, notes Mariah Blake, "Bad Reactors," *Washington Monthly* (January/February 2009), building large numbers of reactors will be expensive and may take much longer than proponents say; expanded nuclear power just may not become available soon enough to do much good. The cost issue is also stressed by Aaron M. Cohen, "Cost May Threaten Nuclear Power's Future," *The Futurist* (May/June 2009).

Internet References . . .

The Population Council

Established in 1952, the Population Council "is an international, nonprofit institution that conducts research on three fronts: biomedical, social science, and public health. This research—and the information it produces—helps change the way people think about problems related to reproductive health and population growth." Many of the Council's publications are available online.

http://www.popcouncil.org/

United Nations Population Division

The United Nations Population Division is responsible for monitoring and appraising a broad range of areas in the field of population. This site offers a wealth of recent data and links.

http://www.un.org/esa/population/unpop.htm

The Agriculture Network Information Center

The Agriculture Network Information Center is a guide to quality agricultural (including biotechnology) information on the Internet as selected by the National Agricultural Library, Land-Grant Universities, and other institutions.

http://www.agnic.org/

Agriculture: Genetic Resources and GMOs

The Agriculture portion of the European Union's portal Web site (EUROPA) provides information on different subjects associated with the genetic base for agricultural activities.

http://www.europa.eu/agriculture/res/index_en.htm

EarthTrends

The World Resources Institute offers data on biodiversity, fisheries, agriculture, population, and a great deal more.

http://earthtrends.wri.org

USDA National Organic Program

The USDA's National Organic Standards Board, comprised of farmers, environmentalists, scientists, and consumer advocates, helps the USDA develop standards for substances used in "organic" farming.

http://www.ams.usda.gov/nop/indexNet.htm

Organic Farming Research Foundation

The Organic Farming Research Foundation sponsors research related to organic farming practices, disseminates research results, and educates the public and decision makers about organic farming issues.

http://www.ofrf.org/

Food and Population

*T*o many, "sustainability" means arranging things so that the natural world—plants and animals, forests and coral reefs, fresh water and landscapes—can continue to exist more or less (mostly less) as it did before human beings multiplied, developed technology, and began to cause extinctions, air and water pollution, soil erosion, desertification, climate change, and so on. To others, "sustainability" means arranging things so that humankind can continue to survive and thrive, even keeping up its history of growth, technological development, and energy use—as if the environment and its resources were infinite.

The two visions of "sustainability" are logically incompatible. Many fear that we are on a collision course with limited resources, with Lester Brown (Issue 13), for instance, writing that food shortages could spell the end of civilization and the National Geographic warning of "The End of Plenty" (June 2009). Can we avoid disaster? Must we reduce the numbers of people on the planet? Their use of technology? Their standard of living? If we do, will human well-being be lessened? If we do not, how can we continue to feed everyone? Is genetic engineering the answer (Issue 14)? Is organic farming (Issue 15)? Both of these issues provoke considerable debate.

- Are Improved Aid Policies the Best Way to Improve Global Food Supply and Protect World Population?

- Is Genetic Engineering the Answer to Hunger?

- Can Organic Farming Feed the World?

ISSUE 13

Are Improved Aid Policies the Best Way to Improve Global Food Supply and Protect World Population?

YES: Robert Paarlberg, from "Evaluating, and Improving, America's Response to Global Hunger," testimony before the Senate Committee on Foreign Relations Hearing on "Alleviating Global Hunger: Challenges and Opportunities for U.S. Leadership" (March 24, 2009)

NO: Lester R. Brown, from "Could Food Shortages Bring Down Civilization?" *Scientific American* (May 2009)

ISSUE SUMMARY

YES: Professor Robert Paarlberg argues that global hunger, which afflicts nearly a billion people worldwide, many of them in Africa, calls for increased aid directed toward agricultural education, science, and research, and infrastructure development.

NO: Lester R. Brown argues that the problem is due more to water shortages, soil losses, rising population, and rising temperatures from global warming than to failures of aid policies. What is needed is immediate attention to the world's environmental problems, lacking which the result will be increased hunger, political conflict, and perhaps even the collapse of civilization.

I n 1798 the British economist Thomas Malthus published his *Essay on the Principle of Population.* In it, he pointed with alarm at the way the human population grew geometrically (a hockey-stick curve of increase) and at how agricultural productivity grew only arithmetically (a straight-line increase). It was obvious, he said, that the population must inevitably outstrip its food supply and experience famine. Contrary to the conventional wisdom of the time, population growth was not necessarily a good thing. Indeed, it led inexorably to catastrophe. For many years, Malthus was something of a laughingstock. The doom he forecast kept receding into the future as new lands were opened to agriculture, new agricultural technologies appeared, new ways of preserving food limited the waste of spoilage, and the developed nations underwent a "demographic transition" from high birthrates and high death rates to low birthrates and low death rates.

Demographers initially attributed the demographic transition to increasing prosperity and predicted that as prosperity increased in countries whose populations were rapidly growing, birthrates would surely fall. Later, some scholars analyzed the historical data and concluded that the transition had actually preceded prosperity. The two views have contrasting implications for public policy designed to slow population growth—economic aid or family planning aid—but neither has worked very well. In 1994 the UN Conference on Population and Development, held in Cairo, Egypt, concluded that better results would follow from improving women's access to education and health care.

The world's human population has grown tremendously. In Malthus's time, there were about 1 billion human beings on earth. By 1950 there were a little over 2.5 billion. In 2007 the tally passed 6.7 billion. By 2050, the United Nations expects world population to be over 9 billion (see *World Population Prospects: The 2008 Revision;* http://www.un.org/esa/population/unpop.htm; United Nations, 2009).

While global agricultural production has also increased, it has not kept up with rising demand, and—because of the loss of topsoil to erosion, the exhaustion of aquifers for irrigation water, the high price of energy for making fertilizer, and the expected effects of global warming (among other things)—the prospect of improvement seems exceedingly slim to many observers. Paul R. Ehrlich and Anne H. Ehrlich argue in "The Population Explosion: Why We Should Care and What We Should Do About It," *Environmental Law* (Winter 1997) that "population growth may be the paramount force moving humanity inexorably towards disaster." They therefore maintain that it is essential to reduce the impact of population in terms of both numbers and resource consumption. See also William G. Moseley, "A Population Remedy Is Right Here at Home: U.S. Overconsumption Is a Bigger Issue than Fertility," *The Philadelphia Inquirer* (July 11, 2007).

Was Malthus wrong? Both environmental scientists and many economists now say that if population continues to grow, problems are inevitable. But the problem is not just a mismatch between population growth and agricultural production. More and more, scientists are seeing that this mismatch is exacerbated by loss of resources such as soil and water that are absolutely essential to agriculture. In the following selections, Wellesley College Professor Robert Paarlberg argues that global hunger, which currently afflicts nearly a billion people worldwide, calls for increased aid directed toward agricultural education, science, and research, and infrastructure development. Lester R. Brown argues that the problem is due more to water shortages, soil losses, rising population, and rising temperatures from global warming than to failures of aid policies. We need to stabilize population and eradicate poverty, but lacking immediate attention to the world's environmental problems, we are likely to see increased hunger, political conflict, and perhaps even the collapse of civilization.

YES

Robert Paarlberg

Evaluating, and Improving, America's Response to Global Hunger

Providing international leadership to alleviate global hunger requires our Government to have strong policies in two separate areas:

- Responding to short-term food emergencies, such as the international food price spike we saw in 2008, which temporarily put up to 100 million more people at risk.
- Attacking the persistent poverty that keeps more than 850 million people hungry even when international food prices are low.

In the first of these areas, the United States Government has done a good job, at least a B+. But in the second area the U.S. has done a poor job over the past 25 years, something close to an F. In 2009 America has a chance to correct this second failing grade by directing more development assistance support to help small farmers in Sub-Saharan Africa and South Asia. Until the productivity of these small farmers goes up, poverty and hunger will not go down.

America's Laudable Response to the 2008 World Food Crisis

When the price of food on the world market suddenly surged upward during the first six month of 2008, it was clear that some developing countries heavily dependent on imported food needed help. In April 2008 the World Bank produced an estimate that an additional 100 million people in the developing world were being pushed into effective poverty because of the much higher food import prices. The New York Times called these high prices a "World Food Crisis." The Economist called it a "Silent Tsunami."

    This was a serious crisis for poor countries heavily dependent on food imports, particularly in West Africa and the Caribbean, but not all developing countries fell into that category. Many governments in the developing world have long made it a point not to depend on imports of basic grains (in the name of national food "self-sufficiency"). For example in South Asia only about 6 percent of total grain consumption is imported, and in India

U.S. Senate, March 24, 2009.

specifically only 1 percent of rice consumption is imported. So when the price of rice for export tripled in 2008 it was a shock in Cameroon and Haiti, but it had little effect on most poor people in India.

International food prices spiked as high as they did precisely because so many developing country governments decided not to let higher international prices cross into their own domestic economy. When export prices starting increasing in 2007, one country after another insulated its domestic market from the increase by placing new restrictions on food exports. China imposed export taxes on grains and grain products. Argentina raised export taxes on wheat, corn and soybeans. Russia raised export taxes on wheat. Malaysia and Indonesia imposed export taxes on palm oil. Egypt, Cambodia, Vietnam, and Indonesia eventually banned exports of rice. India, the world's third largest rice exporter, banned exports of rice other than basmati. When so many export restrictions were suddenly set in place, the quantity of food available for export dropped sharply, triggering the large price spike seen in international markets.

The response of the United States Government to this price spike was timely and commendable. America provided essential global leadership, in two important ways.

First, the United States never placed any restrictions on its own exports of agricultural commodities. While others were imposing export taxes or export bans, the United States continued to leave its domestic food supply open to foreign customers. This was not an easy discipline to maintain. America's decision to place no restriction on its own rice exports meant prices inside the U.S. economy spiked upward along with the international price, which led to an interlude of panic buying. In April 2008, Costco Wholesale Corporation and Wal-Mart's Sam's Club had to limit sales of rice to 4 bags per customer per visit. For wheat, the U.S. decision not to restrict exports implied much higher operating costs for America's baking industry, prompting the American Bakers Association early in 2008 to send delegations to Washington to voice loud complaint. Despite these domestic pressures, our Government never restricted export sales.

Second, when international prices spiked in 2008 the United States dramatically increased its budget for international food aid. In April 2008, President Bush announced the release of $200 million worth of commodities from an emergency food aid reserve for Title II PL480, and then in May 2008 the President requested from Congress an additional $770 million as a crisis response, with roughly 80 percent of this intended to help poor importing governments or support short term feeding of vulnerable populations. According to one unofficial calculation, the United States responded to the 2008 crisis by designating an additional $1.4 billion in food aid above already planned funding levels. Total enacted and estimated international food assistance spending from the United States in FY2009 will be roughly $2 billion.

Our policy response to the 2008 food price spike was far from perfect, in part because our food aid programs are unnecessarily expensive. This is because the United States does not allow any significant sourcing of food from outside of the United States and because shipment on more costly U.S.-flag vessels is required for 75 percent of all gross food aid tonnage. As a result an excessive 65 percent of America's food aid spending goes to administrative and transport costs. Some economists calculate that it costs roughly twice as much to deliver a ton of food to a recipient through this U.S. food aid system as it would cost buying the food from a local market. The United States is heavily criticized abroad for operating its food aid programs this way. On the other hand, if America went to a more efficient system based on foreign sourcing, political support for the program here in Congress would suffer, the size of our food aid budget would fall, and food deliveries to some needy recipients abroad might then fall as well.

America was also heavily criticized in 2008 for the alleged impact of its biofuels policies on world food prices. Federal tax credits, import tariffs, and renewable fuel mandates promoted the diversion of American corn into fuel production, driving up international prices for corn used as food or feed. In 2007–08, ethanol production increased to roughly 23 percent of America's total corn use. On the other hand, much of this diversion would have taken place in 2008 even without any U.S. Government tax credits, tariffs, or mandates, simply due to the unusually high oil prices that prevailed at the time. When bad things happen it is not always the government's fault. It was mostly high oil prices, not government policy, that drove up corn use for ethanol in 2008.

America's Less Helpful Response to Persistent Hunger

America has shown far less leadership in its policy response to the long-term problem of chronic malnutrition in developing countries. This hunger problem, linked especially to rural poverty, is roughly eight times larger than the temporary problem linked to the 2007–08 price spike.

Even before international food prices began to increase in 2007, the United Nations Food and Agriculture Organization (FAO) estimated that there were 850 million chronically malnourished people in the world. Even when food was cheap on the world market in 2005, in Sub-Saharan Africa 23 out of 37 countries in that region were consuming less than their nutritional requirements and nearly one third of all citizens there were malnourished. The problem of hunger in these countries derives primarily from persistent poverty, not from price fluctuations on the world market. In Africa more than 60 percent of all citizens work in the countryside growing crops and herding animals, and it is because the productivity of their labor is so low (incomes average only about $1 a day) that so many are chronically malnourished.

To understand the source of these low incomes, pay a visit to a typical farming community in rural Africa. The farmers you will meet, mostly women, do not have any of the things that farmers everywhere else have required to become more productive and escape poverty:

- Few have had access to formal education. Two out of three adults are not able to read or write in any language.
- Two thirds do not have access to seeds improved by scientific plant breeding.

- Most use no nitrogen fertilizer at all, so they fail to replace soil nutrients and their crop yields per hectare are only one fifth to one tenth as high as in the United States or in Europe.
- Only 4 percent have irrigation, so when the rains fail their crops also fail, and they must sell off their animals and household possessions to survive until the next season.
- Almost none have access to electricity, and powered machinery is completely absent. These farmers still work the fields with hand hoes or wooden plows pulled by animals.
- Few have access to veterinary medicine, so their animals are sick, stunted, and weak.
- Most of these farmers are significantly cut off from markets due to remoteness and high transport costs. Roughly 70 percent of African farmers live more than a half-hour walk from the nearest all-weather road, so most household transport is still by foot.

Given such deficits, it is not surprising that agricultural production in Africa has lagged behind population growth for most of the past three decades. Per capita production of maize, Africa's most important food staple, has actually *declined* 14 percent since 1980. Over the same time period population has doubled, so the numbers of people living in deep poverty (less than $1 a day) has doubled as well, up to 300 million. The number of Africans classified as "food insecure" by the U.S. Department of Agriculture increased to 450 million in 2006, and under a business-as-usual scenario this number will grow by another 30 percent over the next ten years, to reach 645 million.

One reason the current business-as-usual scenario is so bleak has been weak leadership from the United States. Instead of taking action to help address these persistent farm productivity deficits in Africa over the past several decades, the United States Government essentially walked away from the problem:

- America's official development assistance to agriculture in Africa, in real 2008 dollars, declined from more than $400 million annually in the 1980s to just $60 million by 2006, a drop of approximately 85 percent.
- Between 1985 and 2008 the number of Africans supported by USAID for post-graduate agricultural study at American universities declined 83 percent, down to a total of just 42 individuals today.
- From the mid-1980s to 2004, USAID funding to support national agricultural research systems (NARS) in the developing countries as a whole fell by 75 percent, and in Sub-Saharan Africa by 77 percent.
- From the mid-1980s to 2008, United States contributions to the core research budget of the Consultative Group on International Agricultural Research (CGIAR), in real 2008 dollars, fell from more than $90 million annually to just $18.5 million.
- USAID spending for collaborative agricultural research through American universities was nearly $45 million in constant 2008 dollars twenty five years ago. As of 2007, this funding was down to just $25 million.
- These cuts were accompanied by severe agricultural de-staffing at USAID. As late as 1990 USAID still employed 181 agricultural specialists. Currently it employs only 22.

So, while Africa's rural poverty and hunger crisis was steadily growing worse, the United States Government was steadily doing less.

Why Did the United States Stop Investing in Agricultural Development?

Beginning in the 1980s, three factors combined to push the United States away from providing adequate assistance for agricultural development:

First, the enormous success of the original Green Revolution on the irrigated lands of Asia in the 1960s and 1970s left a false impression that all of the world's food production problems had been solved. In fact, on the non-irrigated farmlands of Africa, these problems were just beginning to intensify.

Second, it became fashionable among most donors beginning in the 1980s to rely less on the public sector and more on the private sector, under a so-called "Washington Consensus" doctrine developed inside the IMF and the World Bank. According to this new aid doctrine, the job of the state was mostly to stabilize the macro economy and then get out of the way, so private investors and private markets could create wealth. This approach backfired in rural Africa because the basic public goods needed to support markets and attract private investors—roads, power, and an educated workforce—had not yet been provided.

Third, a new fashion also arose in the 1980s among advocates for social justice and environmental protection. These groups began to depict the improved seeds and fertilizers of the original Green Revolution as a problem, not a solution. They argued that only large farmers would profit and that increased chemical use would harm the environment. This perspective was not appropriate in Africa, where nearly all farmers are smallholders with adequate access to land and where fertilizer use is too low rather than too high. In Africa the social and environmental danger isn't too much Green Revolution farming, but far too little.

I have documented the importance of these NGO objections to Green Revolution farming in a book published last year by Harvard University Press. I show in this book that an influential coalition of social justice and environmental advocates from both North America and Europe was able to discourage international support for agricultural development, including in Africa, beginning in the 1980s. They not only opposed the use of modern biotechnology, such as genetic engineering; they also campaigned against conventionally developed modern seeds and nitrogen fertilizers, even though these were precisely the technologies their own farmers had earlier used back home to become more productive and escape poverty. To Africans they instead promoted agroecological or organic farming methods, not using synthetic pesticides or fertilizers.

The irony is that most farmers in Africa today are already de facto organic, because they do not use any GMOs, or any nitrogen fertilizers, or any synthetic pesticides. This has not made them either productive or prosperous. Nor has it provided any protection to Africa's rural environment, where deforestation, soil erosion, and habitat loss caused by the relentless expansion of low-yield farming is a growing crisis.

How to Correct America's Failing Grade in Agricultural Development

Improving America's dismal policy performance in the area of agricultural development does not have to be difficult. We know what to do, we know it can be done at an affordable cost, and the current political climate even provides new space to act.

A consensus now exists among specialists, even at the World Bank, on what to do. An extensive review conducted by the Bank in preparation for its 2008 *World Development Report* concluded that more public sector action was urgently needed: "the visible hand of the state" was now needed to provide the "core public goods" essential to farm productivity growth. Three kinds of public goods are needed today in the African countryside:

- Public investments in rural and agricultural education, including for women and girls.
- Public investments in agricultural science and local agricultural research to improve crops, animals, and farming techniques.
- Public investments in rural infrastructure (roads, electricity, crop storage) to connect farmers to markets.

Governments in Africa today endorse this consensus. At an African Union summit meeting in Mozambique in 2003, Africa's heads of government pledged to increase their own public spending on agriculture up to at least 10 percent of total national spending. International donors, including the United States, should seize upon this constructive pledge, redirecting assistance efforts so as to partner aggressively with African governments willing to re-invest in the productivity of small farms.

We know exactly what this re-directed assistance effort should look like, thanks to the policy roadmap recently provided by two members of this committee plus the supportive recommendations of a prominent independent study group.

The widely endorsed Global Food Security Act of 2009 (S. 3529), known as the Lugar-Casey bill, would authorize significantly larger investments in agricultural education, extension, and research, to take full international advantage of the superior agricultural resources found within of America's own land grant colleges and universities. The increased investments in institutional strengthening and collaborative research authorized in this bill could be funded at $750 million in year one, increasing to an annual cost of $2.5 billion by year five. Fully funding this initiative would require roughly a 10 percent increase in America's annual development assistance budget, a small increase alongside President Obama's own 2008 campaign pledge—which I strongly endorse—to grow that development assistance budget by 100 percent.

A second worthy blueprint strongly parallels the Lugar-Casey bill. This is a menu of 21 separate recommended actions called the Chicago Initiative on Global Agricultural Development, released just one month ago by an independent bipartisan study group convened by the Chicago Council on Global

Affairs, with financial support from the Bill and Melinda Gates Foundation. This substantial report, which I played a role in preparing, recommends twin thrusts in agricultural education and agricultural research, just like Lugar-Casey. It also recommends closer coordination with the World Bank to increase investments in rural infrastructure, plus a substantial upgrade of agricultural staff at USAID. The Chicago Council report estimates that implementing all 21 of its recommended actions would cost $341 million in the first year (an increase over current programs of $255 million), and only $1.03 billion annually by year five (an increase of $950 million over current expenditures). This implies less than a 5 percent increase in our current development assistance budget.

Why Is 2009 the Best Time to Take These Actions?

The danger is not that Congress will debate these proposals and then reject them as too costly. Both of these proposals are well researched and substantively well defended, and the implied costs are not at all large alongside the anticipated humanitarian, economic, and diplomatic gains. The danger instead is that a serious debate over these proposals will never take place, amid the many distractions of the day, and a decision will simply be deferred. This would be a costly mistake. If new action is deferred, the business-as-usual scenario will kick in and numbers of food insecure people in Sub-Saharan Africa will increase by another 30 percent over the next ten years, to reach a total of 645 million. If the new Administration and Congress decide to put off action until 2013 or 2017, the hunger problem will only become more costly to resolve.

Fortunately, two important windows of political opportunity are open in 2009 to support the embrace of a significant policy initiative in this area. First, both the new Administration and Congress are eager to be seen delivering a "real change" in America's policies abroad, not just at home. A decision in 2009 to reverse, at last, the 25-year decline in U.S. support for agricultural development assistance would be a real change, and it would be recognized as such around the world. It would transform America overnight from being the laggard in this area into being the global leader. With its new agricultural development initiative on the table, America could re-introduce itself to governments around the world—especially in Africa—with a convincing message of hope, not fear. The payoff in farmers' fields would not be seen immediately, but the political and diplomatic gain would be immediate.

For those on this committee looking for an affordable way to recast America's approach toward governments in Africa (e.g., in response to China's growing investment presence and political influence in that region), a new agricultural development initiative is actually one of the most cost-effective ways to proceed. The annual budget cost is low because the best way to support agricultural development is not with a massive front-loaded crash program, but instead with small but steady annual outlays developed and managed in close partnership with recipient governments, maintained for a decade or more.

The second window of opportunity was provided by the 2008 food price crisis itself. Memories of this crisis are still sufficiently fresh in 2009 to motivate action on a significant new agricultural development assistance initiative, to complement the strong leadership we already show in emergency relief and food aid.

Both these windows of political opportunity are currently open. They are not likely to remain open for long.

Lester R. Brown  **NO**

Could Food Shortages Bring Down Civilization?

One of the toughest things for people to do is to anticipate sudden change. Typically we project the future by extrapolating from trends in the past. Much of the time this approach works well. But sometimes it fails spectacularly, and people are simply blindsided by events such as today's economic crisis.

For most of us, the idea that civilization itself could disintegrate probably seems preposterous. Who would not find it hard to think seriously about such a complete departure from what we expect of ordinary life? What evidence could make us heed a warning so dire—and how would we go about responding to it? We are so inured to a long list of highly unlikely catastrophes that we are virtually programmed to dismiss them all with a wave of the hand: Sure, our civilization might devolve into chaos—and Earth might collide with an asteroid, too!

For many years I have studied global agricultural, population, environmental and economic trends and their interactions. The combined effects of those trends and the political tensions they generate point to the breakdown of governments and societies. Yet I, too, have resisted the idea that food shortages could bring down not only individual governments but also our global civilization.

I can no longer ignore that risk. Our continuing failure to deal with the environmental declines that are undermining the world food economy—most important, falling water tables, eroding soils and rising temperatures—forces me to conclude that such a collapse is possible.

The Problem of Failed States

Even a cursory look at the vital signs of our current world order lends unwelcome support to my conclusion. And those of us in the environmental field are well into our third decade of charting trends of environmental decline without seeing any significant effort to reverse a single one.

In six of the past nine years world grain production has fallen short of consumption, forcing a steady drawdown in stocks. When the 2008 harvest began, world carryover stocks of grain (the amount in the bin when the new harvest begins) were at 62 days of consumption, a near record low. In response, world grain prices in the spring and summer of last year climbed to the highest level ever.

From *Scientific American*, May 2009. Copyright © 2009 by Scientific American. All rights reserved. Reprinted by permission. www.sciam.com

As demand for food rises faster than supplies are growing, the resulting food-price inflation puts severe stress on the governments of countries already teetering on the edge of chaos. Unable to buy grain or grow their own, hungry people take to the streets. Indeed, even before the steep climb in grain prices in 2008, the number of failing states was expanding [Purchase the digital edition to see related sidebar]. Many of their problems stem from a failure to slow the growth of their populations. But if the food situation continues to deteriorate, entire nations will break down at an ever increasing rate. We have entered a new era in geopolitics. In the 20th century the main threat to international security was superpower conflict; today it is failing states. It is not the concentration of power but its absence that puts us at risk.

States fail when national governments can no longer provide personal security, food security and basic social services such as education and health care. They often lose control of part or all of their territory. When governments lose their monopoly on power, law and order begin to disintegrate. After a point, countries can become so dangerous that food relief workers are no longer safe and their programs are halted; in Somalia and Afghanistan, deteriorating conditions have already put such programs in jeopardy.

Failing states are of international concern because they are a source of terrorists, drugs, weapons and refugees, threatening political stability everywhere. Somalia, number one on the 2008 list of failing states, has become a base for piracy. Iraq, number five, is a hotbed for terrorist training. Afghanistan, number seven, is the world's leading supplier of heroin. Following the massive genocide of 1994 in Rwanda, refugees from that troubled state, thousands of armed soldiers among them, helped to destabilize neighboring Democratic Republic of the Congo.

Our global civilization depends on a functioning network of politically healthy nation-states to control the spread of infectious disease, to manage the international monetary system, to control international terrorism and to reach scores of other common goals. If the system for controlling infectious diseases—such as polio, SARS or avian flu—breaks down, humanity will be in trouble. Once states fail, no one assumes responsibility for their debt to outside lenders. If enough states disintegrate, their fall will threaten the stability of global civilization itself.

A New Kind of Food Shortage

The surge in world grain prices in 2007 and 2008—and the threat they pose to food security—has a different, more troubling quality than the increases of the past. During the second half of the 20th century, grain prices rose dramatically several times. In 1972, for instance, the Soviets, recognizing their poor harvest early, quietly cornered the world wheat market. As a result, wheat prices elsewhere more than doubled, pulling rice and corn prices up with them. But this and other price shocks were event-driven—drought in the Soviet Union, a monsoon failure in India, crop-shrinking heat in the U.S. Corn Belt. And the rises were short-lived: prices typically returned to normal with the next harvest.

In contrast, the recent surge in world grain prices is trend-driven, making it unlikely to reverse without a reversal in the trends themselves. On the demand side, those trends include the ongoing addition of more than 70 million people a year; a growing number of people wanting to move up the food chain to consume highly grain-intensive livestock products [see "The Greenhouse Hamburger," by Nathan Fiala; *Scientific American,* February 2009]; and the massive diversion of U.S. grain to ethanol-fuel distilleries.

The extra demand for grain associated with rising affluence varies widely among countries. People in low-income countries where grain supplies 60 percent of calories, such as India, directly consume a bit more than a pound of grain a day. In affluent countries such as the U.S. and Canada, grain consumption per person is nearly four times that much, though perhaps 90 percent of it is consumed indirectly as meat, milk and eggs from grain-fed animals.

The potential for further grain consumption as incomes rise among low-income consumers is huge. But that potential pales beside the insatiable demand for crop-based automotive fuels. A fourth of this year's U.S. grain harvest—enough to feed 125 million Americans or half a billion Indians at current consumption levels—will go to fuel cars. Yet even if the entire U.S. grain harvest were diverted into making ethanol, it would meet at most 18 percent of U.S. automotive fuel needs. The grain required to fill a 25-gallon SUV tank with ethanol could feed one person for a year.

The recent merging of the food and energy economies implies that if the food value of grain is less than its fuel value, the market will move the grain into the energy economy. That double demand is leading to an epic competition between cars and people for the grain supply and to a political and moral issue of unprecedented dimensions. The U.S., in a misguided effort to reduce its dependence on foreign oil by substituting grain-based fuels, is generating global food insecurity on a scale not seen before.

Water Shortages Mean Food Shortages

What about supply? The three environmental trends I mentioned earlier—the shortage of freshwater, the loss of topsoil and the rising temperatures (and other effects) of global warming—are making it increasingly hard to expand the world's grain supply fast enough to keep up with demand. Of all those trends, however, the spread of water shortages poses the most immediate threat. The biggest challenge here is irrigation, which consumes 70 percent of the world's freshwater. Millions of irrigation wells in many countries are now pumping water out of underground sources faster than rainfall can recharge them. The result is falling water tables in countries populated by half the world's people, including the three big grain producers—China, India and the U.S.

Usually aquifers are replenishable, but some of the most important ones are not: the "fossil" aquifers, so called because they store ancient water and are not recharged by precipitation. For these—including the vast Ogallala Aquifer that underlies the U.S. Great Plains, the Saudi aquifer and the deep aquifer under the North China Plain—depletion would spell the end of pumping. In arid regions such a loss could also bring an end to agriculture altogether.

In China the water table under the North China Plain, an area that produces more than half of the country's wheat and a third of its corn, is falling fast. Overpumping has used up most of the water in a shallow aquifer there, forcing well drillers to turn to the region's deep aquifer, which is not replenishable. A report by the World Bank foresees "catastrophic consequences for future generations" unless water use and supply can quickly be brought back into balance.

As water tables have fallen and irrigation wells have gone dry, China's wheat crop, the world's largest, has declined by 8 percent since it peaked at 123 million tons in 1997. In that same period China's rice production dropped 4 percent. The world's most populous nation may soon be importing massive quantities of grain.

But water shortages are even more worrying in India. There the margin between food consumption and survival is more precarious. Millions of irrigation wells have dropped water tables in almost every state. As Fred Pearce reported in *New Scientist*:

> Half of India's traditional hand-dug wells and millions of shallower tube wells have already dried up, bringing a spate of suicides among those who rely on them. Electricity blackouts are reaching epidemic proportions in states where half of the electricity is used to pump water from depths of up to a kilometer [3,300 feet].

A World Bank study reports that 15 percent of India's food supply is produced by mining groundwater. Stated otherwise, 175 million Indians consume grain produced with water from irrigation wells that will soon be exhausted. The continued shrinking of water supplies could lead to unmanageable food shortages and social conflict.

Less Soil, More Hunger

The scope of the second worrisome trend—the loss of topsoil—is also startling. Topsoil is eroding faster than new soil forms on perhaps a third of the world's cropland. This thin layer of essential plant nutrients, the very foundation of civilization, took long stretches of geologic time to build up, yet it is typically only about six inches deep. Its loss from wind and water erosion doomed earlier civilizations.

In 2002 a U.N. team assessed the food situation in Lesotho, the small, landlocked home of two million people embedded within South Africa. The team's finding was straightforward: "Agriculture in Lesotho faces a catastrophic future; crop production is declining and could cease altogether over large tracts of the country if steps are not taken to reverse soil erosion, degradation and the decline in soil fertility."

In the Western Hemisphere, Haiti—one of the first states to be recognized as failing—was largely self-sufficient in grain 40 years ago. In the years since, though, it has lost nearly all its forests and much of its topsoil, forcing the country to import more than half of its grain.

The third and perhaps most pervasive environmental threat to food security—rising surface temperature—can affect crop yields everywhere. In many countries crops are grown at or near their thermal optimum, so even a minor temperature rise during the growing season can shrink the harvest. A study published by the U.S. National Academy of Sciences has confirmed a rule of thumb among crop ecologists: for every rise of one degree Celsius (1.8 degrees Fahrenheit) above the norm, wheat, rice and corn yields fall by 10 percent.

In the past, most famously when the innovations in the use of fertilizer, irrigation and high-yield varieties of wheat and rice created the "green revolution" of the 1960s and 1970s, the response to the growing demand for food was the successful application of scientific agriculture: the technological fix. This time, regrettably, many of the most productive advances in agricultural technology have already been put into practice, and so the long-term rise in land productivity is slowing down. Between 1950 and 1990 the world's farmers increased the grain yield per acre by more than 2 percent a year, exceeding the growth of population. But since then, the annual growth in yield has slowed to slightly more than 1 percent. In some countries the yields appear to be near their practical limits, including rice yields in Japan and China.

Some commentators point to genetically modified crop strains as a way out of our predicament. Unfortunately, however, no genetically modified crops have led to dramatically higher yields, comparable to the doubling or tripling of wheat and rice yields that took place during the green revolution. Nor do they seem likely to do so, simply because conventional plant-breeding techniques have already tapped most of the potential for raising crop yields.

Jockeying for Food

As the world's food security unravels, a dangerous politics of food scarcity is coming into play: individual countries acting in their narrowly defined self-interest are actually worsening the plight of the many. The trend began in 2007, when leading wheat-exporting countries such as Russia and Argentina limited or banned their exports, in hopes of increasing locally available food supplies and thereby bringing down food prices domestically. Vietnam, the world's second-biggest rice exporter after Thailand, banned its exports for several months for the same reason. Such moves may reassure those living in the exporting countries, but they are creating panic in importing countries that must rely on what is then left of the world's exportable grain.

In response to those restrictions, grain importers are trying to nail down long-term bilateral trade agreements that would lock up future grain supplies. The Philippines, no longer able to count on getting rice from the world market, recently negotiated a three-year deal with Vietnam for a guaranteed 1.5 million tons of rice each year. Food-import anxiety is even spawning entirely new efforts by food-importing countries to buy or lease farmland in other countries.

In spite of such stopgap measures, soaring food prices and spreading hunger in many other countries are beginning to break down the social order.

In several provinces of Thailand the predations of "rice rustlers" have forced villagers to guard their rice fields at night with loaded shotguns. In Pakistan an armed soldier escorts each grain truck. During the first half of 2008, 83 trucks carrying grain in Sudan were hijacked before reaching the Darfur relief camps.

No country is immune to the effects of tightening food supplies, not even the U.S., the world's breadbasket. If China turns to the world market for massive quantities of grain, as it has recently done for soybeans, it will have to buy from the U.S. For U.S. consumers, that would mean competing for the U.S. grain harvest with 1.3 billion Chinese consumers with fast-rising incomes—a nightmare scenario. In such circumstances, it would be tempting for the U.S. to restrict exports, as it did, for instance, with grain and soybeans in the 1970s when domestic prices soared. But that is not an option with China. Chinese investors now hold well over a trillion U.S. dollars, and they have often been the leading international buyers of U.S. Treasury securities issued to finance the fiscal deficit. Like it or not, U.S. consumers will share their grain with Chinese consumers, no matter how high food prices rise.

Plan B: Our Only Option

Since the current world food shortage is trend-driven, the environmental trends that cause it must be reversed. To do so requires extraordinarily demanding measures, a monumental shift away from business as usual—what we at the Earth Policy Institute call Plan A—to a civilization-saving Plan B.

Similar in scale and urgency to the U.S. mobilization for World War II, Plan B has four components: a massive effort to cut carbon emissions by 80 percent from their 2006 levels by 2020; the stabilization of the world's population at eight billion by 2040; the eradication of poverty; and the restoration of forests, soils and aquifers.

Net carbon dioxide emissions can be cut by systematically raising energy efficiency and investing massively in the development of renewable sources of energy. We must also ban deforestation worldwide, as several countries already have done, and plant billions of trees to sequester carbon. The transition from fossil fuels to renewable forms of energy can be driven by imposing a tax on carbon, while offsetting it with a reduction in income taxes.

Stabilizing population and eradicating poverty go hand in hand. In fact, the key to accelerating the shift to smaller families is eradicating poverty—and vice versa. One way is to ensure at least a primary school education for all children, girls as well as boys. Another is to provide rudimentary, village-level health care, so that people can be confident that their children will survive to adulthood. Women everywhere need access to reproductive health care and family-planning services.

The fourth component, restoring the earth's natural systems and resources, incorporates a worldwide initiative to arrest the fall in water tables by raising water productivity: the useful activity that can be wrung from each drop. That implies shifting to more efficient irrigation systems and to more water-efficient crops. In some countries, it implies growing (and eating) more wheat and less rice, a water-intensive crop. And for industries and cities, it implies doing what some are doing already, namely, continuously recycling water.

At the same time, we must launch a worldwide effort to conserve soil, similar to the U.S. response to the Dust Bowl of the 1930s. Terracing the ground, planting trees as shelterbelts against windblown soil erosion, and practicing minimum tillage—in which the soil is not plowed and crop residues are left on the field—are among the most important soil-conservation measures.

There is nothing new about our four interrelated objectives. They have been discussed individually for years. Indeed, we have created entire institutions intended to tackle some of them, such as the World Bank to alleviate poverty. And we have made substantial progress in some parts of the world on at least one of them—the distribution of family-planning services and the associated shift to smaller families that brings population stability.

For many in the development community, the four objectives of Plan B were seen as positive, promoting development as long as they did not cost too much. Others saw them as humanitarian goals—politically correct and morally appropriate. Now a third and far more momentous rationale presents itself: meeting these goals may be necessary to prevent the collapse of our civilization. Yet the cost we project for saving civilization would amount to less than $200 billion a year, a sixth of current global military spending. In effect, Plan B is the new security budget.

Time: Our Scarcest Resource

Our challenge is not only to implement Plan B but also to do it quickly. The world is in a race between political tipping points and natural ones. Can we close coal-fired power plants fast enough to prevent the Greenland ice sheet from slipping into the sea and inundating our coastlines? Can we cut carbon emissions fast enough to save the mountain glaciers of Asia? During the dry season their meltwaters sustain the major rivers of India and China—and by extension, hundreds of millions of people. Can we stabilize population before countries such as India, Pakistan and Yemen are overwhelmed by shortages of the water they need to irrigate their crops?

It is hard to overstate the urgency of our predicament. [For the most thorough and authoritative scientific assessment of global climate change, see "Climate Change 2007. Fourth Assessment Report of the Intergovernmental Panel on Climate Change," available at www.ipcc.ch]. Every day counts. Unfortunately, we do not know how long we can light our cities with coal, for instance, before Greenland's ice sheet can no longer be saved. Nature sets the deadlines; nature is the timekeeper. But we human beings cannot see the clock.

We desperately need a new way of thinking, a new mind-set. The thinking that got us into this bind will not get us out. When Elizabeth Kolbert, a writer for the *New Yorker,* asked energy guru Amory Lovins about thinking outside the box, Lovins responded: "There is no box."

There is no box. That is the mind-set we need if civilization is to survive.

POSTSCRIPT

Are Improved Aid Policies the Best Way to Improve Global Food Supply and Protect World Population?

Global hunger is a much more enormous problem than most people realize. In a special report on the food crisis, Joel K. Bourne, Jr., "The End of Plenty," *National Geographic* (June 2009), says that we need another green revolution, and in half the time it took to implement the first one. The human population faces a major shortage of essential resources needed to satisfy its need for food, including soil (see Charles Mann, "Our Good Earth: Can We Save It?" *National Geographic,* September 2008) and water (see David Molden, Charlotte de Fraiture, and Frank Rijsberman, "Water Scarcity: The Food Factor," *Issues in Science and Technology,* Summer 2007), and "Perhaps somewhere deep in his crypt . . . Malthus is quietly wagging a bony finger and saying, "Told you so."

P. M. Schmitz and A. Kavallari, "Crop Plants versus Energy Plants—On the International Food Crisis," *Bioorganic & Medicinal Chemistry* (June 2009), argue that demand-driven food shortages can be overcome through proper design of agricultural policy. Testifying at the same Senate hearing as Robert Paarlberg, Dan Glickman and Catherine Bertini describe the Chicago Council on Global Affairs' Global Agricultural Development Project and its report, *Renewing American Leadership in the Fight Against Global Hunger and Poverty* (February 2009). Like Paarlberg, Glickman and Bertini emphasize supporting agricultural education and research and infrastructure development, as well as strengthening aid institutions and improving aid policy. They do not mention environmental problems or suggest addressing them. Sir Gordon Conway, "The Food Crisis," *The Geographical Journal* (vol. 174, No. 3, 2008), pays more attention to water shortages and rising temperatures when he calls for a "Doubly Green Revolution" that would bring to the world's farms drought-tolerant, pest-resistant, disease-resistant crop plants.

Aid, education, research, policy, and technology are all going to be needed. But as Lester R. Brown notes, population is also an important factor. Resources and population come together in the concept of "carrying capacity," defined very simply as the size of the population that the environment can support, or "carry," indefinitely, through both good years and bad. It is not the size of the population that can prosper in good times alone, for such a large population must suffer catastrophically when droughts, floods, hurricanes, or blights arrive or the climate warms or cools. It is a long-term concept, where "long-term" means not decades or generations, nor even centuries, but millennia or more. See Mark Nathan Cohen, "Carrying Capacity," *Free Inquiry* (August/September 2004). It is worth noting that many see the threat of global warming and its accompanying changes in rainfall patterns, sea level, and disease distribution as the beginning of a long run of bad years. See Sarah DeWeerdt, "Climate Change, Coming Home," *World Watch* (May/June 2007), and Jeffrey D. Sachs, "Climate Change Refugees," *Scientific American* (June 2007).

Elizabeth Leahy and Sean Peoples, "Projecting Population," *World Watch* (May/June 2009), remind us that it is difficult to predict future population levels, for our projections depend on assumptions and can be very wrong. Among those assumptions must be counted our estimates of future availability of water, soil, energy, and other resources. It is even harder to estimate Earth's carrying capacity for human beings. It is surely impossible to set a precise figure on the number of human beings the world can support for the long run. As Joel E. Cohen discusses in *How Many People Can the Earth Support?* (W. W. Norton, 1995), estimates of Earth's carrying capacity range from under a billion to over a trillion. The precise number depends on our choices of diet, standard of living, level of technology, willingness to share with others at home and abroad, and desire for an intact physical, chemical, and biological environment, as well as on whether or not our morality permits restraint in reproduction and our political or religious ideology permits educating and empowering women. The key, Cohen stresses, is human choice, and the choices are ones we must make within the next 50 years. See also Joel E. Cohen, "Human Population Grows Up," *Scientific American* (September 2005). Phoebe Hall, "Carrying Capacity," *E Magazine* (March/April 2003), notes that even countries with large land areas and small populations, such as Australia and Canada, can be overpopulated in terms of resource availability. The critical resource appears to be food supply; see Russell Hopfenberg, "Human Carrying Capacity Is Determined by Food Availability," *Population & Environment* (November 2003). Jared Diamond's *Collapse: How Societies Choose to Fail or Succeed* (Viking, 2005) is an excellent presentation of how past societies have run afoul of the need to match resources and population.

Andrew R. B. Ferguson, in "Perceiving the Population Bomb," *World Watch* (July/August 2001), sets the maximum sustainable human population at about 2 billion. Sandra Postel, in the Worldwatch Institute's *State of the World 1994* (W. W. Norton, 1994), says, "As a result of our population size, consumption patterns, and technology choices, we have surpassed the planet's carrying capacity. This is plainly evident by the extent to which we are damaging and depleting natural capital" (including soil and water).

According to current population projections, the population growth rate is decreasing and global population may level off and even begin to decline by the end of this century. But the question of carrying capacity remains. Most estimates of carrying capacity put it at well below the current world population size, and it will take a long time for global population to fall far enough to reach such levels. We seem to be moving in the right direction, but it remains an open question whether our numbers will decline far enough soon enough, i.e., before environmental problems become critical. On the other hand, Jeroen Van den Bergh and Piet Rietveld, "Reconsidering the Limits to World Population: Meta-analysis and Meta-prediction," *Bioscience* (March 2004), set their best estimate of human global carrying capacity at 7.7 billion, which is distinctly reassuring.

ISSUE 14

Is Genetic Engineering the Answer to Hunger?

YES: Gerald D. Coleman, from "Is Genetic Engineering the Answer to Hunger?" *America* (February 21, 2005)

NO: Sean McDonagh, from "Genetic Engineering Is Not the Answer," *America* (May 2, 2005)

ISSUE SUMMARY

YES: Gerald D. Coleman argues that genetically engineered crops are useful, healthful, and nonharmful, and although caution may be justified, such crops can help satisfy the moral obligation to feed the hungry.

NO: Sean McDonagh argues that those who wish to feed the hungry would do better to address land reform, social inequality, lack of credit, and other social issues.

In the early 1970s scientists first discovered that it was technically possible to move genes—the biological material that determines a living organism's physical traits—from one organism to another and thus (in principle) to give bacteria, plants, and animals new features. Most researchers in molecular genetics were excited by the potentialities that suddenly seemed within their reach. However, a few researchers—as well as many people outside the field—were disturbed by the idea; they thought that genetic mix-and-match games might spawn new diseases, weeds, and pests. Researchers in support of genetic experimentation responded by declaring a moratorium on their own work until suitable safeguards (in the form of government regulations) could be devised.

A 1987 National Academy of Sciences report said that genetic engineering posed no unique hazards. And, despite continuing controversy, by 1989 the technology had developed tremendously: researchers could obtain patents for mice with artificially added genes ("transgenic" mice); firefly genes had been added to tobacco plants to make them glow (faintly) in the dark; and growth hormone produced by genetically engineered bacteria was being used to grow low-fat pork and increase milk production in cows. The growing biotechnology industry promised more productive crops that made their own fertilizer and pesticide. Proponents argued that genetic engineering was in no significant way different from traditional selective breeding. Critics argued that genetic engineering was unnatural and violated the rights of both plants and animals to their "species integrity"; that expensive, high-tech, tinkered animals gave the competitive advantage to big agricultural corporations and drove small farmers out of business; and that putting human genes into animals, plants, or bacteria was downright offensive. Trey Popp, "God and the New Foodstuffs," *Science & Spirit* (March/April 2006), discusses objections to genetically modified rice containing human genes. For a summary of events related to the development of agricultural biotechnology, see "Biotechnology Timeline: Chronology of Key Events," *International Debates* (March 2006).

In 1992 the U.S. Office of Science and Technology issued guidelines to bar regulations that are based on the assumption that genetically engineered crops pose greater risks than similar crops produced by traditional breeding methods. The result was the rapid commercial introduction of crops that were genetically engineered to make the bacterial insecticide Bt and to resist herbicides and disease, among other things. Rice has now been engineered for herbicide and stress resistance, attempts have been made to improve nutrient content and yield, and "Transgenic rice can serve as a biofactory for the production of molecules of pharmaceutical and industrial utility. The drive to apply transgenic rice for public good as well as commercial gains has fueled research to an all time high"; see Hitesh Kathuria, et al., "Advances in Transgenic Rice Biotechnology," *Critical Reviews in Plant Sciences* (2007). In 2008, some 70 engineered crop varieties were grown on over 125 million hectares (308 million acres) in 25 countries (see GMO-Compass.org, http://www.gmo-compass.org/eng/agri_biotechnology/gmo_planting/). Commercial GM crops are chiefly soybeans, corn, cotton, and canola. Sales of genetically engineered crop products are expected to reach $210 billion by 2014.

Skepticism about the benefits remains, but in 2000, the national academies of science of the United States, the United Kingdom, China, Brazil, India, Mexico, and the third world recognized that though the use of genetically modified crops has some worrisome potentials that deserve further research and continuing caution, those crops hold the potential to feed the world during the twenty-first century while also protecting the environment (Royal Society of London, et al., "Transgenic Plants and World Agriculture," July 2000; http://books.nap.edu/catalog.php?record_id=9889). Genetic engineering is also poised to improve turfgrass and help produce biofuels (see Derek Burke, "Biofuels—Is There a Role for GM?" *Biologist*, February 2007).

The following selections illustrate the different current perspectives from within the Roman Catholic Church. Gerald D. Coleman argues that genetically engineered crops are useful, healthful, and nonharmful, and though caution may be justified, such crops can help satisfy the moral obligation to feed the hungry. Sean McDonagh, a priest who writes frequently on environmental matters, argues that ethical and environmental problems connected to the implementation of genetically engineered crops stand in the way of their promise. Those who wish to feed the hungry would do better to address land reform, social inequality, lack of credit, and other social issues.

YES

Gerald D. Coleman

Is Genetic Engineering the Answer to Hunger?

Both the developed and developing worlds are facing a critical moral choice in the controversial issue of genetically modified food, also known as genetically modified organisms and genetically engineered crops. Critics of these modifications speak dismissively of biotech foods and genetic pollution. On the other hand, proponents like Nina Federoff and Nancy Marie Brown, authors of *Mendel in the Kitchen: A Scientist's View of Genetically Modified Foods* (2004), promote genetically modified organisms (GMs or GMOs) as "the miracle of seed science and fertilizers."

To mark the 20th anniversary of U.S. diplomatic relations with the Holy See, the U.S. Embassy to the Holy See, in cooperation with the Pontifical Academy of Sciences, hosted a conference last fall at Rome's Gregorian University on "Feeding a Hungry World: The Moral Imperative of Biotechnology." Archbishop Renato Martino, who heads the Pontifical Council for Justice and Peace and has been a strong and outspoken proponent of GMOs, told Vatican Radio: "The problem of hunger involves the conscience of every man. For this reason the Catholic Church follows with special interest and solicitude every development in science to help the solution of a plight that affects . . . humanity."

Americans have grown accustomed, perhaps unwittingly, to GMO products. In the United States, for example, 68 percent of the soybeans, 70 percent of the cotton crop, 26 percent of corn and 55 percent of canola are genetically engineered. GMOs represent an estimated 60 percent of all American processed foods. A recent study by the National Center for Food and Agriculture found that farmers in the United States investing in biotech products harvested 5.3 billion additional pounds of crops and realized $22 billion in increased income. Most of the world's beer and cheese is made with GMOs, as are hundreds of medications. In an article published last October, James Nicholson, then U.S. Ambassador to the Vatican and an aggressive promoter of U.S. policy in Vatican circles, wrote that "millions of Americans, Canadians, Australians, Argentines and other people have been eating genetically modified food for nearly a decade—without one proven case of an illness, allergic reaction or even the hiccups. . . . Mankind has been genetically altering food throughout human history." And according to its supporters, biotechnology helps the environment by reducing the use of pesticides and tilling.

The World Health Association recently reported that more than 3.7 billion people around the world are now malnourished, the largest number in history. To this, opponents of GMOs reply that the "real problems" causing hunger, especially in the developing world, are poverty, lack of education and training, unequal land distribution and lack of access to markets. The moral point they advance is that distribution, not production, is the key to solving hunger.

Another significant moral issue relates to "intellectual property policies" and the interest of companies in licensing potentially valuable discoveries. The Rev. Giulio Albanese, head of the Missionary News Agency, insists that unless the problem of intellectual property is resolved in favor of the poor, it represents a "provocation" to developing countries: "The concern of many in the missionary world over the property rights to GM seeds . . . cannot but accentuate the dependence of the poor nations on the rich ones." In response to this concern, a proposal was made recently (reported in *Science* magazine on March 19) that research universities cooperate to seek open licensing provisions that would allow them to share their intellectual property through a "developing-country license." Universities would still retain rights for research and education and maintain negotiating power with the biotechnology and pharmaceutical industries. Catholic social ethics would support this type of proposal, since it places the good of people over amassing profit.

Three moral paths suggest themselves:

1. *Favor the use of GMOs.* Nobel Prize winner Norman Borlaug, who developed the Green Revolution wheat and rice strains, recently wrote: "Biotechnology absolutely should be part of Africa's agricultural reform. African leaders would be making a grievous error if they turn their back on it." Proponents at the Rome conference agreed, arguing that the use of GMOs decreases pesticide use, creates more nutrient-filled crops that require less water and have greater drought resistance, produces more food at a lower cost and uses less land. One small-scale South African farmer concluded, "We need this technology. We don't want always to be fed food aid. We want access to this technology so that one day we can also become commercial farmers."

   This position concludes that the use of GMOs amounts to a moral obligation.

2. *Condemn the use of GMOs.* Many Catholic bishops take an opposing stance. Perhaps the clearest statement comes from the National Conference of Bishops of Brazil and their Pastoral Land Commission. Their argument is threefold: the use of GMOs involves potential risks to human health; a small group of large corporations will be the greatest beneficiaries, with grave damage to the family farmers; and the environment will be gravely damaged.

   The bishops of Botswana, South Africa and Swaziland agree: "We do not believe that agro-companies or gene technologies will help our farmers to produce the food that is needed in the 21st century." Roland Lesseps and Peter Henriot, two Jesuits working in Zambia who are experts on agriculture in the developing world, state their opposition on principle: "Nature is not just useful to us as humans, but is valued and loved in itself, for itself, by God in Christ. . . . The right to use other creatures does not give us the right to abuse them."

In a similar but distinct criticism, the executive director of the U.S. National Catholic Rural Conference, David Andrews, C.S.C., feels that "the Pontifical Academy of Sciences has allowed itself to be subordinated to the U.S. government's insistent advocacy of biotechnology and the companies which market it." Sean McDonagh states: "With patents [on genetically engineered food], farmers will never own their own food. . . ." He believes that "corporate greed" is at the heart of the GMO controversy. Biowatch's Elfrieda Pschorn-Strauss agrees: "With GM crops, small-scale farmers will become completely reliant on and controlled by big foreign companies for their food supply."

This position concludes that the use of GMOs is morally irresponsible.

3. *Approach the use of GMOs with caution.* Two years ago Pope John Paul II declared that GMO agriculture could not be judged solely on the basis of "short-term economic interests," but needed to be subject to "a rigorous scientific and ethical process of verification." This cautionary stance has been adopted by the Catholic Bishops Conference of the Philippines in urging its government to postpone authorization of GMO corn until comprehensive studies have been made: "We have to be careful because, once it is there, how can we remedy its consequences?"

In 2003 the Rural Life Committee of the North Dakota Conference of Churches also called for "rigorous examination" to understand fully the outcomes of the use of GMOs. This document endorses the "Precautionary Principle" formulated in 1992 by the United Nations Conference on Environment and Development in order to avoid "potential harm and unforeseen and unintended consequences."

This view mandates restraint and places the fundamental burden on demonstrating safety. The arguments are based on three areas of concern: the impact on the natural environment, the size of the benefit to the small farmer if the owners and distributors are giant companies like Bristol-Myers and Monsanto and the long-term effects of GMOs on human and animal health and nutrition.

This position concludes that the use of GMOs should be approached with caution.

While the "Precautionary Principle" seems prudent, there is simultaneously a strong moral argument that a war on hunger is a grave, universal need. Last year, 10 million people died of starvation. Every 3.6 seconds someone dies from hunger—24,000 people each day. Half of sub-Saharan Africans are malnourished, and this number is expected to increase to 70 percent by 2010. It was a moral disgrace that in 2002 African governments gave in to GMO opponents and returned to the World Food Program tons of GMO corn simply because it was produced in the U.S. by biotechnology.

The Roman conference gives solid reasons that GMOs are useful, healthful and nonharmful. After all, organisms have been exchanging genetic information for centuries. The tomato, corn and potato would not exist today if human engineering had not transferred genes between species.

The *Catechism of the Catholic Church* teaches that we have a duty to "make accessible to each what is needed to lead a truly human life." The very first example given is food. In *Populorum Progressio* (1967), *Sollicitude Rei Socialis* (1987) and *Centesimus Annus* (1991), Paul VI and later John Paul II forcefully insisted that rich countries have an obligation to help the poor, just as global economic interdependence places us on a moral obligation to be in solidarity with poor nations. Likewise, *The Challenge of Faithful Citizenship,* published by the U.S. bishops in 2004, argues that the church's preferential option for the poor entails "a moral responsibility to commit ourselves to the common good at all levels."

At the same time, it is critical that farmers in developing countries not become dependent on GMO seeds patented by a small number of companies. Intellectual knowledge must be considered the common patrimony of the entire human family. As the U.S. bishops have stated, "Both public and private entities have an obligation to use their property, including intellectual and scientific property, to promote the good of all people" (*For I Was Hungry and You Gave Me Food, 2003*).

The Catholic Church sees deep sacramental significance in wheat and bread, and insists on the absolute imperative to feed and care for the poor of the world. A vital way to promote and ensure the dignity of every human being is to enable them to have their daily bread.

Sean McDonagh  **NO**

Genetic Engineering
Is Not the Answer

In 1992 the then-chief executive of Monsanto, Robert Shapiro, told the *Harvard Business Review* that genetically modified crops will be necessary to feed a growing world population. He predicted that if population levels were to rise to 10 billion, humanity would face two options: either open up new land for cultivation or increase crop yields. Since the first choice was not feasible, because we were already cultivating marginal land and in the process creating unprecedented levels of soil erosion, we would have to choose genetic engineering. This option, Shapiro argued, was merely a further improvement on the agricultural technologies that gave rise to the Green Revolution that saved Asia from food shortages in the 1960s and 1970s.

Genetically engineered crops might seem an ideal solution. Yet both current data and past examples show problems and provoke doubts as to their necessity.

The Green Revolution

The Green Revolution involved the production of hybrid seeds by crossing two genetically distant parents, which produced an offspring plant that gave increased yield. Critics of genetic engineering question the accepted wisdom that its impact has been entirely positive. Hybrid seeds are expensive and heavily reliant on fertilizers and pesticides. And because they lose their vigor after the first planting, the farmer must purchase new seeds for each successive planting.

In his book *Geopolitics and the Green Revolution*, John H. Perkins describes the environmentally destructive and socially unjust aspects of the Green Revolution. One of its most important negative effects, he says, is that it has contributed to the loss of three-quarters of the genetic diversity of major food crops and that the rate of erosion continues at close to 2 percent per year. The fundamental importance of genetic diversity is illustrated by the fact that when a virulent fungus began to destroy wheat fields in the United States and Canada in 1950, plant breeders staved off disaster by cross-breeding five Mexican wheat varieties with 12 imported ones. In the process they created a new strain that was able to resist so-called "stem rust." The loss of these varieties would have been a catastrophe for wheat production globally.

The Terminator Gene

The development by a Monsanto-owned company of what is benignly called a Technology Protection System—a more apt name is terminator technology—is another reason for asserting that the feed-the-world argument is completely spurious. Genetically engineered seeds that contain the terminator gene self-destruct after the first crop. Once again, this forces farmers to return to the seed companies at the beginning of each planting season. If this technology becomes widely used, it will harm the two billion subsistence farmers who live mainly in the poor countries of the world. Sharing seeds among farmers has been at the very heart of subsistence farming since the domestication of staple food crops 11,000 years ago. The terminator technology will lock farmers into a regime of buying genetically engineered seeds that are herbicide tolerant and insect resistant, tethering them to the chemical treadmill.

On an ethical level, a technology that, according to Professor Richard Lewontin of Harvard University, "introduces a 'killer' transgene that prevents the germ of the harvested grain from developing" must be considered grossly immoral. It is a sin against the poor, against previous generations who freely shared their knowledge of plant life with us, against nature itself and finally against the God of all creativity. To set out deliberately to create seeds that self-destruct is an abomination no civilized society should tolerate. Furthermore, there is danger that the terminator genes could spread to neighboring crops and to the wild and weedy relatives of the plant that has been engineered to commit suicide. This would jeopardize the food security of many poor people.

The current situation promoting genetically modified organisms also means supporting the patenting of living organisms—both crops and animals. I find it difficult to understand the support that Cardinal Renato Martino, prefect of the Pontifical Council for Justice and Peace, seems to be giving to genetically modified organisms, given the Catholic Church's strong pro-life position. In my book *Patenting Life? Stop!* I argue that "patenting life is a fundamental attack on the understanding of life as interconnected, mutually dependent and a gift of God to be shared with everyone. Patenting opts for an atomized, isolated understanding of life." The Indian scientist and activist Dr. Vandana Shiva believes that patented crops will lead to food dictatorship by a handful of northern transnational corporations. This would certainly be a recipe for hunger and starvation—in conflict with Catholic social teaching on food and agriculture.

No Higher Yield, No Reduction in Chemicals

Early in 2003 a researcher at the Institute of Development Studies at Sussex University in England published an analysis of the GMO crops that biotech companies are developing for Africa. Among the plants studied were cotton, maize and the sweet potato. The GMO research on the sweet potato is now approaching its 12th year and has involved the work of 19 scientists; to date it has cost $6 million. Results indicated that yield has increased by 18 percent. On the other hand, conventional sweet potato breeding, working with a small budget, has produced a virus-resistant variety with a 100 percent yield increase.

Claims that GMOs lead to fewer chemicals in agriculture are also being challenged. A comprehensive study using U.S. government data on the use of chemicals on genetically engineered crops was carried out by Charles Benbrook, head of the Northwest Science and Environmental Policy Center in Sandpoint, Idaho. He found that when GMOs were first introduced, they needed 25 percent fewer chemicals for the first three years. But in 2001, 5 percent more chemicals were sprayed compared with conventional crop varieties. Dr. Benbrook stated: "The proponents of biotechnology claim GMO varieties substantially reduce pesticide use. While true in the first few years of widespread planting, it is not the case now. There's now clear evidence that the average pound of herbicide applied per acre planted to herbicide-tolerant varieties have increased compared to the first few years."

Toward a Solution

Hunger and famine around the world have more to do with the absence of land reform, social inequality, bias against women farmers and the scarcity of cheap credit and basic agricultural tools than with lack of agribusiness super-seeds. This fact was recognized by those who attended the World Food Summit in Rome in November 1996. People are hungry because they do not have access to food production processes or the money to buy food. Brazil, for example, is the third largest exporter of food in the world, yet one-fifth of its population, over 30 million people, do not have enough food to eat. Clearly hunger there is not due to lack of food but to the unequal distribution of wealth and the fact that a huge number of people are landless.

Do the proponents of genetically engineered food think that agribusiness companies will distribute such food free to the hungry poor who have no money? There was food in Ireland during the famine in the 1840s, for example, but those who were starving had no access to it or money to buy it.

As a Columban missionary in the Philippines, I saw something similar during the drought caused by El Niño in 1983. There was a severe food shortage among the tribal people in the highlands of Mindanao. The drought destroyed their cereal crops, and they could no longer harvest food in the tropical forest because it had been cleared during the previous decades. Even during the height of the drought, an agribusiness corporation was exporting tropical fruit from the lowlands. There was also sufficient rice and corn in the lowlands, but the tribal people did not have the money to buy it. Had it not been for food aid from nongovernmental organizations, many of the tribal people would have starved.

In 1990 the World Food Program at Brown University calculated that if the global food harvests over the previous few years were distributed equitably among all the people of the world, it could provide a vegetarian diet for over 6 billion people. In contrast, a meat-rich diet, favored by affluent countries and currently available to the global elite, could feed only 2.6 billion people.

Human society is going to be faced with the option of getting protein from plants or from animals. If we opt for animal protein, the consequence will be a much less equitable world, with increasing levels of human misery.

Those who wish to banish hunger should address the social and economic inequalities that create poverty and not claim that a magic-bullet technology will solve all the problems.

POSTSCRIPT

Is Genetic Engineering the Answer to Hunger?

Kathleen Hart, in *Eating in the Dark* (Pantheon, 2002), expresses horror at the fact that "Frankenfood" is not labeled and that U.S. consumers are not as alarmed by genetically modified foods as European consumers are. The worries—and the scientific evidence to support them—are summarized by Kathryn Brown, in "Seeds of Concern," and Karen Hopkin, in "The Risks on the Table," *Scientific American* (April 2001). In the same issue, Sasha Nemecek poses the question "Does the World Need GM Foods?" to two prominent figures in the debate: Robert B. Horsch, vice president of the Monsanto Corporation and recipient of the 1998 National Medal of Technology for his work on modifying plant genes, says yes; Margaret Mellon, of the Union of Concerned Scientists, says no, adding that much more work needs to be done with regard to safety. The May 2002 U.S. General Accounting Office Report to Congressional Requesters, "Genetically Modified Foods: Experts View Regimen of Safety Tests as Adequate, but FDA's Evaluation Process Could Be Enhanced," urges more attention to verifying safety testing performed by biotechnology companies. Carl F. Jordan, in "Genetic Engineering, the Farm Crisis, and World Hunger," *Bioscience* (June 2002), says that a major problem is already apparent in the way agricultural biotechnology is widening the gap between the rich and the poor. In a special report on the food crisis, Joel K. Bourne, Jr., "The End of Plenty," *National Geographic* (June 2009), says that the need for improved food production is critical. We need another green revolution, and in half the time it took to implement the first one fifty years ago. Sir Gordon Conway, "The Food Crisis," *The Geographical Journal* (vol. 174, No. 3, 2008), pays more attention to water shortages and rising temperatures when he calls for a "Doubly Green Revolution" that would bring to the world's farms drought-tolerant, pest-resistant, disease-resistant crop plants. However, he emphasizes selective breeding and hybridization over genetic engineering. The food crisis is particularly acute in Africa, where traditional farming techniques simply cannot provide enough food for the growing population. David King, chief scientific adviser to the British government, believes that Africa will not be able to feed itself because it has so far failed to adopt genetically modified (GM) crops; see Khadja Sharife, "Is GM Food the Future for Africa?" *New African* (January 2009).

Charles Mann, in "Biotech Goes Wild," *Technology Review* (July/August 1999), discusses the continuing "lack of a rigorous regulatory framework to sort out the risks inherent in agricultural biotech." Margaret Kriz, in "Global Food Fight," *National Journal* (March 4, 2000), describes the January 2000 Montreal meeting, in which representatives of 130 countries reached "an agreement that requires biotechnology companies to ask permission before importing genetically altered seeds [and] forces food companies to clearly identify all commodity shipments that may contain genetically altered grain." Also see "Environmental Effects of Transgenic Plants: The Scope and Adequacy of Regulation," a report of the National Research Council's Committee on Environmental Impacts Associated With Commercialization of Transgenic Crops (National Academy Press, 2002).

Gregory Conko and C. S. Prakash, in "The Attack on Plant Biotechnology," in Ronald Bailey, ed., *Global Warming and Other Eco-Myths: How the Environmental Movement Uses False Science to Scare Us to Death* (Prima, 2002), say that genetically engineered crops have successfully increased yields and decreased pesticide usage, have not had notable bad environmental side effects, and will be essential for feeding the world's growing population. Jerry D. Glover, Cindy M. Cox, and John P. Reganold, "Future Farming: A Return to Roots?" *Scientific American* (August 2007), argue that current work aimed at using biotechnology to turn present annual crops such as corn, wheat, and rice into perennial crops will boost the food supply while greatly reducing soil erosion and water pollution. Lee Silver, "Why GM Is Good for Us," *Newsweek (Atlantic Edition)* (March 20, 2006), reports that genetically modified foods may actually have fewer allergy-related problems than organic foods. On the other hand, Jeffrey M. Smith, "Frankenstein Peas," *The Ecologist* (April 2006), reports on a study that genetically modified peas fed to mice caused serious immune system problems. Similar concerns are expressed by Artemis Dona and Ioannis S. Arvanitoyannis, "Health Risks of Genetically Modified Foods," *Critical Reviews in Food Science & Nutrition* (February 2009). See also Javier A. Magana-Gomez and Ana M. Calderon de la Barca, "Risk Assessment of Genetically Modified Crops for Nutrition and Health," *Nutrition Reviews* (January 2009).

The UN Food and Agriculture Organization's 2004 annual report, *The State of Food and Agriculture 2003–2004*, FAO Agriculture Series No. 35 (Rome, 2004), maintains that the biggest problem with genetically engineered crops is that the technology has focused so far on crops of interest to large commercial firms. GM crops have not spread fast enough to small farmers, although where they have been introduced into developing countries, they have yielded economic gains and reduced the use of toxic chemicals. The report concludes that there have been no adverse health or environmental consequences so far. Continued safety will require more research and governmental regulation and monitoring. Jerry Cayford notes in "Breeding Sanity into the GM Food Debate," *Issues in Science and Technology* (Winter 2004) that the issue is one of social justice as much as it is one of science. Who will control the world's food supply? Which philosophy—democratic competition or technocratic monopoly—will prevail? Theodoros H. Varzakas, et al., "The Politics and Science Behind GMO Acceptance," *Critical Reviews in Food Science & Nutrition* (June 2007), notes that in Europe a number of food scares (including Mad Cow disease) have alarmed consumers, and GMOs "despite the intense reactions from Non Governmental Organizations and consumer organizations have entered our lives with inadequate legislative measures to protect consumers from their consumption."

ISSUE 15

Can Organic Farming Feed the World?

YES: Catherine Badgley, et al., "Organic Agriculture and the Global Food Supply," *Renewable Agriculture & Food Systems* (June 2007)

NO: John J. Miller, from "The Organic Myth," *National Review* (February 9, 2004)

ISSUE SUMMARY

YES: Catherine Badgley, et al., argue that organic methods could produce enough food to sustain a global human population that is even larger than today's, and without requiring additional farmland. Organic agriculture would also decrease the undesirable environmental effects of conventional farming.

NO: John J. Miller argues that organic farming is not productive enough to feed today's population, much less larger future populations; it is prone to dangerous biological contamination; and it is not sustainable.

Introduction

There was a time when all farming was organic. Fertilizer was compost and manure. Fields were periodically left fallow (unfarmed) to recover soil moisture and nutrients. Crops were rotated to prevent nutrient exhaustion. Pesticides were nonexistent. And farmers were at the mercy of periodic droughts (despite irrigation) and insect infestations.

As the population grew, so did the demand for food. In Europe and America, the concomitant demand for fertilizer led in the nineteenth century to a booming trade in guano mined from Caribbean and Pacific islands where deposits of seabird dung could be a hundred and fifty feet thick. When the guano deposits were exhausted, there was an agricultural crisis that was relieved only by the invention of synthetic nitrogen-containing fertilizers early in the twentieth century. See Jimmy Skaggs, *The Great Guano Rush: Entrepreneurs and American Overseas Expansion* (St. Martin's, 1994), and G. J. Leigh, *The World's Greatest Fix: A History of Nitrogen and Agriculture* (Oxford, 2004). Unfortunately, synthetic fertilizers do not maintain the soil's content of organic

274

matter (humus). This deficit can be amended by tilling in sewage sludge, but the public is not usually very receptive to the idea, partly because of the "yuck factor," but also because sewage sludge may contain human pathogens and chemical contaminants.

Synthetic pesticides, beginning with DDT, came into use in the 1940s. Their history is nicely outlined in Keith S. Delaplane, *Pesticide Usage in the United States: History, Benefits, Risks, and Trends* (University of Georgia Extension Bulletin 1121). When they turned out to have problems—target species quickly became resistant, and when the chemicals reached human beings and wildlife on food and in water, they proved to be toxic—some people sought alternatives. These alternatives to synthetic fertilizers and pesticides (among other things) comprise what is usually meant by "organic farming." Proponents of organic farming have called it holistic, biodynamic, ecological, and natural and claimed a number of advantages for its practice. They say it both preserves the health of the soil and provides healthier food for people. They also argue that it should be used more, even to the point of replacing chemical-based "industrial" agriculture. Because proponents of chemicals hold that fertilizers and pesticides are essential to produce food in the quantities that a world population of nearly 7 billion people requires, and to hold food prices down to affordable levels, one strand of debate has been over whether organic agriculture can do the job.

In the following selections, University of Michigan professor Catherine Badgley, et al., argue that organic methods could produce enough food to sustain a global human population even larger than that of today, and without needing more farmland. Organic agriculture would also decrease the undesirable environmental effects of conventional farming. John J. Miller argues that organic farming is not productive enough to feed today's population, much less larger future populations; it is prone to dangerous biological contamination; and it is not sustainable. "Wishful thinking is at the heart of the organic-food movement."

YES

Catherine Badgley, et al.

Organic Agriculture and the Global Food Supply

Introduction

Ever since Malthus, the sufficiency of the global food supply to feed the human population has been challenged. One side of the current debate claims that green-revolution methods—involving high-yielding plant and animal varieties, mechanized tillage, synthetic fertilizers and biocides, and now transgenic crops—are essential in order to produce adequate food for the growing human population. Green-revolution agriculture has been a stunning technological achievement. Even with the doubling of the human population in the past 40 years, more than enough food has been produced to meet the caloric requirements for all of the world's people, if food were distributed more equitably. Yet Malthusian doubts remain about the future. Indeed, given the projection of 9 to 10 billion people by 2050 and the global trends of increased meat consumption and decreasing grain harvests per capita, advocates argue that a more intensified version of green-revolution agriculture represents our only hope of feeding the world. Another side of the debate notes that these methods of food production have incurred substantial direct and indirect costs and may represent a Faustian bargain. The environmental price of green-revolution agriculture includes increased soil erosion, surface and groundwater contamination, release of greenhouse gases, increased pest resistance, and loss of biodiversity. Advocates on this side argue that more sustainable methods of food production are essential over the long term.

If the latter view is correct, then we seem to be pursuing a short-term solution that jeopardizes long-term environmental sustainability. A central issue is the assertion that alternative forms of agriculture, such as organic methods, are incapable of producing as much food as intensive conventional methods do. A corollary is that organic agriculture requires more land to produce food than conventional agriculture does, thus offsetting any environmental benefits of organic production. Additionally, critics have argued that there is insufficient organically acceptable fertilizer to produce enough organic food without substantially increasing the land area devoted to agriculture.

Here, we evaluate the potential contribution of organic agriculture to the global food supply. Specifically, we investigate the principal objections against organic agriculture making a significant contribution—low yields and

From *Renewable Agriculture and Food Systems*, June 2007, pp. 86–108. Copyright © 2007 by Cambridge University Press. Reprinted by permission.

insufficient quantities of organic nitrogen fertilizers. The term 'organic' here refers to farming practices that may be called agroecological, sustainable, or ecological; utilize natural (non-synthetic) nutrient-cycling processes; exclude or rarely use synthetic pesticides; and sustain or regenerate soil quality. These practices may include cover crops, manures, compost, crop rotation, intercropping, and biological pest control. We are not referring to any particular certification criteria and include non-certified organic examples in our data.

Methods

We compiled data from the published literature about the current global food supply, comparative yields between organic and non-organic production methods, and biological nitrogen fixation by leguminous crops. These data were the basis for estimating the global food supply that could be grown by organic methods and the amount of nitrogen that could become available through increased use of cover crops as green manures.

Estimation of the Global Food Supply

Estimation of the global food supply grown by organic methods involved compiling data about current global food production, deriving ratios of the yields obtained from organic versus non-organic production methods, and applying these yield ratios to current global production values.

Global Food Production

Summary data from the Food and Agricultural Organization (FAO) for 2001 document the current global food supply—grown primarily by conventional methods in most of the developed world and primarily by low-intensive methods in most of the developing world. The FAO provides estimates of the current food supply in 20 general food categories which we modified for our study. We combined three pairs of categories (into sugars and sweeteners, vegetable oils and oil-crops, meat and offals). We omitted from consideration three categories (spices, stimulants, and "miscellaneous"), because they contribute few calories and little nutritional value to the daily diet and lack comparative data for organic versus non-organic production. In addition, we reported data for seafood and "other aquatic products" but did not estimate yield ratios for these categories, since most of these foods are currently harvested from the wild. Alcoholic beverages were reported since they contribute significantly to the average daily caloric intake, but no assessment of organic yields was made. The data presented for yield ratios pertain to ten categories covering the major plant and animal components of human diets.

Food-production data of the FAO include both commercial and domestic production and exclude losses during harvest. Pre-harvest crop losses are not included in the estimates; these losses may be substantial but are not necessarily more serious for organic production, since a host of methods is available for managing pests. For each country or region, the FAO data for the food supply available for human consumption take into account food production, exports, imports, and stocks, as well as losses of production to become livestock feed, seed, or waste. "Waste" refers to post-harvest loss during storage, transport, and processing. We compiled this information for the world, for developed countries, and for developing countries, following the FAO classification of countries as developed or developing.

Deriving Yield Ratios

We estimated the global organic food supply by multiplying the amount of food in the current (2001) food supply by a ratio comparing average organic: non-organic yields. Comparisons of organic to non-organic production are available for many plant foods and a few animal foods. For each of 293 comparisons of organic or semi-organic production to locally prevalent methods under field conditions, the yield ratio is the ratio of organic : non-organic production. A ratio of 0.96, for example, signifies that the organic yield is 96% that of the conventional yield for the same crop. The comparisons include 160 cases with conventional methods and 133 cases with low-intensive methods. Most examples are from the peer-reviewed, published literature; a minority come from conference proceedings, technical reports, or the Web site of an agricultural research station. Like Stanhill's 1990 survey of organic and conventional production, our data include numerous comparisons from paired farms and controlled experiments at research stations. The studies range in observation length from a single growing season to over 20 years. Despite the observation that yields following conversion from conventional to organic production initially decline and then may increase with time, we included studies regardless of duration. All of Stanhill's examples (which are included here) were from the developed world, whereas our dataset also includes diverse examples from the developing world. No attempt was made to bias the results in favor of organic yields; many examples from developed and developing countries exhibit low comparative yields. We avoided generalizations based on countrywide or regional average yields by organic or conventional methods. Some examples are based on yields before and after conversion to organic methods on the same farm.

We grouped examples into ten general food categories and determined the average yield ratio for all cases in each food category. . . . If no data were available (e.g., tree nuts) for estimating global organic production, then we used the average yield ratio for all plant foods, or all animal foods where relevant. For individual studies in which several yield ratios were reported for a single crop (e.g., 0.80–2.00) grown under the same treatment, we took the average as the value for the study. When different treatments were described, we listed a value for each treatment. Averaging the yield ratios across each general food category reduced the effects of unusually high or low yield ratios from individual studies. As these studies come from many regions in developed and developing countries, the average yield ratios are based on a broad range of soils and climates. The average yield ratio is not intended as a predictor of the yield difference for a specific crop or region but as a general indicator of the potential yield performance of organic relative to other methods of production.

Studies in the global south usually demonstrate increases in yields following conversion to organic methods, but these studies are not comparable with those in the developed world. At present, agriculture in developing countries is generally less intensive than in the developed world. Organic production is often compared with local, resource-poor methods of subsistence farming, which may exhibit low yields because of limited access by farmers to natural resources, purchased inputs, or extension services. While adoption of green-revolution methods has typically increased yields, so has intensification by organic methods. Such methods more often result in non-certified than in certified organic production, since most food produced is for local consumption where certification is not at issue. Data from these studies are relevant for our inquiry, which seeks quantitative comparisons between organic production and prior methods, whether by conventional or subsistence practices, since both prevailing methods contribute to global food production.

Estimating the Global Food Supply

Using the average yield ratio for each food category, we estimated the amount of food that could be grown organically by multiplying the amount of food currently produced times the average yield ratio. Following the FAO methodology, this estimate was then proportionally reduced for imports, exports, and losses to give the estimated organic food supply after losses, which is the food supply available for human consumption. We assumed that all food currently produced is grown by non-organic methods, as the global area of certified organic agriculture is only 0.3%.

We constructed two models of global food production grown by organic methods. Model 1 applied the organic : non-organic (conventional) yield ratios derived from studies in developed countries to the entire agricultural land base. This model effectively assumes that, if converted to organic production, the low-intensity agriculture present in much of the developing world would have the same or a slight reduction in yields that has been reported for the developed world, where green-revolution methods now dominate. Model 2 applied the yield ratios derived from studies in the developed world to food production in the developed world, and the yield ratios derived from studies in the developing world to food production in the developing world. The sum of these separate estimates provides the global estimate. . . .

Additional Model Assumptions

Both models were based on the pattern of food production and the amount of land devoted to crops and pasture in 2001. The models estimate the kinds and relative amounts of food that are currently produced and consumed, including the same pattern of total and per-capita consumption of meat, sugars, and alcoholic beverages. Additional assumptions include (1) the same proportion of foods grown for animal feed (e.g., 36% of global grain production), (2) the same proportion of food wasted (e.g., 10% of starchy roots), and (3) the same nutritional value of food (e.g., for protein and fat content in each food category), even though changes in some of these practices would benefit human or environmental health. Finally, we made no assumptions about food distribution and availability, even though changes in accessibility are necessary to achieve global food security. These assumptions establish the boundary conditions for the models but are not intended as an assessment of the sustainability of the current global food system.

Calories per Capita

The calories per capita resulting from Models 1 and 2 were estimated by multiplying the average yield ratios (organic : non-organic) in each food category by the FAO estimate of per-capita calories currently available in that food category.

Nitrogen Availability with Cover Crops

The main limiting macronutrient for agricultural production is biologically available nitrogen (N) in most areas, with phosphorus limiting in certain tropical regions. For phosphorus and potassium, the raw materials for fertility in organic and conventional systems come largely from mineral sources and are not analyzed here.

Nitrogen amendments in organic farming derive from crop residues, animal manures, compost, and biologically fixed N from leguminous plants. A common practice in temperate regions is to grow a leguminous cover crop during the winter fallow period, between food crops, or as a relay crop during the growing season. Such crops are called green manures when they are not harvested but plowed back into the soil for the benefit of the subsequent crop. In tropical regions, leguminous cover crops can be grown between plantings of other crops and may fix substantial amounts of N in just 46–60 days. To estimate the amount of N that is potentially available for organic production, we considered only what could be derived from leguminous green manures grown between normal cropping periods. Nitrogen already derived from animal manure, compost, grain legume crops, or other methods was excluded from the calculations, as we assumed no change in their use. The global estimate of N availability was determined from the rates of N availability or N-fertilizer equivalency reported in 77 studies—33 for temperate regions and 44 for tropical regions, including three studies from arid regions and 18 studies of paddy rice. N availability values in kg ha^{-1} were obtained from studies as either 'fertilizer replacement value,' determined as the amount of N fertilizer needed to achieve equivalent yields to those obtained using N from cover crops, or calculated as 66% of N fixed by a cover crop becoming available for plant uptake during the growing season following the cover crop. . . . We estimated the total amount of N available for plant uptake by multiplying the area currently in crop production (but not already in leguminous forage production—large-scale plantings of perennial legume systems) by the average amount (kg ha^{-1}) of N available to the subsequent crop from leguminous crops during winter fallow or between crops.

Results and Discussion

Estimates of Food and Caloric Production under Organic Agriculture

. . . According to Model 1, the estimated organic food supply is similar in magnitude to the current food supply for most food categories (grains, sweeteners, tree nuts, oil crops and vegetable oils, fruits, meat, animal fats, milk, and eggs). This similarity occurs because the average yield ratios for these categories range from 0.93 to 1.06. For other food categories (starchy roots, legumes, and vegetables), the average yield ratios range from 0.82 to 0.89, resulting in somewhat lower production levels. The average yield ratio for all 160 examples from developed countries is 0.92, close to Stanhill's average relative yield of 0.91. According to Model 2, the estimated organic food supply exceeds the current food supply in all food categories, with most estimates over 50% greater than the amount of food currently produced. The higher estimates in Model 2 result from the high average yield ratios of organic versus current methods of production in the developing world. The average yield ratio for the 133 examples from the developing world is 1.80. We consider Model 2 more realistic because it uses average yield ratios specific to each region of the world. These two models likely bracket the best estimate of global organic food production. Model 1 may underestimate the potential yield ratios of organic to conventional production, since many agricultural soils in developed countries have been degraded by years of tillage, synthetic fertilizers, and pesticide residues. Conversion to organic methods on such soils typically results in an initial decrease in yields, relative to conventional methods, followed by an increase in yields as soil quality is restored. Model 2 may overestimate the yield ratios for the developing world to the extent that green-revolution methods are practiced.

Both models suggest that organic methods could sustain the current human population, in terms of daily caloric intake. The current world food supply after losses provides 2786 kcal person day. The average caloric requirement for a healthy adult is between 2200 and 2500 kcal day^{-1}. Model 1 yielded 2641 kcal person^{-1} day^{-1}, which is above the recommended value, even if slightly less than the current availability of calories. Model 2 yielded 4381 kcal person^{-1} day^{-1}, which is 57% greater than current availability. This estimate suggests that organic production has the potential to support a substantially larger human population than currently exists. Significantly, both models have high yields of grains, which constitute the major caloric component of the human diet. Under Model 1, the grain yield is 93% that of current production. Under Model 2, the grain yield is 145% that of current production.

The most unexpected aspect of this study is the consistently high yield ratios from the developing world. These high yields are obtained when farmers incorporate intensive agroecological techniques, such as crop rotation, cover cropping, agroforestry, addition of organic fertilizers, or more efficient water management. In some instances, organic-intensive methods resulted in higher yields than conventional methods for the same crop in the same setting (e.g., the system of rice intensification (SRI) in ten developing countries). Critics have argued that some of these examples exceed the intrinsic yield limits set by crop genetics and the environmental context. (Such controversy surrounds the 'SRI' and our data include studies from both sides of this controversy.) Yet alternative agricultural methods may elicit a different pathway of gene expression than conventional methods do. Thus, yield limits for conventionally grown crops may not predict the yield limits under alternative methods.

Crop Rotation and Yield-time Adjustment

Organic grain production frequently uses a different rotation system than conventional production. For example, it is common in organic systems to have a three or four-year rotation (with legumes or other crops) for corn, while the conventional rotation often involves planting corn every other year. In situations like this, it is difficult to make yield comparisons between organic and conventional systems without some sort of time adjustment. Although the high variation among rotation systems worldwide makes it impossible to provide a general time–yield adjustment, evaluating potential differences in performance is important. A thorough evaluation of the rotation effect requires knowledge of the plot-to-plot yield differences between organic and conventional production and the rate of decline of both organic and conventional production as a function of the rotation sequence—information that has not yet been experimentally demonstrated. While rotations would undoubtedly differ under a global organic production system, we have no basis for concluding that this system would be unable to provide enough grain to feed the world.

Organic Nitrogen Fertilizer

In 2001, the global use of synthetic N fertilizers was 82 million Mg (metric ton). Our global estimate of N fixed by the use of additional leguminous crops as fertilizer is 140 million Mg, which is 58 million Mg greater than the amount of synthetic N currently in use. Even in the US, where substantial amounts of synthetic N are used in agriculture, the estimate shows a surplus of available N through the additional use of leguminous cover crops between normal cropping periods. The global estimate is based on an average N availability or N-fertilizer equivalency of 102.8 kg N ha^{-1}. For temperate regions, the average is 95.1 kg Nha^{-1} and for tropical regions, the average is 108.6 kg N ha^{-1}. These rates of biological N fixation and release can match N availability with crop uptake and achieve yields equivalent to those of high-yielding conventionally grown crops. In temperate regions, winter cover crops grow well in fall after harvest and in early spring before planting of the main food crop. Research at the Rodale Institute (Pennsylvania, USA) showed that red clover and hairy vetch as winter covers in an oat/wheat–corn–soybean rotation with no additional fertilizer inputs achieved yields comparable to those in conventional controls. Even in arid and semi-arid tropical regions, where water is limiting between periods of crop production, drought-resistant green manures, such as pigeon peas or groundnuts, can be used to fix N. Use of cover crops in arid regions has been shown to increase soil moisture retention, and management of dry season fallows commonly practiced in dry African savannas can be improved with the use of N-fixing cover crops for both N-fixation and weed control. Areas in sub-Saharan Africa which currently use only very small amounts of N fertilizer (9 kg ha^{-1}, much of it on non-food crops) could easily fix more N with the use of green manures, leading to an increase in N availability and yields in these areas. In some agricultural systems, leguminous cover crops not only contribute to soil fertility but also delay leaf senescence and reduce the vulnerability of plants to disease.

Our estimates of N availability from leguminous cover crops do not include other practices for increasing biologically fixed N, such as intercropping, alley cropping with leguminous trees, rotation of livestock with annual crops, and inoculation of soil with free-living N-fixers—practices that may add considerable N fertility to plant and animal production. In addition, rotation of food-crop legumes, such as pulses, soy, or groundnuts, with grains can contribute as much as 75 kg N ha^{-1} to the grains that follow the legumes.

These methods can increase the N-use efficiency by plants. Since biologically available N is readily leached from soil or volatilized if not taken up quickly by plants, N use in agricultural systems can be as low as 50%. Organic N sources occur in more stable forms in carbon-based compounds, which build soil organic matter and increase the amount of N held in the soil. Consequently, the amount of N that must be added each year to maintain yields may actually decrease, because the release of organic N fixed in one season occurs over several years.

These results imply that, in principle, no additional land area is required to obtain enough biologically available N to replace the current use of synthetic N fertilizers. Although this scenario of biological N fixation is simple, it provides an assessment, based on available data, for one method of organic N-fertility production that is widely used by organic farmers and is fairly easy to implement on a large scale. This scenario is not intended to be prescriptive for any particular rotation or location, but to demonstrate the possibility of this type of cover-cropping system to fix large quantities of N without displacing food crops or expanding land area. The Farm Systems Trial at the Rodale Institute uses legume cover crops grown between main crops every third year as the only source of N fertility and reports comparable grain yields to those of conventionally managed systems, while using non-legume winter cover crops in other years to maintain soil quality and fertility and to suppress weeds. In practice, a range of methods acceptable in organic agriculture provides critical flexibility in N-management, including many sources other than cover crops. Although some environmental and economic circumstances pose challenges to reliance on leguminous fertilizers, the full potential of leguminous cover crops in agriculture is yet to be utilized. Implementation of existing knowledge could increase the use of green manures in many regions of the world. Future selection for crop varieties and green manures that have higher rates of N fixation, especially in arid or semi-arid regions, and perform well under N-limiting conditions, as well as for improved strains of N-fixing symbionts, combined with reductions in the amount of N lost from legume-based production systems, and increases in the planting of legumes, hold great promise for increasing the role of biological N-fixation in fertility management. The capacity for increased reliance on legume fertilizers would be even greater with substantive changes in the food system, such as reduction of food waste and feeding less grain to livestock.

Prospects for More Sustainable Food Production

Our results suggest that organic methods of food production can contribute substantially to feeding the current and future human population on the current agricultural land base, while maintaining soil fertility. In fact, the models suggest the possibility that the agricultural land base could eventually be reduced if organic production methods were employed, although additional intensification via conventional methods in the tropics would have the same effect. Our calculations probably underestimate actual output on many organic farms. Yield ratios were reported for individual crops, but many organic farmers use polycultures and multiple cropping systems, from which the total production per unit area is often substantially higher than for single crops. Also, there is scope for increased production on organic farms, since most agricultural research of the past 50 years has focused on conventional methods. Arguably, comparable efforts focused on organic practices would lead to further improvements in yields as well as in soil fertility and pest management. Production per unit area is greater on small farms than on large farms in both developed and developing countries; thus, an increase in the number of small farms would also enhance food production. Finally, organic production on average requires more hand labor than does conventional production, but the labor is often spread out more evenly over the growing season. This requirement has the potential to alleviate rural unemployment in many areas and to reduce the trend of shantytown construction surrounding many large cities of the developing world.

The Millennium Ecosystem Assessment recommends the promotion of agricultural methods that increase food production without harmful tradeoffs from excessive use of water, nutrients, or pesticides. Our models demonstrate that organic agriculture can contribute substantially to a more sustainable system of food production. They suggest not only that organic agriculture, properly intensified, could produce much of the world's food, but also that developing countries could increase their food security with organic agriculture. The results are not, however, intended as forecasts of instantaneous local or global production after conversion to organic methods. Neither do we claim that yields by organic methods are routinely higher than yields from green-revolution methods. Rather, the results show the potential for serious alternatives to green-revolution agriculture as the dominant mode of food production.

In spite of our optimistic prognosis for organic agriculture, we recognize that the transition to and practice of organic agriculture contain numerous challenges—agronomically, economically, and educationally. The practice of organic agriculture on a large scale requires support from research institutions dedicated to agroecological methods of fertility and pest management, a strong extension system, and a committed public. But it is time to put to rest the debate about whether or not organic agriculture can make a substantial contribution to the food supply. It can, both locally and globally. The debate should shift to how to allocate more resources for research on agroecological methods of food production and how to enhance the incentives for farmers and consumers to engage in a more sustainable production system. Finally, production methods are but one component of a sustainable food system. The economic viability of farming methods, land tenure for farmers, accessibility of markets, availability of water, trends in food consumption, and alleviation of poverty are essential to the assessment and promotion of a sustainable food system.

John J. Miller  **NO**

The Organic Myth: A Food Movement Makes a Pest of Itself

Somewhere in the cornfields of Britain, a hungry insect settled on a tall green stalk and decided to have a feast. It chewed into a single kernel of corn, filled its little belly, and buzzed off—leaving behind a tiny hole that was big enough to invite a slow decay. The agent of the decomposition was a fungus known to biologists as *Fusarium*. Farmers have a much simpler name for it: corn ear rot.

As the mold spread inside the corn, it left behind a cancer causing residue called fumonisin. This sequence repeated itself thousands and thousands of times until the infested corn was harvested and sold last year as Fresh and Wild Organic Maize Meal, Infinity Foods Organic Maize Meal, and several other products.

Consuming trace amounts of fumonisin is harmless, but large doses can be deadly. Last fall, the United Kingdom's Food Standards Agency detected alarming concentrations of the toxin in all six brands of organic corn meal subjected to testing—for a failure rate of 100 percent. The average level of contamination was almost 20 times higher than the safety threshold Europeans have set for fumonisin. The tainted products were immediately recalled from the food chain. In contrast, inspectors determined that 20 of the 24 non-organic corn meal products they examined were unquestionably safe to eat.

Despite this, millions of people continue to assume that organic foods are healthier than non-organic ones, presumably because they grow in pristine settings free from icky chemicals and creepy biotechnology. This has given birth to an energetic political movement. In 2002, activists in Oregon sponsored a ballot initiative that essentially would have required the state to slap biohazard labels on anything that wasn't produced in ways deemed fit by anti-biotech agitators. Voters rejected it, but the cause continues to percolate. Hawaiian legislators are giving serious thought to banning biotech crop tests in their state. In March, California's Mendocino County may outlaw biotech plantings altogether.

Beneath it all lurks the belief that organic food is somehow better for us. In one poll, two-thirds of Americans said that organic food is healthier. But they're wrong: It's no more nutritious than food fueled by industrial fertilizers, sprayed with synthetic pesticides, and genetically altered in science labs. And the problem isn't limited to the fungal infections that recently cursed organic corn meal in Britain; bacteria are a major source of disease in organic food as well. To complicate matters further, organic farming is incredibly inefficient. If its appeal ever grew beyond the boutique, it would pose serious threats to the environment. Consumers who go shopping for products emblazoned with the USDA's "organic" seal of approval aren't really helping themselves or the planet—and they're arguably hurting both.

No Fear

Here's the good news: At no point in human history has food been safer than it is today, despite occasional furors like the recent one over an isolated case of mad-cow disease here in the U.S. People still get sick from food—each year, about 76 million Americans pick up at least a mild illness from what they put in their mouths—but modern agricultural methods have sanitized our fare to the point where we may eat without fear. This is true for all food, organic or otherwise.

And that raises a semantic question: What is it about organic food that makes it "organic"? The food we think of as nonorganic isn't really *in*organic, as if it were composed of rocks and minerals. In truth, everything we eat is organic—it's just not "organic" the way the organic-food movement has come to define the word.

About a decade ago, the federal government decided to wade into this semantic swamp. There was no compelling reason for this, but Congress nonetheless called for the invention of a National Organic Rule. It became official in 2002. Organic food, said the bureaucrats, is produced without synthetic fertilizers, conventional pesticides, growth hormones, genetic engineering, or germ-killing radiation. There are also varying levels of organic-ness: Special labels are available for products that are "made with organic ingredients" (which means the food is 70 percent organic), "organic" (which means 95 percent organic), and "100 percent organic." It's not at all clear what consumers are supposed to do with this information. As the Department of Agriculture explains on its website, the "USDA makes no claims that organically produced food is safer or more nutritious than conventionally produced food."

It doesn't because it can't: There's no scientific evidence whatsoever showing that organic food is healthier. So why bother with a National Organic Rule? When the thing was in development, the Clinton administration's secretary of agriculture, Dan Glickman, offered an answer: "The organic label is a marketing tool. It is not a statement about food safety." In other words, those USDA labels are intended to give people warm fuzzies for buying pricey food.

And herein lies one of the dirty secrets of organic farming: It's big business. Although the organic movement has humble origins, today most of its food isn't produced on family farms in quaint villages or even on hippie communes in Vermont. Instead, the industry has come to be dominated by large corporations that are normally the dreaded bogeymen in the minds of many organic consumers. A single company currently controls about 70 percent of the market in organic milk. California grows about $400 million per year in organic produce—and about half of it comes from just five farms. The membership list of the Organic Trade Association includes the biggest names in agribusiness, such as Archer Daniels Midland, Gerber, and Heinz. Even Nike is a member. When its capitalist slavedrivers aren't exploiting child labor in Third World sweatshops (as they do in the fevered imaginations of campus protesters), they're promoting Nike Organics, a clothing line made from organic cotton.

The Yum Factor

There are, in fact, good reasons to eat organic food. Often it's yummier—though this has nothing to do with the fact that it's "organic." If an organic tomato tastes better than a non-organic one, the reason is usually that it has been grown locally, where it has ripened on the vine and taken only a day or two to get from the picking to the selling. Large-scale farming operations that ship fruits and vegetables across the country or the world can't compete with this kind of homegrown quality, even though they do make it possible for people in Minnesota to avoid scurvy by eating oranges in February. Conventional produce is also a good bargain because organic foods can be expensive—the profit margins are quite high, relative to the rest of the food industry.

Unfortunately, money isn't always the sole cost. Although the overwhelming majority of organic foods are safe to eat, they aren't totally risk-free. Think of it this way: Organic foods may be fresh, but they're also fresh from the manure fields.

Organic farmers aren't allowed to enrich their soils the way most non-organic farmers do, which is with nitrogen fertilizers produced through an industrial process. In their place, many farmers rely on composted manure. When they spread the stuff in their fields, they create luscious breeding grounds for all kinds of nasty microbes. Take the dreaded *E. coli*, which is capable of killing people who ingest it. A study by the Center for Global Food Issues found that although organic foods make up about 1 percent of America's diet, they also account for about 8 percent of confirmed *E. coli* cases. Organic food products also suffer from more than eight times as many recalls as conventional ones.

Some of this problem would go away if organic farmers used synthetic sprays—but this, too, is off limits. Conventional wisdom says that we should avoid food that's been drenched in herbicides, pesticides, and fungicides. Half a century ago, there was some truth in this: Sprays were primitive and left behind chemical deposits that often survived all the way to the dinner table. Today's sprays, however, are largely biodegradable. They do their job in the field and quickly break down into harmless molecules. What's more, advances in biotechnology have reduced the need to spray. About one-third of America's corn crop is now genetically modified. This corn includes a special gene that produces a natural toxin that's safe for every living creature to eat except caterpillars with alkaline guts, such as the European corn borer, a moth larva that can ravage whole harvests. This kind of biotech innovation has helped farmers reduce their reliance on pesticides by about 50 million pounds per year.

Organic farmers, of course, don't benefit from any of this. But they do have some recourse against the bugs, weeds, and fungi that can devastate a crop: They spray their plants with "natural" pesticides. These are less effective than synthetic ones and they're certainly no safer. In rat tests, rotenone—an insecticide extracted from the roots of tropical plants—has been shown to cause the symptoms of Parkinson's disease. The Environmental Protection Agency has described pyrethrum, another natural bug killer, as a human carcinogen. Everything is lethal in massive quantities, of course, and it takes huge doses of pyrethrum to pose a health hazard. Still, the typical organic farmer has to douse his crops with it as many as seven times to have the same effect as one or two applications of a synthetic compound based on the same ingredients. Then there's one of the natural fungicides preferred by organic coffee growers in Guatemala: fermented urine. Think about that the next time you're tempted to order the "special brew" at your local organic java hut.

St. Anthony's Fire, etc.

Fungicides are worth taking seriously—and not just because they might have prevented Britain's corn meal problem a few months ago. Before the advent of modern farming, when all agriculture was necessarily "organic," food-borne fungi were a major problem. One of the worst kinds was ergot, which affects rye, wheat, and other grains. During the Middle Ages, ergot poisoning caused St. Anthony's Fire, a painful contraction of blood vessels that often led to gangrene in limb extremities and sometimes death. Hallucinations also were common, as ergot contains lysergic acid, which is the crucial component of LSD. Historians of medieval Europe have documented several episodes of towns eating large batches of ergot-polluted food and falling into mass hysteria. There is some circumstantial evidence suggesting that ergot was behind the madness of the Salem witches: The warm and damp weather just prior to the infamous events of 1692 would have been ideal for an outbreak. Today, however, chemical sprays have virtually eradicated this affliction.

The very worst thing about organic farming requires the use of a word that doomsaying environmentalists have practically trademarked: It's not *sustainable*. Few activities are as wasteful as organic farming. Its yields are about half of what conventional farmers expect at harvest time. Norman Borlaug, who won the Nobel Peace Prize in 1970 for his agricultural innovations, has said, "You couldn't feed more than 4 billion people" on an all-organic diet.

If organic-food consumers think they're making a political statement when they eat, they're correct: They're declaring themselves to be not only friends of population control, but also enemies of environmental conservation. About half the world's land area that isn't covered with ice or sand is devoted to food production. Modern farming techniques have enabled this limited supply to produce increasing quantities of food. Yields have fattened so much in the last few decades that people refer to this phenomenon as the "Green Revolution," a term that has nothing to do with enviro-greenies and everything to do with improvements in breeding, fertilization, and irrigation. Yet even greater challenges lie ahead, because demographers predict that world population will rise to 9 billion by 2050. "The key is to produce more food," says Alex Avery of CGFI. "Growing more per acre leaves more land for nature."

The alternative is to chop down rainforests so that we may dine on organic soybeans.

There's one more important reason that organics can't feed the world: There just isn't enough cow poop to go around. For fun, pretend that U.N. secretary-general Kofi Annan chowed on some ergot rye, decreed that all of humanity must eat nothing but organic food, and that all of humanity responded by saying, "What the heck, we'll give it a try." Forget about the population boom ahead. The immediate problem would be generating enough manure to fertilize all the brand-new, low-yield organic crop fields. There are a little more than a billion cattle in the world today, and each bovine needs between 3 and 30 acres to support it. Conservative estimates say it would take around 7 or 8 billion cattle to produce sufficient heaps of manure to sustain our all-organic diets. The United States alone would need about a billion head (or rear, to be precise). The country would be made up of nothing but cities and manure fields—and the experiment would give a whole new meaning to the term "fruited plains."

This is the sort of future the organic-food movement envisions—and its most fanatical advocates aren't planning to win any arguments on the merits or any consumers on the quality of organic food. In December, when a single U.S. animal was diagnosed with mad-cow disease, nobody was more pleased than Ronnie Cummins of the Organic Consumers Association, who has openly hoped for a public scare that would spark a "crisis of confidence" in American food. No such thing happened, but Cummins should be careful about what he wishes for: Germany's first case of mad-cow disease surfaced at a slaughter-house that specializes in organic beef.

But then wishful thinking is at the heart of the organic-food movement. Its whole market rationale depends on the misperception that organic foods are somehow healthier for both consumers and Mother Earth. Just remember: Nature's Valley can't be found on any map. It's a state of mind.

POSTSCRIPT

Can Organic Farming Feed the World?

It is worth noting that Alex Avery, author of *The Truth about Organic Foods* (Henderson, 2006) and director of research and education for the Hudson Institute's Center for Global Food Issues, argues in "'Organic Abundance' Report: Fatally Flawed," Center for Global Food Issues, Hudson Institute (September 2007), that Badgley, et al., are guilty of misreporting data, inflating averages by counting high organic yields multiple times, and counting as organic farming clearly nonorganic methods.

Quantitative yield comparisons are not the only organic-related topics that prompt debate. Is organic food better or safer for the consumer than nonorganic food? Faidon Magkos, et al., "Organic Food: Buying More Safety or Just Peace of Mind? A Critical Review of the Literature," *Critical Reviews in Food Science & Nutrition* (January 2006), report that the quality and safety of organic food are largely a matter of perception. "Relevant scientific evidence . . . is scarce, while anecdotal reports abound." Pesticide and herbicide residues may be lower, but even on non-organic food they are low. Environmental contaminants are likely to affect both organic and non-organic foods. "'Organic' does not automatically equal 'safe.'"

Is organic farming better for the environment? Soil fertility is in decline in many parts of the world; see Alfred E. Herteminck, "Assessing Soil Fertility Decline in the Tropics Using Soil Chemical Data," *Advances in Agronomy* (2006). In Africa, the situation is extraordinarily serious. According to the International Center for Soil Fertility and Agricultural Development (http://www.ifdc .org/New_Design/Whats_New/africasfailingagriculture033006.pdf), "About 75 percent of the farmland in sub-Saharan Africa is plagued by severe degradation, losing basic soil nutrients needed to grow the crops that feed Africa, according to a new report . . . on the precipitous decline in African soil health from 1980 to 2004. Africa's crisis in food production and battle with hunger are largely rooted in this 'soil health crisis.'" Proponents of organic farming argue that organic methods are essential to relieving the crisis, but one study of changes in soil fertility, as indicated by crop yield, earthworm numbers, and soil properties, after converting from conventional to organic practices, found that different soils responded differently, with some improving and some not; see Anne Kjersti Bakken, et al., "Soil Fertility in Three Cropping Systems after Conversion from Conventional to Organic Farming," *Acta Agriculturae Scandinavica: Section B, Soil & Plant Science* (June 2006). Richard Wood, et al., "A Comparative Study of Some Environmental Impacts of Conventional and Organic Farming in Australia," *Agricultural Systems* (September 2006), find in a comparison of organic and conventional farms "that direct energy use, energy related emissions, and greenhouse gas emissions are higher for the" former.

Paul Collier, "The Politics of Hunger," *Foreign Affairs* (November/December 2008), says that part of the problem is "the middle- and upper-class love affair with peasant [organic] agriculture." What the world needs is more commercial agriculture and more science, as well as an end to subsidies for biofuels production. According to Catherine M. Cooney, "Sustainable Agriculture Delivers the Crops," *Environmental Science & Technology* (February 15, 2006), "Sustainable agriculture, such as crop rotation, organic farming, and genetically modified seeds, increased crop yields by an average of 79 percent" while also improving the lives of farmers in developing countries. Cong Tu, et al., "Responses of Soil Microbial Biomass and N Availability to Transition Strategies from Conventional to Organic Farming Systems," *Agriculture, Ecosystems & Environment* (April 2006), note that a serious barrier to changing from conventional to organic farming, despite soil improvements, is an initial reduction in yield and increase in pests. David Pimentel, et al., "Environmental, Energetic, and Economic Comparisons of Organic and Conventional Farming Systems," *Bioscience* (July 2005), find that organic farming uses less energy, improves soil, and has yields comparable to those of conventional farming but crops probably cannot be grown as often (because of fallowing), which thus reduces long-term yields. However, "because organic foods frequently bring higher prices in the marketplace, the net economic return per [hectare] is often equal to or higher than that of conventionally produced crops." This is one of the factors that prompts Craig J. Pearson to argue in favor of shifting "from conventional open or leaky systems to more closed, regenerative systems" in "Regenerative, Semiclosed Systems: A Priority for Twenty-First-Century Agriculture," *Bioscience* (May 2007).

Better economic return means that organic farming is currently good for the organic farmer. However, the advantage would disappear if the world converted to organic farming. Initial declines in yield mean the conversion would be difficult, but the longer we wait and the more population grows, the more difficult it will be. Whether the conversion is essential depends on the availability of alternative solutions to the problem, and it is worth noting that high energy prices make chemical fertilizers increasingly expensive. Sadly, Stacey Irwin, "Battle High Fertilizer Costs," *Farm Industry News* (January 2006), does not mention the possibility of using organic methods.

Internet References . . .

The Pesticide Action Network

The Pesticide Action Network North America (PANNA) challenges the global proliferation of pesticides.

http://www.panna.org/

e.hormone

e.hormone is hosted by the Center for Bioenvironmental Research at Tulane and Xavier Universities in New Orleans. It provides accurate, timely information about environmental hormones and their impacts.

http://e.hormone.tulane.edu/

The Silicon Valley Toxics Coalition

The Silicon Valley Toxics Coalition (SVTC) was formed to engage in research, advocacy, and organizing associated with environmental and human problems caused by the rapid growth of the high-tech electronics industry.

http://svtc.etoxics.org/site/PageServer

Superfund

The U.S. Environmental Protection Agency provides a great deal of information on the Superfund program, including material on environmental justice.

http://www.epa.gov/superfund/

The La Hague Nuclear Reprocessing Plant

The AREVA NC La Hague site, located on the western tip of the Cotentin Peninsula in Normandy, reprocesses spent power reactor fuel to recycle reusable energy materials—uranium and plutonium—and to condition the waste into suitable final form.

http://www.lahague.areva-nc.com/scripts/areva-nc/publigen/content/templates/Show.asp?P=13&L=EN

Toxic Chemicals

A *great many of today's environmental issues have to do with industrial development, which expanded greatly during the twentieth century. Just since World War II, many thousands of synthetic chemicals— pesticides, plastics, and antibiotics—have flooded the environment. We have become dependent on the production and use of energy, particularly in the form of fossil fuels. We have discovered that industrial processes generate huge amounts of waste, much of it toxic. Air and water pollution have become global problems. And we have discovered that our actions may change the world for generations to come. The following issues by no means exhaust the possible topics for study and debate.*

- Should DDT Be Banned Worldwide?

- Do Environmental Hormone Mimics Pose a Potentially Serious Health Threat?

- Is the Superfund Program Successfully Protecting Human Health from Hazardous Materials?

- Should the United States Reprocess Spent Nuclear Fuel?

ISSUE 16

Should DDT Be Banned Worldwide?

YES: Anne Platt McGinn, from "Malaria, Mosquitoes, and DDT," *World Watch* (May/June 2002)

NO: Donald R. Roberts, from Statement before the U.S. Senate Committee on Environment & Public Works, Hearing on the Role of Science in Environmental Policy-Making (September 28, 2005)

ISSUE SUMMARY

YES: Anne Platt McGinn, a senior researcher at the Worldwatch Institute, argues that although DDT is still used to fight malaria, there are other, more effective and less environmentally harmful methods. She maintains that DDT should be banned or reserved for emergency use.

NO: Donald R. Roberts argues that the scientific evidence regarding the environmental hazards of DDT has been seriously misrepresented by anti-pesticide activists. The hazards of malaria are much greater and, properly used, DDT can prevent them and save lives.

DDT is a crucial element in the story of environmentalism. The chemical was first synthesized in 1874. Swiss entomologist Paul Mueller was the first to notice that DDT has insecticidal properties, which, it was quickly realized, implied that the chemical could save human lives. It had long been known that more soldiers died during wars because of disease than because of enemy fire. During World War I, for example, some 5 million lives were lost to typhus, a disease carried by body lice. DDT was first deployed during World War II to halt a typhus epidemic in Naples, Italy. It was a dramatic success, and DDT was soon used routinely as a dust for soldiers and civilians. During and after the war, DDT was also deployed successfully against the mosquitoes that carry malaria and other diseases. In the United States cases of malaria fell from 120,000 in 1934 to 72 in 1960, and cases of yellow fever dropped from 100,000 in 1878 to none. In 1948 Mueller received the Nobel Prize for medicine and physiology because DDT had saved so many civilian lives.

DDT was by no means the first pesticide. But its predecessors—arsenic, strychnine, cyanide, copper sulfate, and nicotine—were all markedly toxic to humans. DDT was not only more effective as an insecticide, it was also less hazardous to users. It is therefore not surprising that DDT was seen as a beneficial substance. It was soon applied routinely to agricultural crops and used to control mosquito populations in American suburbs. However, insects quickly became resistant to the insecticide; see James E. McWilliams, *American Pests: The Losing War on Insects from the Colonial Times to DDT* (Columbia University Press, 2008). (In any population of insects, some will be more resistant than others; when the insecticide kills the more vulnerable members of the population, the resistant ones are left to breed and multiply. This is an example of natural selection.) In *Silent Spring* (Houghton Mifflin, 1962), marine scientist Rachel Carson demonstrated that DDT was concentrated in the food chain and affected the reproduction of predators such as hawks and eagles. In 1972 the U.S. Environmental Protection Agency banned almost all uses of DDT (it could still be used to protect public health). Other developed countries soon banned it as well, but developing nations, especially those in the tropics, saw it as an essential tool for fighting diseases such as malaria. Roger Bate, director of Africa Fighting Malaria, argues in "A Case of the DDTs," *National Review* (May 14, 2001) that DDT remains the cheapest and most effective way to combat malaria and that it should remain available for use.

It soon became apparent that DDT is by no means the only pesticide or organic toxin with environmental effects. As a result, on May 24, 2001, the United States joined 90 other nations in signing the Stockholm Convention on Persistent Organic Pollutants (POPs). This treaty aims to eliminate from use the entire class of chemicals to which DDT belongs, beginning with the "dirty dozen," pesticides DDT, aldrin, dieldrin, endrin, chlordane, heptachlor, mirex, and toxaphene, and the industrial chemicals polychlorinated biphenyls (PCBs), hexachlorobenzene (HCB), dioxins, and furans. Since then, 59 countries, not including the United States and the European Union (EU), have formally ratified the treaty. It took effect in May 2004. Fiona Proffitt, in "U.N. Convention Targets Dirty Dozen Chemicals," *Science* (May 21, 2004), notes, "About 25 countries will be allowed to continue using DDT against malaria-spreading mosquitoes until a viable alternative is found." These countries do, however, recognize the problems inherent in using pesticides such as DDT; see e.g., P. C. Abhilash and Nandita Singh, "Pesticide Use and Application: An Indian Scenario," *Journal of Hazardous Materials* (June 2009).

In the following selection, Anne Platt McGinn, granting that malaria remains a serious problem in the developing nations of the tropics, especially Africa, contends that although DDT is still used to fight malaria in these nations, it is far less effective than it used to be. She argues that the environmental effects are also serious concerns and that DDT should be banned or reserved for emergency use. In the second selection, Professor Donald R. Roberts argues that the scientific evidence regarding the environmental hazards of DDT has been seriously misrepresented by anti-pesticide activists. The hazards of malaria are much greater and, properly used, DDT can prevent them and save lives. Efforts to prevent the use of DDT have produced a "global humanitarian disaster."

YES

Anne Platt McGinn

Malaria, Mosquitoes, and DDT

This year, like every other year within the past couple of decades, uncountable trillions of mosquitoes will inject malaria parasites into human blood streams billions of times. Some 300 to 500 million full-blown cases of malaria will result, and between 1 and 3 million people will die, most of them pregnant women and children. That's the official figure, anyway, but it's likely to be a substantial underestimate, since most malaria deaths are not formally registered, and many are likely to have escaped the estimators. Very roughly, the malaria death toll rivals that of AIDS, which now kills about 3 million people annually.

But unlike AIDS, malaria is a low-priority killer. Despite the deaths, and the fact that roughly 2.5 billion people (40 percent of the world's population) are at risk of contracting the disease, malaria is a relatively low public health priority on the international scene. Malaria rarely makes the news. And international funding for malaria research currently comes to a mere $150 million annually. Just by way of comparison, that's only about 5 percent of the $2.8 billion that the U.S. government alone is considering for AIDS research in fiscal year 2003.

The low priority assigned to malaria would be at least easier to understand, though no less mistaken, if the threat were static. Unfortunately it is not. It is true that the geographic range of the disease has contracted substantially since the mid-20th century, but over the past couple of decades, malaria has been gathering strength. Virtually all areas where the disease is endemic have seen drug-resistant strains of the parasites emerge—a development that is almost certainly boosting death rates. In countries as various as Armenia, Afghanistan, and Sierra Leone, the lack or deterioration of basic infrastructure has created a wealth of new breeding sites for the mosquitoes that spread the disease. The rapidly expanding slums of many tropical cities also lack such infrastructure; poor sanitation and crowding have primed these places as well for outbreaks—even though malaria has up to now been regarded as predominantly a rural disease.

What has current policy to offer in the face of these threats? The medical arsenal is limited; there are only about a dozen antimalarial drugs commonly in use, and there is significant malaria resistance to most of them. In the absence of a reliable way to kill the parasites, policy has tended to focus on killing the mosquitoes that bear them. And that has led to an abundant use of synthetic pesticides, including one of the oldest and most dangerous: dichlorodiphenyl trichloroethane, or DDT.

DDT is no longer used or manufactured in most of the world, but because it does not break down readily, it is still one of the most commonly detected pesticides in the milk of nursing mothers. DDT is also one of the "dirty dozen" chemicals included in the 2001 Stockholm Convention on Persistent Organic Pollutants [POPs]. The signatories to the "POPs Treaty" essentially agreed to ban all uses of DDT except as a last resort against disease-bearing mosquitoes. Unfortunately, however, DDT is still a routine option in 19 countries, most of them in Africa. (Only 11 of these countries have thus far signed the treaty.) Among the signatory countries, 31—slightly fewer than one-third—have given notice that they are reserving the right to use DDT against malaria. On the face of it, such use may seem unavoidable, but there are good reasons for thinking that progress against the disease is compatible with *reductions* in DDT use.

Malaria is caused by four protozoan parasite species in the genus *Plasmodium.* These parasites are spread exclusively by certain mosquitoes in the genus *Anopheles.* An infection begins when a parasite-laden female mosquito settles onto someone's skin and pierces a capillary to take her blood meal. The parasite, in a form called the *sporozoite,* moves with the mosquito's saliva into the human bloodstream. About 10 percent of the mosquito's lode of sporozoites is likely to be injected during a meal, leaving plenty for the next bite. Unless the victim has some immunity to malaria—normally as a result of previous exposure—most sporozoites are likely to evade the body's immune system and make their way to the liver, a process that takes less than an hour. There they invade the liver cells and multiply asexually for about two weeks. By this time, the original several dozen sporozoites have become millions of *merozoites*— the form the parasite takes when it emerges from the liver and moves back into the blood to invade the body's red blood cells. Within the red blood cells, the merozoites go through another cycle of asexual reproduction, after which the cells burst and release millions of additional merozoites, which invade yet more red blood cells. The high fever and chills associated with malaria are the result of this stage, which tends to occur in pulses. If enough red blood cells are destroyed in one of these pulses, the result is convulsions, difficulty in breathing, coma, and death.

As the parasite multiplies inside the red blood cells, it produces not just more merozoites, but also *gametocytes,* which are capable of sexual reproduction. This occurs when the parasite moves back into the mosquitoes; even as they inject sporozoites, biting mosquitoes may ingest gametocytes if they are feeding on a person who is already infected. The gametocytes reproduce in the insect's gut and the resulting eggs move into the gut cells. Eventually, more sporozoites emerge from the gut and penetrate the mosquito's salivary glands, where they await a chance to enter another human bloodstream, to begin the cycle again.

Of the roughly 380 mosquito species in the genus *Anopheles,* about 60 are able to transmit malaria to people. These malaria vectors are widespread throughout the tropics and warm temperate zones, and they are very efficient at spreading the disease. Malaria is highly contagious, as is apparent from a measurement that epidemiologists call the "basic reproduction number," or BRN. The BRN indicates, on average, how many new cases a single infected person is likely to cause. For example, among the nonvectored diseases (those in which the pathogen travels directly from person to person without an intermediary like a mosquito), measles is one of the most contagious. The BRN for measles is 12 to 14, meaning that someone with measles is likely to infect 12 to 14 other people. (Luckily, there's an inherent limit in this process: as a pathogen spreads through any particular area, it will encounter fewer and fewer susceptible people who aren't already sick, and the outbreak will eventually subside.) HIV/AIDS is on the other end of the scale: it's deadly, but it burns through a population slowly. Its BRN is just above 1, the minimum necessary for the pathogen's survival. With malaria, the BRN varies considerably, depending on such factors as which mosquito species are present in an area and what the temperatures are. (Warmer is worse, since the parasites mature more quickly.) But malaria can have a BRN in excess of 100: over an adult life that may last about a week, a single, malaria-laden mosquito could conceivably infect more than 100 people.

Seven Years, Seven Months

"Malaria" comes from the Italian "mal'aria." For centuries, European physicians had attributed the disease to "bad air." Apart from a tradition of associating bad air with swamps—a useful prejudice, given the amount of mosquito habitat in swamps—early medicine was largely ineffective against the disease. It wasn't until 1897 that the British physician Ronald Ross proved that mosquitoes carry malaria.

The practical implications of Ross's discovery did not go unnoticed. For example, the U.S. administration of Theodore Roosevelt recognized malaria and yellow fever (another mosquito-vectored disease) as perhaps the most serious obstacles to the construction of the Panama Canal. This was hardly a surprising conclusion, since the earlier and unsuccessful French attempt to build the canal—an effort that predated Ross's discovery—is thought to have lost between 10,000 and 20,000 workers to disease. So the American workers draped their water supplies and living quarters with mosquito netting, attempted to fill in or drain swamps, installed sewers, poured oil into standing water, and conducted mosquito-swatting campaigns. And it worked: the incidence of malaria declined. In 1906, 80 percent of the workers had the disease; by 1913, a year before the Canal was completed, only 7 percent did. Malaria could be suppressed, it seemed, with a great deal of mosquito netting, and by eliminating as much mosquito habitat as possible. But the labor involved in that effort could be enormous.

That is why DDT proved so appealing. In 1939, the Swiss chemist Paul Müller discovered that this chemical was a potent pesticide. DDT was first used during World War II, as a delousing agent. Later on, areas in southern Europe, North Africa, and Asia were fogged with DDT, to clear malaria-laden mosquitoes from the paths of invading Allied troops. DDT was cheap and it seemed to be harmless to anything other than insects. It was also long-lasting: most other insecticides lost their potency in a few days, but in the early years of its use, the effects of a single dose of DDT could last for up to six months. In 1948, Müller won a Nobel Prize for his work and DDT was hailed as a chemical miracle.

A decade later, DDT had inspired another kind of war—a general assault on malaria. The "Global Malaria Eradication Program," launched in 1955, became one of the first major undertakings of the newly created World Health Organization [WHO]. Some 65 nations enlisted in the cause. Funding for DDT factories was donated to poor countries and production of the insecticide climbed.

The malaria eradication strategy was not to kill every single mosquito, but to suppress their populations and shorten the lifespans of any survivors, so that the parasite would not have time to develop within them. If the mosquitoes could be kept down long enough, the parasites would eventually disappear from the human population. In any particular area, the process was expected to take three years—time enough for all infected people either to recover or die. After that, a resurgence of mosquitoes would be merely an annoyance, rather than a threat. And initially, the strategy seemed to be working. It proved especially effective on islands—relatively small areas insulated from reinfestation. Taiwan, Jamaica, and Sardinia were soon declared malaria-free and have remained so to this day. By 1961, arguably the year at which the program had peak momentum, malaria had been eliminated or dramatically reduced in 37 countries.

One year later, Rachel Carson published *Silent Spring,* her landmark study of the ecological damage caused by the widespread use of DDT and other pesticides. Like other organochlorine pesticides, DDT bioaccumulates. It's fat soluble, so when an animal ingests it—by browsing contaminated vegetation, for example—the chemical tends to concentrate in its fat, instead of being excreted. When another animal eats that animal, it is likely to absorb the prey's burden of DDT. This process leads to an increasing concentration of DDT in the higher links of the food chain. And since DDT has a high chronic toxicity—that is, long-term exposure is likely to cause various physiological abnormalities—this bioaccumulation has profound implications for both ecological and human health.

With the miseries of malaria in full view, the managers of the eradication campaign didn't worry much about the toxicity of DDT, but they were greatly concerned about another aspect of the pesticide's effects: resistance. Continual exposure to an insecticide tends to "breed" insect populations that are at least partially immune to the poison. Resistance to DDT had been reported as early as 1946. The campaign managers knew that in mosquitoes, regular exposure to DDT tended to produce widespread resistance in four to seven years. Since it took three years to clear malaria from a human population, that didn't leave a lot of leeway for the eradication effort. As it turned out, the logistics simply couldn't be made to work in large, heavily infested areas with high human populations, poor housing and roads, and generally minimal infrastructure. In 1969, the campaign was abandoned. Today, DDT resistance is widespread in *Anopheles,* as is resistance to many more recent pesticides.

Undoubtedly, the campaign saved millions of lives, and it did clear malaria from some areas. But its broadest legacy has been of much more dubious value. It engendered the idea of DDT as a first resort against mosquitoes and it established the unstable dynamic of DDT resistance in *Anopheles* populations. In mosquitoes, the genetic mechanism that confers resistance to DDT does not usually come at any great competitive "cost"—that is, when no DDT is being sprayed, the resistant mosquitoes may do just about as well as nonresistant mosquitoes. So once a population acquires resistance, the trait is not likely to disappear even if DDT isn't used for years. If DDT is reapplied to such a population, widespread resistance will reappear very rapidly. The rule of thumb among entomologists is that you may get seven years of resistance-free use the first time around, but you only get about seven months the second time. Even that limited respite, however, is enough to make the chemical an attractive option as an emergency measure—or to keep it in the arsenals of bureaucracies committed to its use.

Malaria Taxes

In December 2000, the POPs Treaty negotiators convened in Johannesburg, South Africa, even though, by an unfortunate coincidence, South Africa had suffered a potentially embarrassing setback earlier that year in its own POPs policies. In 1996, South Africa had switched its mosquito control programs from DDT to a less persistent group of pesticides known as pyrethroids. The move seemed solid and supportable at the time, since years of DDT use had greatly reduced *Anopheles* populations and largely eliminated one of the most troublesome local vectors, the appropriately named *A. funestus* ("funestus" means deadly). South Africa seemed to have beaten the DDT habit: the chemical had been used to achieve a worthwhile objective; it had then been discarded. And the plan worked—until a year before the POPs summit, when malaria infections rose to 61,000 cases, a level not seen in decades. *A. funestus* reappeared as well, in KwaZulu-Natal, and in a form resistant to pyrethroids. In early 2000, DDT was reintroduced, in an indoor spraying program. (This is now a standard way of using DDT for mosquito control; the pesticide is usually applied only to walls, where mosquitoes alight to rest.) By the middle of the year, the number of infections had dropped by half.

Initially, the spraying program was criticized, but what reasonable alternative was there? This is said to be the African predicament, and yet the South African situation is hardly representative of sub-Saharan Africa as a whole.

Malaria is considered endemic in 105 countries throughout the tropics and warm temperate zones, but by far the worst region for the disease is sub-Saharan Africa. The deadliest of the four parasite species, *Plasmodium falciparum,* is widespread throughout this region, as is one of the world's most effective malaria vectors, *Anopheles gambiae*. Nearly half the population of sub-Saharan Africa is at risk of infection, and in much of eastern and central Africa, and pockets of west Africa, it would be difficult to find anyone who has not been exposed to the parasites. Some 90 percent of the world's malaria infections and deaths occur in sub-Saharan Africa, and the disease now accounts for 30 percent of African childhood mortality. It is true that malaria is a grave problem in many parts of the world, but the African experience is misery on a very different order of magnitude. The average Tanzanian suffers more infective bites each *night* than the average Thai or Vietnamese does in a year.

As a broad social burden, malaria is thought to cost Africa between $3 billion and $12 billion annually. According to one economic analysis, if the disease had been eradicated in 1965, Africa's GDP would now be 35 percent higher than it currently is. Africa was also the gaping hole in the global eradication program: the WHO planners thought there was little they could do on the continent and limited efforts to Ethiopia, Zimbabwe, and South Africa, where eradication was thought to be feasible.

But even though the campaign largely passed Africa by, DDT has not. Many African countries have used DDT for mosquito control in indoor spraying programs, but the primary use of DDT on the continent has been as an agricultural insecticide. Consequently, in parts of west Africa especially, DDT resistance is now widespread in *A. gambiae.* But even if *A. gambiae* were not resistant, a full-bore campaign to suppress it would probably accomplish little, because this mosquito is so efficient at transmitting malaria. Unlike most *Anopheles* species, *A. gambiae* specializes in human blood, so even a small population would keep the disease in circulation. One way to get a sense for this problem is to consider the "transmission index"—the threshold number of mosquito bites necessary to perpetuate the disease. In Africa, the index overall is 1 bite per person per month. That's all that's necessary to keep malaria in circulation. In India, by comparison, the TI is 10 bites per person per month.

And yet Africa is not a lost cause—it's simply that the key to progress does not lie in the general suppression of mosquito populations. Instead of spraying, the most promising African programs rely primarily on "bednets"—mosquito netting that is treated with an insecticide, usually a pyrethroid, and that is suspended over a person's bed. Bednets can't eliminate malaria, but they can "deflect" much of the burden. Because *Anopheles* species generally feed in the evening and at night, a bednet can radically reduce the number of infective bites a person receives. Such a person would probably still be infected from time to time, but would usually be able to lead a normal life.

In effect, therefore, bednets can substantially reduce the disease. Trials in the use of bednets for children have shown a decline in malaria-induced mortality by 25 to 40 percent. Infection levels and the incidence of severe anemia also declined. In Kenya, a recent study has shown that pregnant women who use bednets tend to give birth to healthier babies. In parts of Chad, Mali, Burkina Faso, and Senegal, bednets are becoming standard household items. In the tiny west African nation of The Gambia, somewhere between 50 and 80 percent of the population has bednets.

Bednets are hardly a panacea. They have to be used properly and retreated with insecticide occasionally. And there is still the problem of insecticide resistance, although the nets themselves are hardly likely to be the main cause of it. (Pyrethroids are used extensively in agriculture as well.) Nevertheless, bednets can help transform malaria from a chronic disaster to a manageable public health problem—something a healthcare system can cope with.

So it's unfortunate that in much of central and southern Africa, the nets are a rarity. It's even more unfortunate that, in 28 African countries, they're taxed or subject to import tariffs. Most of the people in these countries would have trouble paying for a net even without the tax. This problem was addressed in the May 2000 "Abuja Declaration," a summit agreement on infectious diseases signed by 44 African countries. The Declaration included a pledge to do away with "malaria taxes." At last count, 13 countries have actually acted on the pledge, although in some cases only by reducing rather than eliminating the taxes. Since the Declaration was signed, an estimated 2 to 5 million Africans have died from malaria.

This failure to follow through with the Abuja Declaration casts the interest in DDT in a rather poor light. Of the 31 POPs treaty signatories that have reserved the right to use DDT, 21 are in Africa. Of those 21, 10 are apparently still taxing or imposing tariffs on bednets. (Among the African countries that have *not* signed the POPs treaty, some are almost certainly both using DDT and taxing bednets, but the exact number is difficult to ascertain because the status of DDT use is not always clear.) It is true that a case can be made for the use of DDT in situations like the one in South Africa in 1999—an infrequent flare-up in a context that lends itself to control. But the routine use of DDT against malaria is an exercise in toxic futility, especially when it's pursued at the expense of a superior and far more benign technology.

Learning to Live with the Mosquitoes

A group of French researchers recently announced some very encouraging results for a new anti-malarial drug known as G25. The drug was given to infected aotus monkeys, and it appears to have cleared the parasites from their systems. Although extensive testing will be necessary before it is known whether the drug can be safely given to people, these results have raised the hope of a cure for the disease.

Of course, it would be wonderful if G25, or some other new drug, lives up to that promise. But even in the absence of a cure, there are opportunities for progress that may one day make the current incidence of malaria look like some dark age horror. Many of these opportunities have been incorporated into an initiative that began in 1998, called the Roll Back Malaria (RBM) campaign, a collaborative effort between WHO, the World Bank, UNICEF, and the UNDP [United Nations Development Programme]. In contrast to the earlier WHO eradication program, RBM grew out of joint efforts between WHO and various African governments specifically to address African malaria. RBM focuses on household- and community-level intervention and it emphasizes apparently modest changes that could yield major progress. Below are four "operating principles" that are, in one way or another, implicit in RBM or likely to reinforce its progress.

1. Do away with all taxes and tariffs on bednets, on pesticides intended for treating bednets, and on antimalarial drugs. Failure to act on this front

certainly undercuts claims for the necessity of DDT; it may also undercut claims for antimalaria foreign aid.

2. Emphasize appropriate technologies. Where, for example, the need for mud to replaster walls is creating lots of pothole sized cavities near houses— cavities that fill with water and then with mosquito larvae—it makes more sense to help people improve their housing maintenance than it does to set up a program for squirting pesticide into every pothole. To be "appropriate," a technology has to be both affordable and culturally acceptable. Improving home maintenance should pass this test; so should bednets. And of course there are many other possibilities. In Kenya, for example, a research institution called the International Center for Insect Physiology and Ecology has identified at least a dozen native east African plants that repel *Anopheles gambiae* in lab tests. Some of these plants could be important additions to household gardens.

3. Use existing networks whenever possible, instead of building new ones. In Tanzania, for example, an established healthcare program (UNICEF's Inte-grated Management of Childhood Illness Program) now dispenses antimalarial drugs—and instruction on how to use them. The UNICEF program was already operating, so it was simple and cheap to add the malaria component. Reported instances of severe malaria and anemia in infants have declined, apparently as a result. In Zambia, the government is planning to use health and prenatal clinics as the network for a coupon system that subsidizes bednets for the poor. Qualifying patients would pick up coupons at the clinics and redeem them at stores for the nets.

4. Assume that sound policy will involve action on many fronts. Malaria is not just a health problem—it's a social problem, an economic problem, an environmental problem, an agricultural problem, an urban planning problem. Health officials alone cannot possibly just make it go away. When the disease flares up, there is a strong and understandable temptation to strap on the spray equipment and douse the mosquitoes. But if this approach actually worked, we wouldn't be in this situation today. Arguably the biggest opportunity for progress against the disease lies, not in our capacity for chemical innovation, but in our capacity for *organizational innovation*—in our ability to build an awareness of the threat across a broad range of policy activities. For example, when government officials are considering loans to irrigation projects, they should be asking: has the potential for malaria been addressed? When foreign donors are designing antipoverty programs, they should be asking: do people need bednets? Routine inquiries of this sort could go a vast distance to reducing the disease.

Where is the DDT in all of this? There isn't any, and that's the point. We now have half a century of evidence that routine use of DDT simply will not prevail against the mosquitoes. Most countries have already absorbed this les-son, and banned the chemical or relegated it to emergency only status. Now the RBM campaign and associated efforts are showing that the frequency and intensity of those emergencies can be reduced through systematic attention to the chronic aspects of the disease. There is less and less justification for DDT, and the futility of using it as a matter of routine is becoming increasingly apparent: in order to control a disease, why should we poison our soils, our waters, and ourselves?

Donald R. Roberts  **NO**

Statement before the U.S. Senate Committee on Environment & Public Works, Hearing on the Role of Science in Environmental Policy-Making

Thank you, Chairman Inhofe, and distinguished members of the Committee on Environment and Public Works, for the opportunity to present my views on the misuse of science in public policy. My testimony focuses on misrepresentations of science during decades of environmental campaigning against DDT.

Before discussing how and why DDT science has been misrepresented, you first must understand why this misrepresentation has not helped, but rather harmed, millions of people every year all over the world. Specifically you need to understand why the misrepresentation of DDT science has been and continues to be deadly. By way of explanation, I will tell you something of my experience.

I conducted malaria research in the Amazon Basin in the 1970s. My Brazilian colleague—who is now the Secretary of Health for Amazonas State—and I worked out of Manaus, the capitol of Amazonas State. From Manaus we traveled two days to a study site where we had sufficient numbers of cases for epidemiological studies. There were no cases in Manaus, or anywhere near Manaus. For years before my time there and for years thereafter, there were essentially no cases of malaria in Manaus. However, in the late 1980s, environmentalists and international guidelines forced Brazilians to reduce and then stop spraying small amounts of DDT inside houses for malaria control. As a result, in 2002 and 2003 there were over 100,000 malaria cases in Manaus alone.

Brazil does not stand as the single example of this phenomenon. A similar pattern of declining use of DDT and reemerging malaria occurs in other countries as well, Peru for example. Similar resurgences of malaria have occurred in rural communities, villages, towns, cities, and countries around the world. As illustrated by the return of malaria in Russia, South Korea, urban areas of the Amazon Basin, and increasing frequencies of outbreaks in the United States, our malaria problems are growing worse. Today there are 1 to 2 million malaria deaths each year and hundreds of millions of cases. The poorest of the world's people are at greatest risk. Of these, children and pregnant women are the ones most likely to die.

U.S. Senate Committee on Environment and Public Works, September 28, 2005.

We have long known about DDT's effectiveness in curbing insect-borne disease. Othmar Zeidler, a German chemistry student, first synthesized DDT in 1874. Over sixty years later in Switzerland, Paul Müller discovered the insecticidal property of DDT. Allied forces used DDT during WWII, and the new insecticide gained fame in 1943 by successfully stopping an epidemic of typhus in Naples, an unprecedented achievement. By the end of the war, British, Italian, and American scientists had also demonstrated the effectiveness of DDT in controlling malaria-carrying mosquitoes. DDT's proven efficacy against insect-borne diseases, diseases that had long reigned unchecked throughout the world, won Müller the Nobel Prize for Medicine in 1948. After WWII, the United States conducted a National Malaria Eradication Program, commencing operations on July 1, 1947. The spraying of DDT on internal walls of rural homes in malaria endemic counties was a key component of the program. By the end of 1949, the program had sprayed over 4,650,000 houses. This spraying broke the cycle of malaria transmission, and in 1949 the United States was declared free of malaria as a significant public health problem. Other countries had already adopted DDT to eradicate or control malaria, because wherever malaria control programs sprayed DDT on house walls, the malaria rates dropped precipitously. The effectiveness of DDT stimulated some countries to create, for the first time, a national malaria control program. Countries with pre-existing programs expanded them to accommodate the spraying of houses in rural areas with DDT. Those program expansions highlight what DDT offered then, and still offers now, to the malaria endemic countries. As a 1945 U.S. Public Health Service manual explained about the control of malaria: "Drainage and larviciding are the methods of choice in towns of 2,500 or more people. But malaria is a rural disease. Heretofore there has been no economically feasible method of carrying malaria control to the individual tenant farmer or sharecropper. Now, for the first time, a method is available—the application of DDT residual spray to walls and ceilings of homes." Health workers in the United States were not the only ones to recognize the particular value of DDT. The head of malaria control in Brazil characterized the changes that DDT offered in the following statement: "Until 1945–1946, preventive methods employed against malaria in Brazil, as in the rest of the world, were generally directed against the aquatic phases of the vectors (draining, larvicides, destruction of bromeliads, etc. . . .). These methods, however, were only applied in the principal cities of each state and the only measure available for rural populations exposed to malaria was free distribution of specific drugs."

DDT was a new, effective, and exciting weapon in the battle against malaria. It was cheap, easy to apply, long-lasting once sprayed on house walls, and safe for humans. Wherever and whenever malaria control programs sprayed it on house walls, they achieved rapid and large reductions in malaria rates. Just as there was a rush to quickly make use of DDT to control disease, there was also a rush to judge how DDT actually functioned to control malaria. That rush to judgment turned out to be a disaster. At the heart of the debate—to the extent there was a debate—was a broadly accepted model that established a mathematical framework for using DDT to kill mosquitoes and eradicate malaria. Instead of studying real data to see how DDT actually worked in controlling malaria, some scientists settled upon what they thought was a logical conclusion: DDT worked solely by killing mosquitoes. This conclusion was based on their belief in the model. Scientists who showed that DDT did not function by killing mosquitoes were ignored. Broad acceptance of the mathematical model led to strong convictions about DDT's toxic actions. Since they were convinced that DDT worked only by killing mosquitoes, malaria control specialists became very alarmed when a mosquito was reported to be resistant to DDT's toxic actions. As a result of concern about DDT resistance, officials decided to make rapid use of DDT before problems of resistance could eliminate their option to use DDT to eradicate malaria. This decision led to creation of the global malaria eradication program. The active years of the global malaria eradication program were from 1959 to 1969. Before, during, and after the many years of this program, malaria workers and researchers carried out their responsibilities to conduct studies and report their research. Through those studies, they commonly found that DDT was functioning in ways other than by killing mosquitoes. In essence, they found that DDT was functioning through mechanisms of repellency and irritancy. Eventually, as people forgot early observations of DDT's repellent actions, some erroneously interpreted new findings of repellent actions as the mosquitoes' adaptation to avoid DDT toxicity, even coining a term, "behavioral resistance," to explain what they saw. This new term accommodated their view that toxicity was DDT's primary mode of action and categorized behavioral responses of mosquitoes as mere adaptations to toxic affects. However this interpretation depended upon a highly selective use of scientific data. The truth is that toxicity is not DDT's primary mode of action when sprayed on house walls. Throughout the history of DDT use in malaria control programs there has always been clear and persuasive data that DDT functioned primarily as a spatial repellent. Today we know that there is no insecticide recommended for malaria control that rivals, much less equals, DDT's spatial repellent actions, or that is as long-acting, as cheap, as easy to apply, as safe for human exposure, or as efficacious in the control of malaria as DDT. . . . The 30 years of data from control programs of the Americas plotted . . . illustrate just how effective DDT is in malaria control. The period 1960s through 1979 displays a pattern of malaria controlled through house spraying. In 1979 the World Health Organization (WHO) changed its strategy for malaria control, switching emphasis from spraying houses to case detection and treatment. In other words, the WHO changed emphasis from malaria prevention to malaria treatment. Countries complied with WHO guidelines and started to dismantle their spray programs over the next several years. . . .

I find it amazing that many who oppose the use of DDT describe its earlier use as a failure. Our own citizens who suffered under the burden of malaria, especially in the rural south, would hardly describe it thus.

Malaria was a serious problem in the United States and for some localities, such as Dunklin County, Missouri, it was a very serious problem indeed. For four counties in Missouri, the average malaria mortality from 1910 to 1914 was 168.8 per 100,000 population. For Dunklin County, it was 296.7 per 100,000, a rate almost equal to malaria deaths in Venezuela and actually greater than the mortality rate for Freetown, Sierra Leone. Other localities in other states were equally as malarious. Growing wealth and improved living conditions were gradually reducing malaria rates, but cases resurged during WWII. The advent of DDT, however, quickly eradicated malaria from the United States.

DDT routed malaria from many other countries as well. The Europeans who were freed of malaria would hardly describe its use as a failure. After DDT was introduced to malaria control in Sri Lanka (then Ceylon), the number of malaria cases fell from 2.8 million in 1946 to just 110 in 1961. Similar spectacular decreases in malaria cases and deaths were seen in all the regions that began to use DDT. The newly formed Republic of China (Taiwan) adopted DDT use in malaria control shortly after World War II. In 1945 there were over 1 million cases of malaria on the island. By 1969 there were only 9 cases and shortly thereafter the disease was eradicated from the island and remains so to this day. Some countries were less fortunate. South Korea used DDT to eradicate malaria, but without house spray programs, malaria has returned across the demilitarized zone with North Korea. As DDT was eliminated and control programs reduced, malaria has returned to other countries such as Russia and Argentina. Small outbreaks of malaria are even beginning to appear more frequently in the United States.

These observations have been offered in testimony to document first that there were fundamental misunderstandings about how DDT functioned to exert control over malaria. Second, that regardless of systematic misunderstandings on the part of those who had influence over malaria control strategies and policies, there was an enduring understanding that DDT was the most cost-effective compound yet discovered for protecting poor rural populations from insect-borne diseases like malaria, dengue, yellow fever, and leishmaniasis. I want to emphasize that misunderstanding the mode of DDT action did not lead to the wholesale abandonment of DDT. It took an entirely new dimension in the misuse of science to bring us to the current humanitarian disaster represented by DDT elimination.

The misuse of science to which I refer has found fullest expression in the collection of movements within the environmental movement that seek to stop production and use of specific man-made chemicals. Operatives within these movements employ particular strategies to achieve their objectives. By characterizing and understanding the strategies these operatives use, we can identify their impact in the scientific literature or in the popular press.

The first strategy is to develop and then distribute as widely as possible a broad list of claims of chemical harm. This is a sound strategy because individual scientists can seldom rebut the scientific foundations of multiple and diverse claims. Scientists generally develop expertise in a single, narrow field and are disinclined to engage issues beyond their area of expertise. Even if an authoritative rebuttal of one claim occurs, the other claims still progress. A broad list of claims also allows operatives to tailor platforms for constituencies, advancing one set of claims with one constituency and a different combination for another. Clever though this technique is, a list of multiple claims of harm is hardly sufficient to achieve the objective of a ban. The second strategy then is to mount an argument that the chemical is not needed and propose that alternative chemicals or methods can be used instead. The third strategy is to predict that grave harm will occur if the chemical continues to be used.

The success of Rachel Carson's *Silent Spring* serves as a model for this tricky triad. In *Silent Spring*, Rachel Carson used all three strategies on her primary target, DDT. She described a very large list of potential adverse effects of insecticides, DDT in particular. She argued that insecticides were not really needed and that the use of insecticides produces insects that are insecticide resistant, which only exacerbates the insect control problems. She predicted scary scenarios of severe harm with continued use of DDT and other insecticides. Many have written rebuttals to Rachel Carson and others who have, without scientific justification, broadcast long lists of potential harms of insecticides. . . .

[T]ime and science have discredited most of Carson's claims. Rachel Carson's descriptions of inappropriate uses of insecticides that harmed wildlife are more plausible. However, harm from an inappropriate use does not meet the requirements of anti-pesticide activists. They can hardly lobby for eliminating a chemical because someone used it wrongly. No, success requires that even the proper use of an insecticide will cause a large and systematic adverse effect. However, the proper uses of DDT yield no large and systematic adverse effects. Absent such adverse actions, the activists must then rely on claims about insidious effects, particularly insidious effects that scientists will find difficult to prove one way or the other and that activists can use to predict a future catastrophe.

Rachel Carson relied heavily on possible insidious chemical actions to alarm and frighten the public. Many of those who joined her campaign to ban DDT and other insecticides made extensive use of claims of insidious effects. These claims were amplified by the popular press and became part of the public perception about modern uses of chemicals. For example, four well-publicized claims about DDT were:

1. DDT will cause the obliteration of higher trophic levels. If not obliterated, populations will undergo reproductive failure. Authors of this claim speculated that, even if the use of DDT were stopped, systematic and ongoing obliterations would still occur.
2. DDT causes the death of algae. This report led to speculations that use of DDT could result in global depletion of oxygen.
3. DDT pushed the Bermuda petrel to the verge of extinction and that full extinction might happen by 1978.
4. DDT was a cause of premature births in California sea lions.

Science magazine, the most prestigious science journal in the United States, published these and other phantasmagorical allegations and/or predictions of DDT harm. Nonetheless, history has shown that each and every one of these claims and predictions were false.

1. The obliteration of higher trophic levels did not occur; no species became extinct; and levels of DDT in all living organisms declined precipitously after DDT was de-listed for use in agriculture. How could the prediction have been so wrong? Perhaps it was so wrong

because the paper touting this view used a predictive model based on an assumption of no DDT degradation. This was a startling assertion even at the time as *Science* and other journals had previously published papers that showed DDT was ubiquitously degraded in the environment and in living creatures. It was even more startling that *Science* published a paper that flew so comprehensively in the face of previous data and analysis.

2. DDT's action against algae reportedly occurred at concentrations of 500 parts per billion. But DDT cannot reach concentrations in water higher than about 1.2 parts per billion, the saturation point of DDT in water.

3. Data on the Bermuda petrel did not show a cause-and-effect relationship between low numbers of birds and DDT concentrations. DDT had no affect on population numbers, for populations increased before DDT was de-listed for use in agriculture and after DDT was de-listed as well.

4. Data gathered in subsequent years showed that "despite relatively high concentrations [of DDT], no evidence that population growth or the health of individual California sea lions have been compromised. The population has increased throughout the century, including the period when DDT was being manufactured, used, and its wastes discharged off southern California."

If time and science have refuted all these catastrophic predictions, why do many scientists and the public not know these predictions were false? In part, we do not know the predictions were false because the refutations of such claims rarely appear in the literature.

When scientists hear the kinds of claims described above, they initiate research to confirm or refute the claims. After Charles Wurster published his claim that DDT kills algae and impacts photosynthesis, I initiated research on planktonic algae to quantify DDT's effects. From 1968–1969, I spent a year of honest and demanding research effort to discover that not enough DDT would even go into solution for a measurable adverse effect on planktonic algae. In essence, I conducted a confirmatory study that failed to confirm an expected result. I had negative data, and journals rarely accept negative data for publication. My year was practically wasted. Without a doubt, hundreds of other scientists around the world have conducted similar studies and obtained negative results, and they too were unable to publish their experimental findings. Much in the environmental science literature during the last 20–30 years indicates that an enormous research effort went into proving specific insidious effects of DDT and other insecticides. Sadly, the true magnitude of such efforts will never be known because while the positive results of research find their way into the scientific literature, the negative results rarely do. Research on insidious actions that produce negative results all too often ends up only in laboratory and field notebooks and is forgotten. For this reason, I place considerable weight on a published confirmatory study that fails to confirm an expected result.

The use of the tricky triad continues. A . . . recent paper . . . published in *The Lancet* illustrates the triad's modern application. Two scientists at the National

Institute of Environmental Health Sciences, Walter Rogan and Aimin Chen, wrote this paper, entitled "Health risks and benefits of bis(4-chlorophenyl)-1,1, 1-trichloroethane (DDT)." It is interesting to see how this single paper spins all three strategies that gained prominence in Rachel Carson's *Silent Spring*.

The journal *Emerging Infectious Diseases* had already published a slim version of this paper, which international colleagues and I promptly rebutted. The authors then filled in some parts, added to the claims of harm, and republished the paper in the British journal, *The Lancet*. To get the paper accepted by editors, the authors described studies that support (positive results) as well as studies that do not support (negative results) each claim. Complying with strategy number 1 of the triad, Rogan and Chen produce a long list of possible harms, including the charge that DDT causes cancer in nonhuman primates. The literature reference for Rogan and Chen's claim that DDT causes cancer in nonhuman primates was a paper by Takayama et al. Takayama and coauthors actually concluded from their research on the carcinogenic effect of DDT in nonhuman primates that "the two cases involving malignant tumors of different types are inconclusive with respect to a carcinogenic effect of DDT in nonhuman primates." Clearly, the people who made the link of DDT with cancer were not the scientists who actually conducted the research.

The authors enacted strategy number two of the triad by conducting a superficial review of the role of DDT in malaria control with the goal of discrediting DDT's value in modern malaria control programs. The authors admitted that DDT had been very effective in the past, but then argued that malaria control programs no longer needed it and should use alternative methods of control. Their use of the second strategy reveals, in my opinion, the greatest danger of granting authority to anti-pesticide activists and their writings. As *The Lancet* paper reveals, the NIEHS scientists assert great authority over the topic of DDT, yet they assume no responsibility for the harm that might result from their erroneous conclusions. After many malaria control specialists have expressed the necessity for DDT in malaria control, it is possible for Rogan and Chen to conclude that DDT is not necessary in malaria control only if they have no sense of responsibility for levels of disease and death that will occur if DDT is not used.

Rogan and Chen also employ the third strategy of environmentalism. Their list of potential harms caused by DDT includes toxic effects, neurobehavior effects, cancers, decrements in various facets of reproductive health, decrements in infant and child development, and immunology and DNA damage. After providing balanced coverage of diverse claims of harm, the authors had no option but to conclude they could not prove that DDT caused harm. However, they then promptly negated this honest conclusion by asserting that if DDT is used for malaria control, then great harm might occur. So, in an amazing turn, they conclude they cannot prove DDT causes harm, but still predict severe harm if it is used.

Rogan and Chen end their paper with a call for more research. One could conclude that the intent of the whole paper is merely to lobby for research to better define DDT harm, and what's the harm in that? Surely increasing knowledge is a fine goal. However, if you look at the specific issue of the relative need for research, you will see that the harm of this technique is great. Millions of children and pregnant women die from malaria every year, and the disease sickens hundreds of millions more. This is an indisputable fact: impoverished people engage in real life and death struggles every day with malaria. This also is a fact: not one death or illness can be attributed to an environmental exposure to DDT. Yet, a National Library of Medicine literature search on DDT reveals over 1,300 published papers from the year 2000 to the present, almost all in the environmental literature and many on potential adverse effects of DDT. A search on malaria and DDT reveals only 159 papers. DDT is a spatial repellent and hardly an insecticide at all, but a search on DDT and repellents will reveal only 7 papers. Is this not an egregiously disproportionate research emphasis on non-sources of harm compared to the enormous harm of malaria? Does not this inequity contribute to the continued suffering of those who struggle with malaria? Is it possibly even more than an inequity? Is it not an active wrong?

Public health officials and scientists should not be silent about enormous investments into the research of theoretical risks while millions die of preventable diseases. We should seriously consider our motivations in apportioning research money as we do. For consider this: the U.S. used DDT to eradicate malaria. After malaria disappeared as an endemic disease in the United States, we became richer. We built better and more enclosed houses. We screened our windows and doors. We air conditioned our homes. We also developed an immense arsenal of mosquito control tools and chemicals. Today, when we have a risk of mosquito borne disease, we can bring this arsenal to bear and quickly eliminate risks. And, as illustrated by aerial spray missions in the aftermath of hurricane Katrina, we can afford to do so. Yet, our modern and very expensive chemicals are not what protect us from introductions of the old diseases. Our arsenal responds to the threat; it does not prevent the appearance of old diseases in our midst. What protects us is our enclosed, screened, air-conditioned housing, the physical representation of our wealth. Our wealth is the factor that stops dengue at the border with Mexico, not our arsenal of new chemicals. Stopping mosquitoes from entering and biting us inside our homes is critical in the prevention of malaria and many other insect-borne diseases. This is what DDT does for poor people in poor countries. It stops large proportions of mosquitoes from entering houses. It is, in fact, a form of chemical screening, and until these people can afford physical screening or it is provided for them, this is the only kind of screening they have.

DDT is a protective tool that has been taken away from countries around the world, mostly due to governments acceding to the whims of the anti-pesticide wing of environmentalism, but it is not only the anti-pesticide wing that lobbies against DDT. The activists have a sympathetic lobbying ally in the pesticide industry. As evidence of insecticide industry working to stop countries from using DDT, I am attaching an email message dated 23rd September and authored by a Bayer official. . . . The Bayer official states

> "[I speak] Not only as the responsible manager for the vector control business in Bayer, being the market leader in vector control and pointing out by that we know what we are talking about and have decades of

experiences in the evolution of this very particular market. [but] Also as one of the private sector representatives in the RBM Partnership Board and being confronted with that discussion about DDT in the various WHO, RBM et al circles. So you can take it as a view from the field, from the operational commercial level—but our companies [sic] point of view. I know that all of my colleagues from other primary manufacturers and internationally operating companies are sharing my view."

The official goes on to say that

"DDT use is for us a commercial threat (which is clear, but it is not that dramatical because of limited use), it is mainly a public image threat."

However the most damning part of this message was the statement that

"we fully support EU to ban imports of agricultural products coming from countries using DDT"

[There is] . . . clear evidence of international and developed country pressures to stop poor countries from using DDT to control malaria. This message also shows the complicity of the insecticide industry in those internationally orchestrated efforts.

Pressures to eliminate spray programs, and DDT in particular, are wrong. I say this not based on some projection of what might theoretically happen in the future according to some model, or some projection of theoretical harms, I say this based firmly on what has already occurred. The track record of the anti-pesticide lobby is well documented, the pressures on developing countries to abandon their spray programs are well documented, and the struggles of developing countries to maintain their programs or restart their uses of DDT for malaria control are well documented. The tragic results of pressures against the use of DDT, in terms of increasing disease and death, are quantified and well documented. How long will scientists, public health officials, the voting public, and the politicians who lead us continue policies, regulations and funding that have led us to the current state of a global humanitarian disaster? How long will support continue for policies and programs that favor phantoms over facts?

POSTSCRIPT

Should DDT Be Banned Worldwide?

Professor Roberts is not alone in his disapproval of the efforts to halt the use of DDT. Alexander Gourevitch, "Better Living Through Chemistry," *Washington Monthly* (March 2003), comes close to accusing environmentalists of condemning DDT on the basis of politics or ideology rather than of science. Angela Logomasini comes even closer in "Chemical Warfare: Ideological Environmentalism's Quixotic Campaign Against Synthetic Chemicals," in Ronald Bailey, ed., *Global Warming and Other Eco-Myths: How the Environmental Movement Uses False Science to Scare Us to Death* (Prima, 2002). Her admission that public health demands have softened some environmentalists' resistance to the use of DDT points to a basic truth about environmental debates: over and over again, they come down to what we should do first: Should we meet human needs regardless of whether or not species die and air and water are contaminated? Or should we protect species, air, water, and other aspects of the environment even if some human needs must go unmet? What if this means endangering the lives of children? In the debate over DDT, the human needs are clear, for insect-borne diseases have killed and continue to kill a great many people. Yet the environmental needs are also clear. The question is one of choosing priorities and balancing risks. See John Danley, "Balancing Risks: Mosquitoes, Malaria, Morality, and DDT," *Business and Society Review* (Spring 2002). It is worth noting that John Beard, "DDT and Human Health," *Science of the Total Environment* (February 2006), finds the evidence for the ill effects of DDT more convincing and says that it is still too early to say it does not contribute to human disease.

The serious nature of the malaria threat is well covered in Michael Finkel, "Bedlam in the Blood: Malaria," *National Geographic* (July 2007). Mosquitoes can be controlled in various ways: Swamps can be drained (which carries its own environmental price), and other breeding opportunities can be eliminated. Fish can be introduced to eat mosquito larvae. And mosquito nets can be used to keep the insects away from people, although bednets must be subsidized in the poorest areas and people must be taught how to use them properly; see Phoebe Williams, et al., "Malaria Prevention in Sub-Saharan Africa: A Field Study in Rural Uganda," *Journal of Community Health* (August 2009). See also Jeffrey Marlow, "Malaria: New Methods and New Hope in Battling an Old Scourge," *World Watch* (May/June 2008). But these (and other) alternatives do not mean that there does not remain a place for chemical pesticides. In "Pesticides and Public Health: Integrated Methods of Mosquito Management," *Emerging Infectious Diseases* (January–February 2001), Robert I. Rose, an arthropod biotechnologist with the Animal and Plant Health Inspection Service of the U.S. Department of Agriculture, says, "Pesticides have a role in public health as part of sustainable integrated mosquito management. Other components of such management include surveillance, source reduction or prevention, biological control, repellents, traps, and pesticide-resistance management." "The most effective programs today rely on a range of tools," says McGinn in "Combating Malaria," *State of the World 2003* (W. W. Norton, 2003). Still, some countries see DDT as essential. See Tina Rosenberg, "What the World Needs Now Is DDT," *New York Times Magazine* (April 11, 2004). However when the World Health Organization (WHO) endorsed the use of DDT to fight malaria, there was immediate outcry; see Allan Schapira, "DDT: A Polluted Debate in Malaria Control," *Lancet* (December 16, 2006). So far, however, the outcry seems fruitless; see Apoorva Mandavilli, "DDT Is Back," *Discover* (January 2007).

It has proven difficult to find effective, affordable drugs against malaria; see Ann M. Thayer, "Fighting Malaria," *Chemical and Engineering News* (October 24, 2005), and Claire Panosian Dunavan, "Tackling Malaria," *Scientific American* (December 2005). A great deal of effort has gone into developing vaccines against malaria, but the parasite has demonstrated a persistent talent for evading all attempts to arm the immune system against it. See Z. H. Reed, M. Friede, and M. P. Kieny, "Malaria Vaccine Development: Progress and Challenges," *Current Molecular Medicine* (March 2006). A newer approach is to develop genetically engineered (transgenic) mosquitoes that either cannot support the malaria parasite or cannot infect humans with it; see George K. Christophides, "Transgenic Mosquitoes and Malarial Transmission," *Cellular Microbiology* (March 2005), and M. J. Friedrich, "Malaria Researchers Target Mosquitoes," *JAMA: Journal of the American Medical Association* (May 23, 2007). On the other hand, Nicholas J. White, "Malaria—Time to Act," *New England Journal of Medicine* (November 9, 2006), argues that rather than wait for a perfect solution, we should recognize that present tools—including DDT—are effective enough now that there is no excuse to avoid using them.

It is worth stressing that malaria is only one of several mosquito-borne diseases that pose threats to public health. Two others are yellow fever and dengue. A recent arrival to the United States is West Nile virus, which mosquitoes can transfer from birds to humans. However, West Nile virus is far less fatal than malaria, yellow fever, or dengue, and a vaccine is in development.

It is also worth stressing that global warming means climate changes that may increase the geographic range of disease-carrying mosquitoes. Many climate researchers are concerned that malaria, yellow fever, and other now mostly tropical and subtropical diseases may return to temperate-zone nations and even spread into areas where they have never been known. See Atul A. Khasnis and Mary D. Nettleman, "Global Warming and Infectious Disease," *Archives of Medical Research* (November 2005). Sarah DeWeerdt, "Climate Change, Coming Home," *World Watch* (May/June 2007), notes that "by 2020 human-triggered climate change could kill 300,000 people worldwide every year," in large part because of the spread of diseases such as malaria.

ISSUE 17

Do Environmental Hormone Mimics Pose a Potentially Serious Health Threat?

YES: Michele L. Trankina, from "The Hazards of Environmental Estrogens," *The World & I* (October 2001)

NO: Michael Gough, from "Endocrine Disrupters, Politics, Pesticides, the Cost of Food and Health," *Daily Commentary* (December 15, 1997)

ISSUE SUMMARY

YES: Professor of biological sciences Michele L. Trankina argues that a great many synthetic chemicals behave like estrogen, alter the reproductive functioning of wildlife, and may have serious health effects—including cancer—on humans.

NO: Michael Gough, a biologist and expert on risk assessment and environmental policy, argues that only "junk science" supports the hazards of environmental estrogens.

Following World War II there was an exponential growth in the industrial use and marketing of synthetic chemicals. These chemicals, known as "xenobiotics," were used in numerous products, including solvents, pesticides, refrigerants, coolants, and raw materials for plastics. This resulted in increasing environmental contamination. Many of these chemicals, such as DDT, PCBs, and dioxins, proved to be highly resistant to degradation in the environment; they accumulated in wildlife and were serious contaminants of lakes and estuaries. Carried by winds and ocean currents, these chemicals were soon detected in samples taken from the most remote regions of the planet, far from their points of introduction into the ecosphere.

Until very recently most efforts to assess the potential toxicity of synthetic chemicals to living things, including human beings, focused almost exclusively on their possible role as carcinogens. This was because of legitimate public concern about rising cancer rates and the belief that cancer causation was the most likely outcome of exposure to low levels of synthetic chemicals.

Some environmental scientists urged public health officials to give serious consideration to other possible health effects of xenobiotics. They were generally ignored because of limited funding and the common belief that toxic effects other than cancer required larger exposures than usually resulted from environmental contamination.

In the late 1980s Theo Colborn, a research scientist for the World Wildlife Fund who was then working on a study of pollution in the Great Lakes, began linking together the results of a growing series of isolated studies. Researchers in the Great Lakes region, as well as in Florida, the West Coast, and Northern Europe, had observed widespread evidence of serious and frequently lethal physiological problems involving abnormal reproductive development, unusual sexual behavior, and neurological problems exhibited by a diverse group of animal species, including fish, reptiles, amphibians, birds, and marine mammals. Through Colborn's insights, communications among these researchers, and further studies, a hypothesis was developed that all of these wildlife problems were manifestations of abnormal estrogenic activity. The causative agents were identified as more than 50 synthetic chemical compounds that have been shown in laboratory studies to either mimic the action or disrupt the normal function of the powerful estrogenic hormones responsible for female sexual development and many other biological functions.

Concern that human exposure to these ubiquitous environmental contaminants may have serious health repercussions was heightened by a widely publicized European research study, which concluded that male sperm counts had decreased by 50 percent over the past several decades (a result that is disputed by other researchers) and that testicular cancer rates have tripled. Some scientists have also proposed a link between breast cancer and estrogen disrupters.

In response to the mounting scientific evidence that environmental estrogens may be a serious health threat, the U.S. Congress passed legislation requiring that all pesticides be screened for estrogenic activity and that the Environmental Protection Agency (EPA) develop procedures for detecting environmental estrogenic contaminants in drinking water supplies; see the EPA's Endocrine Disruptor Screening Program Web site at http://www. epa.gov/scipoly/oscpendo/index.htm. In April 2009, the EPA released its list of chemicals for initial screening.

In the following selections, Michele L. Trankina argues that a great many synthetic chemicals behave like estrogen, alter the reproductive functioning of wildlife, and may have serious health effects—including cancer—on humans. She insists that regulatory agencies must minimize public exposure. Michael Gough argues that only "junk science" supports the hazards of environmental estrogens. Expensive testing and regulatory programs can only drive up the cost of food, he says, which will make it harder for the poor to afford fresh fruits and vegetables. Furthermore, health protection will not be increased.

YES

Michele L. Trankina

The Hazards of Environmental Estrogens

What do Barbie dolls, food wrap, and spermicides have in common? And what do they have to do with low sperm counts, precocious puberty, and breast cancer? "Everything," say those who support the notion that hormones are disrupting everything from fish gender to human fertility. "Nothing," counter others who regard the connection as trumped up, alarmist chemophobia. The controversy swirls around the significance of a number of substances that behave like estrogens and appear to be practically everywhere—from plastic toys to topical sunscreens.

Estrogens are a group of hormones produced in both the female ovaries and male testes, with larger amounts made in females than in males. They are particularly influential during puberty, menstruation, and pregnancy, but they also help regulate the growth of bones, skin, and other organs and tissues.

Over the past 10 years, many synthetic compounds and plant products present in the environment have been found to affect hormonal functions in various ways. Those that have estrogenic activity have been labeled as environmental estrogens, ecoestrogens, estrogen mimics, or xenoestrogens (*xenos* means foreign). Some arise as artifacts during the manufacture of plastics and other synthetic materials. Others are metabolites (breakdown products) generated from pesticides or steroid hormones used to stimulate growth in livestock. Ecoestrogens that are produced naturally by plants are called phytoestrogens (*phyton* means plant).

Many of these estrogen mimics bind to estrogen receptors (within specialized cells) with roughly the same affinity as estrogen itself, setting up the potential to wreak havoc on reproductive anatomy and physiology. They have therefore been labeled as disruptors of endocrine function.

Bizarre Changes in Reproductive Systems

Heightened concern about estrogen-mimicking substances arose when several nonmammalian vertebrates began to exhibit bizarre changes in reproductive anatomy and fertility. Evidence that something was amiss came serendipitously in 1994, from observations by reproductive physiologist Louis Guillette of the University of Florida. In the process of studying the decline of alligator populations at Lake Apopka, Florida, Guillette and coworkers noticed that many male alligators had smaller penises than normal. In addition, females superovulated, with multiple nuclei in some of the surplus ova. Closer scrutiny linked these findings to a massive spill of DDT (dichloro-diphenyl-trichloroethane) into Lake Apopka in 1980. Guillette concluded that the declining alligator population was related to the effects of DDT exposure on the animals' reproductive systems.

Although DDT was banned for use in the United States in the early 1970s, it continues to be manufactured in this country and marketed abroad, where it is sprayed on produce that is then sold in U.S. stores. The principal metabolite derived from DDT is called DDE, a xenoestrogen that lingers in fat deposits in the human body for decades. Historically, there have been reports of estrogen-mimicking effects in various fish species, especially in the Great Lakes, where residual concentrations of DDT and PCBs (polychlorinated biphenyls) are high. These effects include feminization and hermaphroditism in males. In fact, fish serve as barometers of the effects of xenoestrogen contamination in bodies of water. An index of exposure is the presence of vitellogenin, a protein specific to egg yolk, in the blood of male fish. Normally, only females produce vitellogenin in their livers, upon stimulation by estrogen from the ovaries.

Ecoestrogens are further concentrated in animals higher up in the food chain. In the Great Lakes region, birds including male herring gulls, terns, and bald eagles exhibit hermaphroditic changes after feeding on contaminated fish. Increased embryo mortality among these avians has also been observed. In addition, evidence from Florida links sterility in male and female panthers to their predation on animals exposed to pesticides with estrogenic activity.

The harmful effects of DDT have also been observed in rodents. Female rodents treated with high concentrations of DDT become predisposed to mammary tumors, while males tend to develop testicular cancer. These observations raise the question of whether pharmacological (that is, low) doses of substances with estrogenic activity translate into physiological effects. Usually they do not, but chronic exposure may be enough to trigger such effects.

Dangers to Humans

If xenoestrogens cause such dramatic reproductive effects in vertebrates, including mammals, what might be the consequences for humans? Nearly a decade ago, Frederick vom Saal, a developmental biologist at the University of Missouri Columbia, cautioned that mammalian reproductive mechanisms are similar enough to warrant concern about the effects of hormone disruptors on humans.

A 1993 article in *Lancet* looked at decreasing sperm counts in men in the United States and 20 other countries and correlated these decreases with the growing concentration of environmental estrogens. The authors—Niels Skakkenbaek, a Danish reproductive endocrinologist, and Richard Sharpe, of the British Medical Research Council Reproductive Biology Unit in Scotland— performed a meta-analysis of 61 sperm-count studies published between 1938 and 1990 to make their connection.

"Nonbelievers" state that this interpretation is contrived. Others suggest alternative explanations. For instance, the negative effects on sperm counts could have resulted from simultaneous increases in the incidence of venereal diseases. Besides, there are known differences in steroid hormone metabolism between lower vertebrates (including nonprimate mammals) and primates, so one cannot always extrapolate from the former group to the latter.

Even so, the incidence of testicular cancer, which typically affects young men in their 20s and 30s, has increased worldwide. Between 1979 and 1991, over 1,100 new cases were reported in England and Wales—a 55 percent increase over previous rates. In Denmark, the rate of testicular cancer increased by 300 percent from 1945 to 1990. Intrauterine exposure to xenoestrogens during testicular development is thought to be the cause.

Supporting evidence comes from Michigan, where accidental contamination of cattle feed with PCBs in 1973 resulted in high concentrations in the breast milk of women who consumed tainted beef. Their sons exhibited defective genitalia. Furthermore, observers in England have noted increased incidences of cryptorchidism (undescended testes), which results in permanent sterility if left untreated, and hypospadias (urethral orifice on the underside of the penis instead of at the tip). The one area of agreement between those who attribute these effects to ecoestrogens and those who deny such a connection is that more research is needed.

Perhaps one of the most disturbing current trends is the alarming increase in breast cancer incidence. Fifty years ago, the risk rate was 1 woman in 20; today it is 1 in 8. Numerous studies have implicated xenoestrogens as the responsible agents. For instance, high concentrations of pesticides, especially DDT, have been found in the breast tissue of breast cancer patients on Long Island. In addition, it has long been known that under certain conditions, estrogen from any source can be tumor promoting, and that most breast cancer cell types have estrogen receptors.

Precocious Puberty

If that were not enough, the growing number of estrogen mimics in the environment has been linked to early puberty in girls. The normal, average age of onset is between 12 and 13. A recent study of 17,000 girls in the United States indicated that 7 percent of white and 27 percent of black girls exhibited physical signs of puberty by age seven. For 10-year-old girls, the percentages increased to 68 and 95, respectively. Studies from the United Kingdom, Canada, and New Zealand have shown similar changes in the age of puberty onset.

It is difficult, however, to elucidate the exact mechanisms that underlie these trends toward precocious puberty. One explanation, which applies especially to cases in the United States, points to the increasing number of children who are overweight or obese as a result of high-calorie diets and lack of regular exercise. Physiologically, an enhanced amount of body fat implies reproductive readiness and signals the onset of puberty in both boys and girls.

For girls, more body fat ensures that there is enough stored energy to support pregnancy and lactation. Young women with low percentages of body fat caused by heavy exercise, sports, or eating disorders usually do not experience the onset of menses until their body composition reflects adequate fat mass.

Many who study the phenomenon of premature puberty attribute it to environmental estrogens in plastics and secondhand exposure through the meat and milk of animals treated with steroid hormones. An alarming increase in the numbers of girls experiencing precocious puberty occurred in the 1970s and '80s in Puerto Rico. Among other effects, breast development occurred in girls as young as one year. Premature puberty was traced to consumption of beef, pork, and dairy products containing high concentrations of estrogen.

Another study from Puerto Rico revealed higher concentrations of phthalate —a xenoestrogen present in certain plastics—in girls who showed signs of early puberty, compared with controls.

It may be that excess body fat and exposure to estrogenic substances operate in concert to hasten puberty. Body fat is one site of endogenous estrogen synthesis. Exposure to environmental estrogens may add just enough exogenous hormone to exert the synergistic effect necessary to bring on puberty, much like the last drop of water that causes the bucket to overflow.

Although most xenoestrogens produce detrimental effects, at least one subgroup—the phytoestrogens—includes substances that can have beneficial effects. Phytoestrogens are generally weaker than natural estrogens. They are found in various foods, such as flax seeds, soybeans and other legumes, some herbs, and many fruits and vegetables. Some studies suggest that soy products may offer protection against certain cancers, including breast, prostate, uterus, and colon cancers. On the other hand, high doses of certain phytoestrogens— such as coumestrol (in sunflower seeds and alfalfa sprouts)—have been found to adversely affect the fertility and reproductive cycles of animals.

Unlike artificially produced xenoestrogens, phytoestrogens are generally not stored in the body but are readily metabolized and excreted. Their health effects should be evaluated on a case-by-case basis, considering such factors as the individual's age, medical and family history, and potential interactions with medications or supplements.

Plastics, Plastics Everywhere

It has been estimated that perhaps 100,000 synthetic chemicals are registered for commercial use in the world today and 1,000 new ones are formulated every year. While many are toxic and carcinogenic, little is known about the chronic effects of the majority of them. And there is growing concern about their potential hormone-disrupting effects. The problem of exposure is complicated by numerous carrier routes, including air, food, water, and consumer products.

Consider certain synthetics that turn up in familiar places, including food and consumer goods. Such seemingly inert products as plastic soda and water bottles, baby bottles, food wrap, Styrofoam, many toys, cosmetics, sunscreens, and even spermicides either contain or break down to yield xenoestrogens. In addition, environmental estrogens are among the byproducts created by such processes as the incineration of biological materials or industrial waste and chlorine bleaching of paper products.

In April 1999, Consumers Union confirmed information previously reported by the Food and Drug Administration regarding 95 percent of baby bottles sold in the United States. The bottles, made of a hard plastic known as polycarbonate, leach out the synthetic estrogen named bisphenol-A, especially when heated or scratched. Studies verifying the estrogenic activity of bisphenol-A were published in *Nature* in the 1930s, but it did not arouse much concern then. A 1993 report published in *Endocrinology* showed that bisphenol-A produced estrogenic effects in a culture of human breast cancer cells.

Additional studies published by vom Saal in 1997 and '98 have shown that bisphenol-A stimulates precocious puberty and obesity in mice. Others have detected leaching of bisphenol-A from polycarbonate products such as plastic tableware, water cooler jugs, and the inside coatings of certain cans (used for some canned foods) and bottle tops. Autoclaving in the canning process causes bisphenol-A to migrate into the liquid in cans.

Spokespersons from polycarbonate manufacturers have stated that they cannot replicate vom Saal's results, but he counters that industry researchers have not done the experiments correctly.

DEHA (di-[2-ethylhexyl] adipate) is a liquid plasticizer added to some plastic food wraps made of polyvinyl chloride (PVC). Scientific studies have shown that the fat-soluble DEHA can migrate into foods, especially luncheon meats, cheese, and other products with a high fat content. For a 45-pound child eating cheese wrapped in such plastic, the limit of safe intake is 1.5 ounces by European standards or 2.5 ounces by Environmental Protection Agency criteria. Studies conducted by Consumers Union indicate that DEHA leaching from commercial plastic wraps is eight times higher than European directives allow. Fortunately, alternative wraps—such as Handi-Wrap™ and Glad Cling Wrap—are made of polyethylene; chemicals in them do not appear to leach into foods.

Barbie dolls manufactured in the 1950s and '60s are made of PVC containing a stabilizer that degrades to a sticky, estrogen-mimicking residue and accumulates on the dolls' bodies. This phenomenon was noted by Danish museum officials in August 2000. Yvonne Shashoua, an expert in materials preservation at the National Museum of Denmark, warns that young children who play with older Barbie and Ken dolls expose themselves to this estrogenic chemical, and she suggests enclosing the dolls with xenoestrogen-free plastic wrap. Storing the dolls in a cool, dark place also helps prevent the harmful stabilizer from oozing out. Who would have thought that these models of glamour would become health hazards?

Another consumer nightmare is a group of chemical plasticizers known as phthalates. They have been associated with various problems, including testicular depletion of zinc, a necessary nutrient for spermatogenesis. Zinc deficiency results in sperm death and consequent infertility. Many products that previously contained phthalates have been reformulated to eliminate them, but phthalates continue to be present in vinyl flooring, medical tubing and bags, adhesives, infants' toys, and ink used to print on food wrap made of plastic and cardboard. They have been detected in fat-soluble foods such as infant formula, cheese, margarine, and chips.

Some environmental estrogens can be found in unusual places, such as contraceptive products containing the spermicide nonoxynol-9. This chemical degrades to nonylphenol, a xenoestrogen shown to stimulate breast cancer cells. Nonylphenol and other alkylphenols have been detected in human umbilical cords, plastic test tubes, and industrial detergents.

Various substances added to lotions (including sunscreens) and cosmetics serve as preservatives. Some of them, members of a chemical family called parabens, were shown to be estrogen mimics by a study at Brunel University in the United Kingdom. The researchers warned that "the safety of these chemicals should be reassessed with particular attention being paid to . . . levels of systemic exposure to humans." Officials of the European Cosmetic Toiletry and Perfumery Association dismissed the Brunel study as "irrelevant," on the grounds that parabens do not enter the systemic circulation. But they ignored the possibility of transdermal introduction.

Additional questions have been raised about the safety—in terms of xenoestrogen content—of recycled materials, especially plastics and paper. Because it is unlikely that a moratorium on chemical synthesis will occur anytime soon, such questions will continue to surface until the public is satisfied that regulatory agencies are doing all they can to minimize exposure. Fortunately, organizations such as the National Institutes of Health, the National Academy of Sciences, the Environmental Protection Agency, the Centers for Disease Control, many universities, and other institutions are involved in efforts to monitor and minimize the effects of environmental estrogens on wildlife and humans.

Michael Gough  **NO**

Endocrine Disrupters, Politics, Pesticides, the Cost of Food and Health

Environmentalists and politicians and federal regulators have added environmental estrogens or endocrine disrupters to the "concerns" or scares that dictate "environmental health policy." That policy, from its beginning, has been based on ideology, not on science. To provide some veneer to the ideology, its proponents have spawned bad science and junk science that claims chemicals in the environment are a major cause of human illness. There is no substance to the claims, but the current policies threaten to cost billions of dollars in wasted estrogen testing programs and to drive some substantial proportion of pesticides from the market.

Rachel Carson, conjuring up a cancer-free, pre-industrial Garden of Eden launched the biggest environmental scare of all in the 1960s. She charged that modern industrial chemicals in the environment caused human cancers. It mattered not at all to her or to her readers that cancers are found in every society, pre-industrial and modern. What mattered were opinions of people such as Umberto Safiotti of the National Cancer Institute, who wrote:

> I consider cancer as a social disease, largely caused by external agents which are derived from our technology, conditioned by our societal lifestyle and whose control is dependent on societal actions and policies.

When Saffioti said "societal actions and policies," he meant government regulations.

By 1968, environmental groups and individuals—including some scientists—appeared on TV and on the floors of the House and Senate to say, over and over again, "The environment causes 90 percent of cancers." They didn't have to say "environment" meant pollution from modern industry and chemicals—especially pesticides—everyone already knew that. Saffioti and others had told them.

In the 1970s, the National Cancer Institute [NCI] released reports that blamed elevated rates for all kinds of cancers on chemicals in the workplace or in the environment. The institute did not have evidence to link those exposures to cancer. It didn't exist then, and it doesn't, except for a limited number

From *Daily Commentary*, December 15, 1997. Copyright © 1997 by Cato Institute. Reprinted by permission.

of high exposures in the workplace, exist now. So what? The reports were gobbled up by the press, politicians, and the public.

In our ignorance of what causes most cancers, the "90 percent" misstatement provided great hope. If the carcinogenic agents in the environment could be identified and eliminated, cancer rates should drop. NCI scientists said so, and they said success was just around the corner if animal tests were used to identify carcinogens. Congress responded. It created the Environmental Protection Agency [EPA] and the Occupational Safety and Health Administration [OSHA]. Both agencies have lots of tasks, but both place an emphasis on controlling exposures to carcinogens. Congress passed and amended law after law. The Clean Air Act, the Clean Water Act, the Safe Drinking Water Act, amendments to the Fungicide, Insecticide, and Rodenticide Act—the euphonious FIFRA—the Toxic Substances Control Act, and the Resource Recovery and Control Act poured forth from Capitol Hill.

And in return, EPA and OSHA, to justify their existence, generated scare after scare. They are aided by all kinds of people eager for explanations about their health problems or for government grants and contracts for research or other work or for money to compensate for health effects or other problems that could be blamed on chemicals.

In 1978, we had the occupational exposure scare. Astoundingly, according to a government report, six workplace substances caused 38 percent of all the cancers in the United States. It was nonsense, of course, and many scientists ridiculed the report, but the government never retracted it. The government scientists who contributed to it never repudiated it.

At about the same time, wastes disposed in Love Canal near Niagara Fall, NY, spewed liquids and gases into a residential community. The chemicals were blamed as the cause of cancers, birth defects, miscarriages, skin diseases, you name it. None of it was true, but waste sites around the country were routinely identified as "another Love Canal" or a "Love Canal in the making," and Congress gave the nation the Superfund Law. Since its passage, Superfund has enriched lawyers and provided secure employment to thousands who wear moon suits and dig up, burn, and rebury wastes, and done nothing for the nation's health. For those who doubt the importance of politics in the environmental health saga, it's worth recalling that every state had two waste sites on the first list of sites slated for priority cleanup under Superfund.

By the 1980s, EPA was chucking out the scares. We had the 2,4,5-T scare, the dioxin scare, the 2,4-D scare, the asbestos in schools scare, the radon in homes scare, the Alar scare, the EMF scare. I've left some out, but the common thread that linked the scares together was cancer. Each scare prompted investigations by affected industries and non-government scientists. Each scare fell apart, revealed as a house of cards jerry-rigged from bad science, worse interpretations of the science, and terrible policy.

In fact, by the late 1970s, there was ample evidence that the much-talked about "cancer epidemic" and the 90 percent statement were simply wrong. Cancer rates were not increasing and rates for some cancers were higher in industrial countries and rates for others were higher in non-industrial countries. The U.S. fell in the middle of countries when ranked by cancer rates.

Sure, there are some carcinogenic substances in the nation's workplaces, but the best estimates are that they cause four percent or less of all cancers, and the percentage is decreasing because the biggest occupational threat, asbestos, is gone. Environmental exposures might cause two or three percent of cancer—on the outside—and they might cause much less.

The research into causes of cancer—not stories designed to bolster the chemicals cause cancer myth—did reveal that there are preventable causes of cancer. Not smoking is a good idea, as is eating lots of fruits and vegetables, not gaining too much weight, restricting the number of sexual partners, and, for those who are fair-skinned, being careful about sun exposures. It's not a lot different from what your mother or grandmother told you.

The government can take a nanny role in urging us to behave, but that's not where the big bucks are. The big bucks are in regulation, and regulation doesn't seem to have much to do with cancer.

In any case, cancer death rates began to fall in 1990, they've fallen since, and the fall is growing steeper. Maybe that information is blunting the cancer scare. I somehow doubt it. I think that the public has become numb to the cancer scare or that it fatalistically accepts the notion that "everything causes cancer." In any case, the environmentalists and the regulators needed a new scare.

The collapse of the cancer scare wasn't good news to everyone. Government bureaucrats and scientists in the anti-carcinogen offices and programs at EPA and elsewhere have secure jobs. Congress easily finds the will to write laws establishing environmental protection activities, but it lacks the will or patience to examine those activities to see if they've accomplished anything. And, let's face it, Congress doesn't eliminate established programs. But the growth of programs slows, and money can become scarce, and that can squeeze researchers who depend on EPA grants and contracts to fund their often senseless surveys and testing programs. Moreover, the fading of scares doesn't benefit environmental organizations that utter shrill cries about scares and coming calamities in their campaigns for contributions.

Here's an example of just how disappointed some people can be that cancer isn't on the rise. Dr. Theo Colborn, a wildlife biologist working for the Conservation Foundation in the late 1980s, was convinced that the chemicals in the Great Lakes were causing human cancer. She set out to prove it by reviewing the available literature about cancer rates in that region. She couldn't. In fact, she found that the rates for some cancers in the Great Lakes region were lower than the rates for the same cancers in other parts of the United States and Canada. . . .

Failing to find cancer slowed her down but didn't stop her. She knew that those chemicals were causing something. All she had to do was find it.

And find it, she did. She collected every paper that described any abnormality in wildlife that live on or around the Great Lakes, and concluded that synthetic chemicals were mimicking the effects of hormones. They were causing every problem in the literature, whether it was homosexual behavior among gulls, crossed bills in other birds, cancer in fish, or increases or decreases in any wildlife population.

The chemicals that have those activities were called "environmental estrogens" or "endocrine disrupters." There was no more evidence to link them to every abnormality in wildlife than there had been in the 1960s to link every human cancer to chemicals. The absence of evidence wasn't much of a problem. Colborn and her colleagues believed that chemicals were the culprit, and the press and much of the public, nutured on the idea that chemicals were bad, didn't require evidence.

Even so, Colborn had a problem that EPA faced in its early days. Soon after EPA was established, the agency leaders realized that protecting wildlife and the environment might be a good thing, but that Congress might not decide to lavish funds on such activities. They were sure, however, that Congress would throw money at programs that were going to protect human health from environmental risks. Whether Colborn knew that history or not, she apparently realized that any real splash for endocrine disrupters depended on tying them to human health effects.

Using the same techniques she'd used to catalogue the adverse effects of endocrine disrupters on wildlife, she reviewed the literature about human health effects that some way or another might be related to disruption of hormone activity. The list was long, including cancers, birth defects, and learning disabilities, but the big hitter on the list was decreased sperm counts. According to Colborn and others' analyses of sperm counts made in different parts of the world under different conditions of nutrition and stress and at different time periods, sperm counts had decreased by 50 percent in the post–World War II period.

If there's anything that catches the attention of Congress, it's risks to males. Congress banned leaded gasoline after EPA released a report that said atmospheric lead was a cause of heart attacks in middle-aged men. The reported decrease in sperm counts leaped up for attention, and attention it got. Congressional hearings were held, magazine articles were written, experts opined about endocrine disrupters and sexual dysfunctions.

And then it fell apart. Scientists found large geographical variations in sperm counts that have not changed over time. Those geographical variations and poor study designs accounted for the reported decrease. That scare went away, but endocrine disrupters were here to stay.

Well-organized and affluent women's groups are convinced that breast cancer is unusually common on Long Island. . . . We know that obesity, estrogen replacement therapy, and late child-bearing or no child-bearing, all of which are more common in affluent women, are associated with breast cancer. Nevertheless, from the very beginning, environmental chemicals have been singled out as the cause of the breast cancer excess. The insecticide aldicarb, which is very resistant to degradation was blamed, but subsequent studies failed to confirm the link. A well-publicized study found a link between DDT and breast cancer, but larger, follow-up studies failed to confirm it. But there're lots of environmental chemicals, and no evidence is required to justify a suggestion of a link between the chemicals and cancer.

Senator Al D'Amato is from Long Island, and he shares his constituents' concerns. During a hearing about the Clean Water Act, Senator D'Amato heard testimony by Dr. Anna Soto from Tufts University about her "E-Screen." According to Dr. Soto, her quick laboratory test could identify chemicals that behave as environmental estrogens or endocrine disrupters for $500 a chemical. Since environmental estrogens seem to some people to be a likely cause of breast cancer, Soto's test appeared to be a real bargain.

Senator D'Amato pushed for an amendment to require E-Screen testing of chemicals that are regulated under the Clean Water Act, but he was unsuccessful. Later in 1995, a senior Senate staffer, Jimmy Powell, took the E-Screen amendment to a very junior Senate staffer and told her to incorporate it into the Safe Drinking Water Act as an "administrative amendment." She did, it passed the Senate, and, for the first time, there was a legislative requirement for endocrine testing.

In the spring of 1996, House Committees were considering legislation to amend the Safe Drinking Water Act and new legislation related to pesticides in food. Aware of the Senate's action, some members of the House committees were eager to include endocrine testing in their legislation, but there was resistance as well. Chemical companies viewed the imposition of yet another test as certain to be an expense, unlikely to cost as little as $500 a chemical, and bound to raise new concerns about chemicals that would require far more extensive tests and research to understand or discount.

Furthermore, so far as food was concerned, there was a general conviction that all the safety factors built into the testing of pesticides and other chemicals that might end up in food provided adequate margins of protection. That conviction was shattered by rumors that reached the House in May 1996. According to the rumors, Dr. John McLachlan and his colleagues at Tulane University had shown that mixtures of pesticides and other environmental chemicals such as PCBs were far more potent in activating estrogen receptors, the first step in estrogen modulation of biochemical pathways than were single chemicals. In the most extreme case, two chemicals at concentrations considered safe by all conventional toxicity tests were 1600 times as potent as estrogen receptor activators as either chemical by itself. The powerful synergy raised new alarms.

In May, everyone concerned about pesticides knew that EPA had a draft of the Tulane paper, and EPA staff were drifting around House offices, but they refused to answer questions about the Tulane results. The silence signaled the expected significance of the paper. A month later, in June, the paper appeared in *Science*. It was a big deal. *Science* ran a news article about the research with a picture of the Tulane researchers. It also ran an editorial by a scientist from the National Institutes of Health who offered some theoretical explanations for how combinations of pesticides at very low levels could affect cells and activate the estrogen receptors. *The New York Times, The Washington Post,* other major newspapers, and news magazines and TV reported the news. If there was ever any doubt that FQPA [Food Quality Protection Act] would require tests for endocrine activity, the flurry of news about the Tulane results erased them.

While the House was drafting the FQPA, Dr. Lynn Goldman, an assistant administrator at EPA, established a committee called the "Endocrine Disrupter

Screening and Testing Advisory Committee" (EDSTAC) [which is now] considering tests for all of the 70,000 chemicals that it estimates are present in commerce, and it's not limiting its recommendations to tests for estrogenic activity. It's adding tests for testosterone and thyroid hormone activity as well as for anti-estrogenic, anti-testosterone, and anti-thyroid activity. The relatively simple E-Screen, which is run on cultured cells, is to be supplemented by some whole animal tests. Tests on single compounds will have to be complemented by tests of mixtures of compounds. The FQPA requires that "valid" tests be used. None of the tests being considered by EDSTAC has been validated; many of them have never been done.

EDSTAC's estimate of 70,000 chemicals in commerce is on the high side—some of those chemicals are used in such small amounts and under such controlled conditions that there's no exposure to them. Dr. Dan Byrd has estimated that 50,000 is a more realistic number. He's also looked at the price lists from commercial testing laboratories to see how much they would charge for a battery of tests something like EDSTAC is considering. Some of the tests haven't been developed, but assuming they can be, Dr. Byrd estimates testing each chemical will cost between $100,000 and $200,000. The total cost would be between $5 and $10 billion. . . .

The Tulane results played some major role in the passage of FQPA, the focus on endocrine disrupters, and Dr. Goldman's establishment of EDSTAC. The Tulane results are wrong. Several groups of scientists tried to replicate the Tulane results. None could. At first, Dr. McLachlan insisted his results were correct. He said that the experiments he reported required expertise and finesse and suggested that the scientists who couldn't repeat his findings were at fault, essentially incompetent. That changed. In July 1997, just 13 months after he published his report, he threw in the towel, acknowledging that neither his laboratory nor anyone else had been able to produce the results that had created such a stir.

Whether the initial results were caused by a series of mistakes or a willful desire to show, once and for all, that environmental chemicals, especially pesticides are bad, bad, bad, we don't know. We do know that the results were wrong.

No matter, EPA now assumes as a matter of policy that synergy occurs. Good science, repeatable science that showed the reported synergy didn't occur has been brushed aside. In its place, we have bad science or junk science. If the Tulane results were the products of honest mistakes, they're bad science; if they flowed from ideology, they're junk science. The effect is the same, but the reasons are different.

The estrogenic disrupter testing under FQPA is going to cost a lot of money and cause a lot of mischief. But the effects of that testing are off somewhere in the future. More immediately, a combination of ideology-driven science and congressional misreading of that science threatens to drive between 50 and 80 percent of all pesticides from the market.

In 1993, a committee of the National Research Council spun together the facts that childhood development takes place at specific times as an infant matures into a toddler and then a child, that infants, toddlers, and children eat, proportionally, far larger amounts of foods such as apple juice and apple sauce and orange juice than do adults, and that pesticides can be present in those processed foods. From those three observations, the committee concluded that an additional safety factor should be included in setting acceptable levels for pesticides in those foods. Left out from the analysis was any evidence that current exposures cause any harm to any infants, toddlers, or children. No matter.

Most people who worry about pesticides expected EPA and the Food and Drug Administration to react to the NRC recommendation by reducing the allowable levels of pesticides on foods that are destined for consumption by children. Maybe they would have. We'll never know. In the FQPA, Congress directed that a new ten-fold safety factor be incorporated into the evaluation of the risks from pesticides.

Safety factors are a fundamental part in the evaluation of pesticide risks. Pesticides are tested in laboratory animals to determine what concentrations to cause effects on the nervous, digestive, endocrine, and other systems. At some, sufficiently low dose that varies from pesticide to pesticide, the chemical does not cause those adverse effects. That dose, called the "No Observed Adverse Effect Level" (or NOAEL), is then divided by 100 to set the acceptable daily limit for human ingestion of the chemical. The FQPA requires division by another factor of 10, so the acceptable daily limit will be the NOAEL divided by 1000 instead of 100. Acceptable limits will be ten-fold less.

Dr. Byrd has estimated that up to 80 percent of all currently permitted uses of pesticides would be eliminated by an across the board application of the 1000-fold safety factor. He cites another toxicologist who estimates that 50 percent of all pesticides would be eliminated from the market. The extent to which these draconian reductions will be forced remains to be seen, but pesticide manufacturers and users can look forward to a period of even-greater limbo as EPA sorts through it new responsibilities and decides how to implement FQPA.

There's no convincing evidence that pesticides in food contribute to cancer causation and none that they cause other adverse health effects. Restrictions on pesticides in food will not have a demonstrable effect on human health. On the other hand, the estrogen testing program and the new safety factor will drive pesticide costs up and pesticide availability down.

Some manufacturers may lose profitable product lines; some may even lose their businesses. Farmers will pay more. They will pass those costs onto middlemen and processors, who, in turn, will pass them onto consumers. Increases in the costs of fruits and vegetables won't change the food purchasing habits of the middle class, but they may and probably will affect the purchases of the poor. The poor are already at greater risks because of poor diets, and the increased costs can be expected to further decrease their consumption of fresh fruits and vegetables.

POSTSCRIPT

Do Environmental Hormone Mimics Pose a Potentially Serious Health Threat?

Stephen H. Safe's "Environmental and Dietary Estrogens and Human Health: Is There a Problem?" *Environmental Health Perspectives* (April 1995) is often cited to support the contention that there is no causative link between environmental estrogens and human health problems. He draws a cautious conclusion, calling the link "implausible" and "unproven." Gough's belief that the battle against environmental estrogens is motivated by environmentalist ideology rather than facts is repeated by Angela Logomasini in "Chemical Warfare: Ideological Environmentalism's Quixotic Campaign Against Synthetic Chemicals," in Ronald Bailey, ed., *Global Warming and Other Eco-Myths: How the Environmental Movement Uses False Science to Scare Us to Death* (Prima, 2002). Some caution is certainly warranted, for the complex and variable manner by which different compounds with estrogenic properties may affect organisms makes projections from animal effects to human effects risky.

Sheldon Krimsky, in "Hormone Disruptors: A Clue to Understanding the Environmental Cause of Disease," *Environment* (June 2001), summarizes the evidence that many chemicals released to the environment affect—both singly and in combination or synergistically—the endocrine systems of animals and humans and may threaten human health with cancers, reproductive anomalies, and neurological effects. He cautions that the regulatory machinery is likely to move very slowly, adding that we cannot wait for scientific certainty about the hazards before we act. See also M. Gochfeld, "Why Epidemiology of Endocrine Disruptors Warrants the Precautionary Principle," *Pure & Applied Chemistry* (December 1, 2003).

Theo Colborn, a senior scientist with the World Wildlife Fund, first drew public attention to the potential problems of environmental estrogens with the book *Our Stolen Future* (Dutton, 1996), coauthored by Dianne Dumanoski and John Peterson Myers. Colborn clearly believes that the problem is real; she finds the evidence that extensive damage is being done to wildlife by synthetic estrogenic chemicals convincing and thinks it likely that humans are experiencing similar health problems. Recent data reinforce her and Krimsky's points; see Rebecca Renner, "Human Estrogens Linked to Endocrine Disruption," *Environmental Science and Technology* (January 1, 1998) and Ted Schettler, et al., *Generations at Risk: Reproductive Health and the Environment* (MIT Press, 1999). Julia R. Barrett, "Phthalates and Baby Boys," *Environmental Health Perspectives*, (August 2005), report effects of phthalate plastic-softeners on the development of the human male reproductive system. Laboratory work has found that at least one phthalate reduces the number of sperm-generating cells in fetal testis tissue; see Romain Lambrot, et al., "Phthalates Impair Germ Cell Development in the Human Fetal Testis in Vitro without Change in Testosterone Production," *Environmental Health Perspectives* (January 2009). When researchers added synthetic estrogen to a Canadian lake, they found that the reproduction of fathead minnows was so severely affected that the population nearly died out; see Karen A. Kidd, et al., "Collapse of a Fish Population After Exposure to a Synthetic Estrogen," *Proceedings of the National Academy of Sciences of the United States of America* (May 22, 2007). In 1999 the National Research Council published *Hormonally Active Agents in the Environment* (National Academy Press), in which the council's Committee on Hormonally Active Agents in the Environment reports on its evaluation of the scientific evidence pertaining to endocrine disruptors. The National Environmental Health Association has called for more research and product testing; see Ginger L. Gist, "National Environmental Health Association Position on Endocrine Disrupters," *Journal of Environmental Health* (January–February 1998).

Elisabete Silva, Nissanka Rajapakse, and Andreas Kortenkamp, in "Something From 'Nothing'—Eight Weak Estrogenic Chemicals Combined at Concentrations Below NOECs Produce Significant Mixture Effects," *Environmental Science and Technology* (April 2002), find synergistic effects of exactly the kind dismissed by Gough. Also, after reviewing the evidence, the U.S. National Toxicology Program found that low-dose effects had been demonstrated in animals (see Ronald Melnick, et al., "Summary of the National Toxicology Program's Report of the Endocrine Disruptors Low-Dose Peer Review," *Environmental Health Perspectives* [April 2002]). Some researchers say there is reason to think endocrine disruptors are linked to the epidemic of obesity in the U.S. and to a decline in the proportion of male births in the U.S. and Japan (Julia R. Barrett, "Shift in the Sexes," *Environmental Health Perspectives*, June 2007). Retha R. Newbold, et al., "Effects of Endocrine Disruptors on Obesity," *International Journal of Andrology* (April 2008), review the literature on the obesity link and conclude that the evidence is strong enough to warrant expanding "the focus on obesity [and consequent diabetes and heart disease] from intervention and treatment to include prevention and avoidance of these chemical modifiers."

The view that environmental hormone mimics or disruptors have potentially serious effects continues to gain support. Evidence continues to accumulate, and changes are beginning to happen. Plastic bottles, which contain the hormone disruptor bisphenol-A, are being replaced by metal bottles; see Nancy Macdonald, "Plastic Bottles Get the Eco-Boot," *Maclean's* (October 15, 2007). Medical plastics, softened by phthalates, are being reformulated; see "Medical Devices to Get Phthalate Substitution Rule," *European Environment & Packaging Law Weekly* (June 1, 2007).

ISSUE 18

Is the Superfund Program Successfully Protecting Human Health from Hazardous Materials?

YES: **Robert H. Harris, Jay Vandeven, and Mike Tilchin**, from "Superfund Matures Gracefully," *Issues in Science & Technology,* Summer 2003, Vol. 19, Issue 4

NO: **Randall Patterson**, from "Not in Their Back Yard," *Mother Jones* (May/June 2007)

ISSUE SUMMARY

YES: Environmental consultants Robert H. Harris, Jay Vandeven, and Mike Tilchin argue that although the Superfund program still has room for improvement, it has made great progress in risk assessment and treatment technologies.

NO: Journalist Randall Patterson argues that the Superfund Program is not applied to some appropriate situations, largely because people resist its application.

T he potentially disastrous consequences of ignoring hazardous wastes and disposing of them improperly burst upon the consciousness of the American public in the late 1970s. The problem was dramatized by the evacuation of dozens of residents of Niagara Falls, New York, whose health was being threatened by chemicals leaking from the abandoned Love Canal, which was used for many years as an industrial waste dump. Awakened to the dangers posed by chemical dumping, numerous communities bordering on industrial manufacturing areas across the country began to discover and report local sites where chemicals had been disposed of in open lagoons or were leaking from disintegrating steel drums. Such esoteric chemical names as dioxins and PCBs have become part of the common lexicon, and numerous local citizens' groups have been mobilized to prevent human exposure to these and other toxins.

The expansion of the industrial use of synthetic chemicals following World War II resulted in the need to dispose of vast quantities of wastes laden with organic and inorganic chemical toxins. For the most part, industry adopted a casual attitude toward this problem and, in the absence of regulatory restraint, chose the least expensive means of disposal available. Little attention was paid to the ultimate fate of chemicals that could seep into surface water or groundwater. Scientists have estimated that less than 10 percent of the waste was disposed of in an environmentally sound manner.

The magnitude of the problem is truly mind-boggling: Over 275 million tons of hazardous waste is produced in the United States each year; as many as 10,000 dump sites may pose a serious threat to public health, according to the federal Office of Technology Assessment; and other government estimates indicate that more than 350,000 waste sites may ultimately require corrective action at a cost that could easily exceed $500 billion.

Congressional response to the hazardous waste threat is embodied in two complex legislative initiatives. The Resource Conservation and Recovery Act (RCRA) of 1976 mandated action by the Environmental Protection Agency (EPA) to create "cradle to grave" oversight of newly generated waste, and the Comprehensive Environmental Response, Compensation, and Liability Act of 1980 (CERCLA), commonly called "Superfund," gave the EPA broad authority to clean up existing hazardous waste sites. The implementation of this legislation has been severely criticized by environmental organizations, citizens' groups, and members of Congress who have accused the EPA of foot-dragging and a variety of politically motivated improprieties. Less than 20 percent of the original $1.6 billion Superfund allocation was actually spent on waste cleanup. In 2009, the U.S. Inspector General issued a report accusing the EPA of mismanagement; see "Inspector General's Report Faults EPA for Management of 819 Superfund Accounts," *BNA's Environmental Compliance Bulletin* (April 6, 2009) (for EPA commentary, visit http://www.epa.gov/oig/reports/2009/20090318-09-P-0119_glance.pdf).

Amendments designed to close RCRA loopholes were enacted in 1984, and the Superfund Amendments and Reauthorization Act (SARA) added $8.6 billion to a strengthened cleanup effort in 1986 and an additional $5.1 billion in 1990. While acknowledging some improvement, both environmental and industrial policy analysts remain very critical about the way that both RCRA and Superfund/SARA are being implemented. Efforts to reauthorize and modify both of these hazardous waste laws have been stalled in Congress since the early 1990s. But the work went on. The Superfund program continued to identify hazardous waste sites that warranted cleanup and to clean up sites; see http://www.epa.gov/superfund/ for the latest news. In 2004, it even declared that the infamous Love Canal site was finally safe.

In the following selections, Robert H. Harris, Jay Vandeven, and Mike Tilchin argue that the Superfund program has had to struggle with unexpectedly large cleanup tasks such as mines, harbors, and the ruins of the World Trade Center in New York; is subject to greater resource demands than ever before; and still has room for improvement. But, they say, the program has made great progress in risk assessment and treatment technologies. Journalist Randall Patterson makes it clear that not all places made dangerous to human health by chemicals are industrial sites. They can be suburbs such as El Dorado Hills, CA, where ordinary construction activities produce hazardous waste of another sort, in this case dust containing asbestos. He argues that El Dorado Hills has not been labeled a Superfund site largely because people deny there is a problem and resist the EPA's intervention.

YES    Robert H. Harris, Jay Vandeven, and Mike Tilchin

Superfund Matures Gracefully

Superfund, one of the main programs used by the Environmental Protection Agency (EPA) to clean up serious, often abandoned, hazardous waste sites, has been improved considerably in recent years. Notably, progress has been made in two important areas: the development of risk assessments that are scientifically valid yet flexible, and the development and implementation of better treatment technologies.

The 1986 Superfund Amendments and Reauthorization Act (SARA) provided a broad refocus to the program. The act included an explicit preference for the selection of remediation technologies that "permanently and significantly reduce the volume, toxicity, or mobility of hazardous substances." SARA also required the revision of the National Contingency Plan (NCP) that sets out EPA's rules and guidance for site characterization, risk assessment, and remedy selection.

The NCP specifies the levels of risk to human health that are allowable at Superfund sites. However, "potentially responsible parties"—companies or other entities that may be forced to help pay for the cleanup—have often challenged the risk assessment methods used as scientifically flawed, resulting in remedies that are unnecessary and too costly. Since SARA was enacted, fundamental changes have evolved in the policies and science that EPA embraces in evaluating health risks at Superfund sites, and these changes have in turn affected which remedies are most often selected. Among the changes are three that collectively can have a profound impact on the selected remedy and attendant costs: EPA's development of land use guidance, its development of guidance on "principal threats," and the NCP requirement for the evaluation of "short-term effectiveness."

Before EPA's issuance in 1995 of land use guidance for evaluating the potential future public health risks at Superfund sites, its risk assessments usually would assume a future residential use scenario at a site, however unrealistic that assumption might be. This scenario would often result in the need for costly soil and waste removal remedies necessary to protect against hypothetical risks, such as those to children playing in contaminated soil or drinking contaminated ground water, even at sites where future residential use was highly improbable. The revised land use guidance provided a basis for selecting more realistic future use scenarios, with projected exposure patterns that may allow for less costly remedies.

Potentially responsible parties also complained that there was little room to tailor remedies to the magnitude of cancer risk at a site, and that the same costly remedies would be chosen for sites where the cancer risks may differ by several orders of magnitude. However, EPA's guidance on principal threats essentially established a risk-based hierarchy for remedy selection. For example, if cancer risks at a site exceed 1 in 1,000, then treatment or waste removal or both might be required. Sites that posed a lower lifetime cancer risk could be managed in other ways, such as by prohibiting the installation of drinking water wells, which likely would be far less expensive than intrusive remedies.

Revisions to the NCP in 1990 not only codified provisions required by the 1986 Superfund amendments, but also refined EPA's evolving remedy-selection criteria. For example, these revisions require an explicit consideration of the short-term effectiveness of a remedy, including the health and safety risks to the public and to workers associated with remedy implementation. EPA had learned by bitter experience that to ignore implementation risks, such as those associated with vapor and dust emissions during the excavation of wastes, could lead to the selection of remedies that proved costly and created unacceptable risks.

Although these changes in risk assessment procedures have brought greater rationality to the evaluation of Superfund sites, EPA still usually insists on the use of hypothetical exposure factors (for example, the length of time that someone may come in contact with the site) that may overstate risks. The agency has been slow in embracing other methodologies, such as probabilistic exposure analysis, that might offer more accurate assessments. Thus, some remedies are still fashioned on risk analyses that overstate risk.

Technological Evolution

Cleanup efforts in Superfund's early years were dominated by containment and excavation-and-disposal remedies. But over the years, cooperative work by government, industry, and academia have led to the development and implementation of improved treatment technologies.

The period from the mid-1980s to the early 1990s was marked by a dramatic increase in the use of source control treatment, reflecting the preference expressed in SARA for "permanent solutions and alternative treatment technologies or resource recovery technologies to the maximum extent practicable." Two types of source control technologies that have been widely used are incineration and soil vapor extraction. Although the use of incineration decreased during the 1990s because of cost and other factors, soil vapor extraction remains a proven technology at Superfund sites.

Just as early source control remedies relied on containment or excavation and disposal offsite, the presumptive remedy for groundwater contamination has historically been "pump and treat." It became widely recognized in the early 1990s that conventional pump-and-treat technologies had significant limitations, including relatively high costs. What emerged to fill the gap was an approach called "monitored natural attenuation" (MNA), which makes use of a variety of technologies, such as biodegradation, dispersion, dilution, absorption, and volatilization. As the name suggests, monitoring the effectiveness of the process is a key element of this technology. And although cleanup times still may be on the order of years, there is evidence that MNA can achieve comparable results in comparable periods and at significantly lower costs than conventional pump-and-treat systems. EPA has taken an active role in promoting this technology, and its use has increased dramatically in recent years.

As suggested by the MNA example, what may prove an even more formidable challenge than selecting specific remedies is the post-remedy implementation phase—that is, the monitoring and evaluation that will be required during coming decades to ensure that the remedy chosen is continuing to protect human health and the environment. Far too few resources have been devoted to this task, which will require not only monitoring and maintaining the physical integrity of the technology used and ensuring the continued viability of institutional controls, but also evaluating and responding to the developing science regarding chemical detection and toxicity.

Coming Challenges

In recent years, the rate at which waste sites are being added to the National Priorities List (NPL) has been decreasing dramatically as compared with earlier years. In fiscal years 1983 to 1991, EPA placed an average of 135 sites on the NPL annually. The rate dropped to an average of 27 sites per year between 1992 and 2001. Although many factors have contributed to this trend, three stand out:

- There was a finite group of truly troublesome sites before Superfund's passage, and after a few years most of those were identified.
- The program's enforcement authority has had a profound impact on how wastes are managed, significantly reducing, although not eliminating, the types of waste management practices that result in the creation of Superfund sites.
- A range of alternative cleanup programs, such as voluntary cleanup programs and those for brownfields, have evolved at both the federal and state levels. No longer is Superfund the only path for cleaning up sites.

But such programmatic changes are about more than just site numbers. In 1988, most NPL sites were in the investigation stage, and the program was widely criticized as being too much about studies and not enough about cleanup. Superfund is now a program predominantly focused on the design and construction of cleanup remedies.

This shift reflects the natural progress of sites through the Superfund pipeline, the changes in NPL listing activity, and a deliberate emphasis on achieving "construction completion," which is the primary measure of achievement for the program as established under the Government Performance and Results Act. It is a truism in regulatory matters that what gets done is what gets measured, and Superfund is no exception.

In the late 1990s, many observers believed that the demands on Superfund were declining and that it would be completed sometime in the middle of the first decade of the new century. But this is not proving to be true. Although expenditures have not been changing dramatically over time, the resource demands on the program are greater today than ever before.

Few people would have predicted, for example, that among the biggest technical and resource challenges facing Superfund at this date would be the cleanup of hard-rock mining sites and of large volumes of sediments from contaminated waterways and ports. These sites tend to be very costly to clean up, with the driver behind these great costs weighted more toward the protection of natural resources than of human health. In mapping the future course of the program, Congress and EPA must address the question of whether Superfund is the most appropriate program for cleaning up these types of sites.

There are other uncertainties, as well. The substantial role that Superfund has played in emergency response in the aftermath of 9/11, the response to the anthrax attacks of October 2001, and the program's role in the recovery of debris from the crash of the space shuttle Columbia were all totally unforeseeable. Although many valuable lessons have been learned over the past 20 years of the program, there remain substantial opportunities for improvement as well as considerable uncertainty about the kinds of environmental problems Superfund will tackle in the coming decade.

Randall Patterson

Not in Their Back Yard

When the EPA discovered asbestos in their Little League fields, the residents of idyllic El Dorado Hills rushed to protect themselves—from reality.

The land in question is a high, rolling hillside, where California's Central Valley slopes into the Sierra foothills. Many have been drawn here, and after the first stampede, for gold in 1849, most left disappointed. Then the land lay virtually empty for many years until, as the state capital grew, its emptiness became valuable. In the late 1970s, as the local Chamber of Commerce tells it, a developer was traveling along U.S. Highway 50 when, 25 miles east of Sacramento, he looked up and had a vision.

Grand homes started to appear, followed by fine schools, lush golf courses, and a central shopping district modeled after a Tuscan village. The entire hillside was transformed into a patchwork of gated communities, and over the past 15 years, El Dorado Hills, as developers dubbed the spot, has been one of the fastest-growing areas in California. "High per capita income along with low crime rates have attracted many new families to this vibrant and desirable destination of the future," says the Chamber of Commerce's website. "As the name 'El Dorado' translates, this is a land of golden opportunity!"

Some residents of the bedroom community like to think of it as the next Silicon Valley, but the president of the Chamber of Commerce admits there are few jobs, and "the one thing very different about El Dorado Hills" is that no one has to be here. Almost no one leaves either, which is surprising when you consider that what lies beneath these hills isn't gold but something potentially deadly.

Housing developers and real estate agents don't like talking about the dark side of paradise, so I had to rely on Terry Trent to give me an unofficial tour. Trent, a 50-ish retired construction consultant, no longer lives in El Dorado County but admits that his former home has become something of an obsession. "There are some really dangerous things in the ground," he said as we drove down the main street, El Dorado Hills Boulevard. Talking nonstop, Trent pointed to the hillside that forms the spine of the town, which, according to his map from the state's Division of Mines and Geology, contains a vast deposit of naturally occurring asbestos.

There are similar deposits all over California; serpentine, a primary source of asbestos, is the state rock. It's not dangerous—unless it becomes airborne dust and gets in your lungs. The trouble is, in El Dorado Hills, dust isn't just a fact of life but a sign of progress. "Here in the mountains, we don't let no rocks stand in our way," Trent said. Over the past 10 years, as the town's population has doubled to around 35,000, hundreds of bulldozers have scraped over this ground, clearing land for thousands of new homes. In that earth is a form of asbestos—amphibole—that's significantly more toxic than the type commonly used commercially.

Trent's map of the amphibole deposits was made 50 years ago. Still, most residents were unaware of the asbestos until 11 years ago, when Trent uncovered a huge, glistening vein of it in his yard. He went to the local papers with the news, and there was a brief outcry. But the construction went on, and most of El Dorado Hills has been built in the years since.

Turning onto Harvard Way, Trent stopped before a grassy embankment that he claims is shot through with asbestos. "You can see how they carved through it to build this road," he said. He recalled watching as construction crews gouged out a spot for a new middle school nearby, crushing the excavated rock and transporting the gravel to become the foundation for the new community center. A half-mile stretch of Harvard Way was covered in dust, he said, for months.

Nine years ago, tests commissioned by the *Sacramento Bee* found high concentrations of amphibole fibers inside Trent's home. He moved soon afterward. To this day, he still can't understand why few of his former neighbors seem concerned about the threat beneath them, or why the town is still fully occupied, its homes still worth hundreds of thousands of dollars. "I don't like to be mean," he said, "but I'm a little cranky with these people. It's like they're in shock. They don't want this stuff to exist; therefore, they put their heads in the sand. They say, 'Well, it's not here, and it's not bad, and here, I'll prove that it's not here and it's not bad.'"

In the fall of 2004, agents from the Environmental Protection Agency descended on El Dorado Hills in respirators and protective suits and headed for a town park. There, they began playing as children would—tossing baseballs, kicking soccer balls, biking, and running. All the while, they took air samples, and all the while they were quietly watched by the citizens of El Dorado Hills.

The civic leaders of El Dorado Hills had spent many months trying to stave off these tests, scrambling to protect the community not from potentially toxic substances, but from the EPA's potentially toxic information. Taking the lead was Vicki Barber, the superintendent of schools. A stout woman with compressed lips and an unwavering gaze, she recently won an award for being "a person who does not accept the word 'no' when it comes to what is good for students." After asbestos was found during the construction of a high school soccer field in 2002, Barber questioned a costly EPA-mandated cleanup. When a citizen formally asked the EPA to test the town's public areas for asbestos in 2003, Barber quickly emerged as the agency's most determined local foe. Before the study was even under way, she began writing to the EPA as well as to senators and congressmen, questioning whether the agency had the "legal and scientific authority" to conduct what she called a "science experiment" with "limited benefit to the residents." At least four state legislators and one congressman responded by putting pressure on the EPA, which in turn agreed not to declare El Dorado Hills a Superfund site, regardless of what it might find there.

In May 2005, the EPA announced its findings: Almost every one of more than 400 air samples collected at the park, as well as along a hiking trail and in three school yards, contained asbestos fibers. Most of the samples contained amphibole fibers known as tremolite—long, thin strands that, when inhaled, tend to wedge in the lungs for much longer and at higher concentrations than other forms of asbestos, even at very low exposure levels.

The correlation between amphibole asbestos and lung disease has been demonstrated around the world, from Canada to Cyprus, and perhaps most convincingly in the mining town of Libby, Montana, where hundreds of residents have been diagnosed with asbestos-related diseases and many have died. After reviewing the EPA's findings in El Dorado Hills, a Canadian epidemiologist told the *Fresno Bee* that the area's exposure levels were comparable to those in towns where mining had gone on for a century. "You can certainly say people are going to die, and there are going to be increased cases of cancer," he said. "I wouldn't live there. I wouldn't want my family to live there."

The EPA, however, avoided making such dire statements. Its report neither specified how toxic the samples were nor quantified the health risk to the residents of El Dorado Hills. Rather, it simply noted that the exposure levels were "of concern." Jere Johnson, who oversaw the study, explained that the agency was trying not to sound alarmist. "We wanted to inform the community without scaring them," she said.

But local officials received the EPA's findings almost as a declaration of war on El Dorado Hills and its way of life. Jon Morgan, the director of the county's environmental management department, thundered to a local television station that the tremolite announcement "may unnecessarily scare the living daylights out of every man, woman, and child in El Dorado County and could possibly devastate the county for years to come."

When I visited El Dorado Hills, more than a year after the EPA had run its tests, I was told over and over that the biggest threat facing the town was big-government intrusion. "The issue is not asbestos," declared James Sweeney, chairman of the county Board of Supervisors. "The issue is the EPA." Superintendent Barber said that El Dorado Hills is like any other community in California, "and yet why is it that the EPA has decided to focus on only here?" One county supervisor claimed that an EPA official had told her he was going to make El Dorado Hills "a poster child for asbestos." "What I heard," town general manager Wayne Lowery said, "is that the Libby, Montana, project is being wrapped up and they have all these employees with training, and they're looking for a place to keep them employed."

What were officials supposed to do with the EPA's information, wondered Debbie Manning, the president of the Chamber of Commerce. "Are you just going to make it a ghost town?" There was no need to rush to judgment until all the facts had been determined. "I'm sure you know that asbestos is a complicated issue," she told me, explaining that there had been a locally led effort to get the information out about asbestos. Fortunately, El Dorado Hills is a highly educated community, she said, and you can trust that "with good information, people can make proper decisions."

One of those struggling to do just that was Vicki Summers, a 47-year-old mom who said that her chief duty was to protect her two children from "any type of health issues." Surrounded by stuffed animals, she sat on a couch in her spacious living room, her eyes darting as she spoke of the dangers they faced. "Every day in the newspaper, there's something new," she said—chemicals in the water, mad cow disease, bird flu. She often felt overwhelmed, especially when experts said different things. Summers had guzzled green tea for its anti-cancer benefits until she heard it caused colon cancer. She had gorged on fish during her first pregnancy, thinking it was "brain food," only to be told, while expecting her second child, that mercury could cause brain damage.

The same sort of thing had happened to her dad. When she was a girl, he had made a point of buying margarine because he wanted to protect his family from cholesterol. "And now he's had three strokes," she explained, "and they say it's the plaque buildup from the hydrogenated vegetable oil." So much conflicting information was almost paralyzing. "It would be better to be ignorant," she said, "because then you wouldn't have to be stressed out by all this."

Summers called her home her "dream house," and said she had long considered El Dorado Hills "heaven on Earth," but when she heard about the asbestos, her first thought was, "If this is Love Canal, and there's going to be a mass exodus, I don't want to be the last one out."

But no exodus from El Dorado Hills ever began. Many residents seemed oddly comforted by the apparent uncertainty of the situation. Just after the EPA report came out, Summers joined a thousand other citizens in the town gym, looking for answers. There, school superintendent Barber reassured the crowd that the town was "deeply committed to maintaining public health and safety." Then she ripped into the EPA report for offering no solid "risk information." She pointed out there had been no abundance of "pulmonary cases" in the area. In other words, with nobody getting sick or dying, there was no real evidence of any hazard, so why worry? "Risk is a part of all of our lives," Barber said. "But we also need to keep it in perspective." She received enthusiastic applause.

Scientists, however, are more certain about the dangers of asbestos. There is no known safe exposure level to asbestos, whether it is in a commercial or a natural form; even low doses can cause malignant mesothelioma, a rare lung cancer. And the evidence keeps mounting. In the summer of 2005, researchers from the University of California-Davis Department of Public Health Sciences announced the findings of a study comparing the addresses of 2,900 Californians suffering from malignant mesothelioma against a geological map of the state. They concluded that the risk of developing lung cancer was directly related to how close the patients had lived to areas of rock associated with naturally occurring tremolite asbestos. (Likewise, the odds of getting mesothelioma drop 1 percent for every mile one moves away from an asbestos source.) As one of the authors explained, "We showed that breathing asbestos in your community is not magically different from breathing asbestos in an industrial setting."

All of this passed over El Dorado Hills like clouds in the blue sky. County supervisor Helen Baumann's reaction to the UC Davis report was that "there are studies counter to that." Local government would "err on the side of public health and safety," she said, "until some better science comes forward."

So began the effort to live in El Dorado Hills without coming into contact with the earth. In the park and the school playing fields where the EPA had collected samples, workers trucked in clean soil to replace two feet of topsoil, as the agency had recommended. The county banned using leaf blowers on town property, "except for emergency situations." Residents were encouraged to take precautions such as removing their shoes before entering their homes, driving slowly with the windows up on unpaved roads, and "limiting time spent on dirt." Home-builders agreed to spray down building sites and rinse off their trucks' tires after work. Any sightings of "fugitive dust" were to be reported to a new "dust enforcement" team.

Life in El Dorado Hills under the new regime was "manageable," said Baumann. That seemed to be the point of the show: If you took a few precautions, there was no need to panic. She proudly pointed out that the dust hot line rarely received a call. Gerri Silva, the interim director of the county department of environmental management, told me the measures had worked so well that in all of El Dorado Hills, "there is no dust."

In the town's main coffee shop, Bella Bru, languor pervaded as well. An engineer named Matt Parisek said that asbestos was "pretty muck a fake issue." "It's pretty obvious [the EPA] selected this county because of our conservative Republican reputation," he said. And, wondered retiree Carole Gilmore, "If asbestos is all over California, I don't know why they zero in on El Dorado Hills."

Real estate broker Charles Hite, who is the current resident of Terry Trent's old house, pointed out that none of his neighbors had died. "If it is such a health hazard," he asked, "why are people still buying and building? Why are real estate prices going up? Why doesn't government shut the city down?"

There was a circular logic at work. The local government had told everyone to stay calm, and so residents weren't afraid. And if no one was scared, then perhaps there was nothing to fear. "The vast, vast majority" trusted that they were safe, said Baumann. But, she added, there remained "a very, very small centralized group that keeps pushing issues and pushing issues to the extreme." She specifically mentioned Terry Trent, who to this day sends regular, dire emails to residents, reporters, and scientists. The group of extremists who "do not represent our community" also included, in Baumann's view, some members of a community group that meets monthly to discuss asbestos.

Vicki Summers had joined the Asbestos Community Advisory Group as part of her effort to educate herself. When someone would mention the low body count as evidence that nothing was wrong, she could now cite the UC Davis study, or note that the latency period for asbestos-related diseases is about 30 years. Who lived in these hills back then, she asked, "a dozen ranchers?" Other parents assumed that if they couldn't see asbestos in the air, it wasn't there; Summers could picture invisible fibers burrowing into her kids' lungs. "It never disappears," she said, wide-eyed. "That's the scary thing. And it's a known carcinogen. It's a known carcinogen. And if we know that, we know it's not good for our kids."

Summers said she was thinking only of her family when, at a community meeting, she passed a note to an EPA official, asking, "Should I move?" After that was reported by a newspaper, she received an answer in an anonymous, late-night phone call: "Vicki Summers, I think you should move!"

Short of evacuating, there are only two ways to solve El Dorado Hills' asbestos problem: Either pave over every asbestos deposit in town—an impossible task—or make the asbestos magically disappear. Soon after the EPA report came out, Superintendent Barber got in touch with the National Stone, Sand and Gravel Association, an industry trade group, which commissioned the R.J. Lee Group to scrutinize the EPA's data. In 2000, it was reported that R.J. Lee's president, Richard Lee, had been paid about $7 million for testifying more than 250 times on behalf of the asbestos industry. When the *Seattle Post-Intelligencer* found asbestos in Crayola crayons, it was Lee who did a study that discovered none. When people began getting sick and dying in Libby, Montana, R.J. Lee found that local asbestos levels had been overstated by the EPA.

As in Libby, R.J. Lee declared that what the EPA had been calling asbestos in El Dorado Hills was, in fact, not asbestos at all. Some of the fibers were not the proper size; others contained a bit too much aluminum. The EPA had, in other words, completely goofed.

Supervisor Baumann hailed "a study that has profound information in it." Barber bustled off to Washington, D.C., at taxpayer expense, carrying news of "this startling scientific development." And Wayne Lowery, the town manager, admitted it was hard to get excited about controlling asbestos "when you've got two different interpretations of the same data."

At a meeting of the Asbestos Community Advisory Group in the fire station last winter, Summers and a handful of other residents met with the EPA's Jere Johnson to try to understand how asbestos could vanish before their eyes. Two builders were also at the table. Baumann seemed to have been thinking of them when, at an earlier county supervisors' meeting, she had issued a "public thank you" to "the very, very smart people attending those meetings, making sure another thought process is heard."

One of the builders introduced R.J. Lee's report and started to expound on "cleavage fragments," mineral composition, microns, and the like. "I don't want to get bogged down with these semantics," a fireman interrupted. "We know we have it here." The group's chairman shushed him, though, saying he didn't want to get bogged down in semantics either, "but I think we almost have to." The builder persisted with his mantra of scientific uncertainty. "We all wish science to be this nice black-and-white thing," he said, "and sometimes it isn't." The debate seemed essentially over by the time the EPA's Johnson clarified the topic: It was not geology but public health. "The body can't tell the difference between a fragment and a fiber," Johnson said. Whatever you called it, it would still get stuck in your lungs.

For Summers, the starkness of the issue had again faded to gray. When she started looking for a safer place to live, she was alarmed at first to realize there were cancer clusters and crime everywhere, and then she had taken comfort in danger. If nowhere on Earth was completely safe, she reasoned, why should she leave El Dorado Hills? Perhaps it was as Vicki Barber had said: Risk is a part of all of our lives. Her family could stay, she decided, as long as she took precautions, such as wiping down her home with damp rags, and kept pushing to get all the facts. And if new studies revealed the asbestos to be "just fragments and dust," well, she admitted, she'd be "thrilled to death."

But she knew otherwise, didn't she? She knew that asbestos was here, and she knew that it was bad? The question silenced her. "I don't want to know it, though," she finally answered. "I don't want to know it. That's when I can't sleep. I mean, I love it here. I love it here. That's why it's hard. Part of me wants to be in denial."

After reviewing R.J. Lee's report on El Dorado Hills, the EPA declared last spring that the company had violated "generally accepted scientific principles." R.J. Lee, in turn, discovered "a number of important differences of opinion as well as factual misstatements" in the EPA's response. The EPA at last asked the U.S. Geological Survey to step in and do its own testing. In December, the USGS announced that its analysis confirmed that "material that can be classified as tremolite asbestos is" in El Dorado Hills. But the geologists were uncomfortable assessing the health risks, and so the controversy continues.

Supervisor Baumann thought the EPA should just leave El Dorado Hills be. Her constituents had "really calmed down, been educated." Why, she wondered, would anyone want to excite them again? The public health debate seemed to have come full circle when Sweeney, the chairman of the Board of Supervisors, griped about the "extreme cost" of dust controls, saying, "We're putting our public at risk by telling them to do things that are absolutely unnecessary."

Up on the ridge, behind the high school, Terry Trent stood beside a driveway with a piece of tremolite in his hand. As he tossed the rock back to the ground, there was the sound of a small engine starting. "Watch this," Trent said, and a smile broke over his face as a gardener passed by with a leaf blower, dust billowing all around.

POSTSCRIPT

Is the Superfund Program Successfully Protecting Human Health from Hazardous Materials?

Superfund cleanups, when those responsible for contaminated sites could not pay or could not be found, were to have been funded by taxes on industry (e.g., the Crude Oil Tax, the Chemical Feedstock Tax, the Toxic Chemicals Importation Tax, and the Corporate Environmental Income Tax). These taxes expired in 1995, and Congress has so far refused to reauthorize them, but the Obama Administration's 2010 budget presumes that the tax will be reinstated soon; see "Obama Budget Would Reinstate Superfund Tax," *Chemical Week* (March 2, 2009). Meanwhile, the Environmental Protection Agency has received $600 million of economic stimulus funds to clean up 50 superfund sites. Some of these sites will also benefit from federal, state, and private funds designed to turn "brownfields" into new industrial and commercial developments; see JoAnn Greco, "Brown Is the New Green," *Planning* (March 2009).

The Competitive Enterprise Institute objects that taxes such as the superfund taxes are an assault on consumer pocketbooks, as is the Comprehensive Environmental Response, Compensation, and Liability Act's (CERCLA's) "joint and several liability" clause, which can make minor contributors to toxics sites liable for large cleanup costs even when they acted according to all laws and regulations in force at the time. However, in May 2009, the U.S. Supreme Court ruled that some contributors may not be liable for cleanup costs; see "US Supreme Court Ruling Limits Superfund Liability," *Oil Spill Intelligence Report* (May 7, 2009).

Meanwhile, the hazardous waste problem takes new forms. Even in the 1990s, an increasingly popular method of disposing of hazardous wastes was to ship them from the United States to "dumping grounds" in developing countries. Iwonna Rummel-Bulska's "The Basel Convention: A Global Approach for the Management of Hazardous Wastes," *Environmental Policy and Law* (vol. 24, no. 1, 1994) describes an international treaty designed to prevent or at least limit such waste dumping. But eight years later, in February 2002, the Basel Action Network and the Silicon Valley Toxics Coalition (http://svtc.org) published *Exporting Harm: The High-Tech Trashing of Asia*. This lengthy report documents the shipping of electronics wastes, including defunct personal computers, monitors, and televisions, as well as circuit boards and other products rich in lead, beryllium, cadmium, mercury, and other toxic materials. Some 50 to 80 percent of the "e-waste" collected for recycling in the western United States is shipped to destinations such as China, India, and Pakistan, where recycling and disposal methods lead to widespread human and environmental contamination. An updated version of this report, *Poison PCs and Toxic TVs: E-Waste Tsunami to Roll Across the US: Are We Prepared?*, was released in February 2004. In August 2005, Greenpeace released K. Brigden, et al., *Recycling of Electronic Wastes in China and India: Workplace and Environmental Contamination* (http://www.e-takeback.org/press_open/greenpeace.pdf). Currently the problem of electronic waste continues to grow; see Elisabeth Jeffries, "E-Wasted," *World Watch* (July/August 2006).

Among the solutions that have been urged to address the hazardous waste problem are "take-back" and "remanufacturing" practices. Gary A. Davis and Catherine A. Wilt, in "Extended Product Responsibility," *Environment* (September 1997), urge such solutions as crucial to the minimization of waste and describe how they are becoming more common in Europe. Brad Stone, "Tech Trash, E-Waste: By Any Name, It's an Issue," *Newsweek* (December 12, 2005), describes their appearance in the United States, where some states are making take-back programs mandatory for computers, televisions, and other electronic devices. After *Exporting Harm* was published and drew considerable attention from the press, some industry representatives hastened to emphasize such practices as Hewlett Packard's recycling of printer ink cartridges. See Doug Bartholomew's "Beyond the Grave," *Industry Week* (March 1, 2002), which also stressed the need to minimize waste by intelligent design. The Institute of Industrial Engineers published in its journal *IIE Solutions* Brian K. Thorn's and Philip Rogerson's "Take It Back" (April 2002), which stressed the importance of designing for reuse or remanufacturing. Anthony Brabazon and Samuel Idowu, in "Costing the Earth" *Financial Management (CIMA)* (May 2001), note that "take-back schemes may [both] provide opportunities to build goodwill and [help] companies to use resources more efficiently."

ISSUE 19

Should the United States Reprocess Spent Nuclear Fuel?

YES: Phillip J. Finck, from Statement before the House Committee on Science, Energy Subcommittee, Hearing on Nuclear Fuel Reprocessing (June 16, 2005)

NO: Charles D. Ferguson, "An Assessment of the Proliferation Risks of Spent Fuel Reprocessing and Alternative Nuclear Waste Management Strategies," from Testimony before the U.S. House of Representatives Committee on Science and Technology Hearing on Advancing Technology for Nuclear Fuel Recycling: What Should Our Research, Development and Demonstration Strategy Be? (June 17, 2009).

ISSUE SUMMARY

YES: Phillip J. Finck argues that by reprocessing spent nuclear fuel, the United States can enable nuclear power to expand its contribution to the nation's energy needs while reducing carbon emissions, nuclear waste, and the need for waste repositories such as Yucca Mountain.

NO: Charles D. Ferguson, Philip D. Reed senior fellow for science and technology at the Council on Foreign Relations, argues that even though reprocessing can help reduce nuclear waste management problems, because as currently practiced it both poses a significant risk that weapons-grade material will fall into the wrong hands and raises the price of nuclear fuel (compared to the once-through fuel cycle), it should not be pursued at present.

$\mathbf{A}$s nuclear reactors operate, the nuclei of uranium-235 atoms split, releasing neutrons and nuclei of smaller atoms called fission products, which are themselves radioactive. Some of the neutrons are absorbed by uranium-238, which then becomes plutonium. The fission product atoms eventually accumulate to the point where the reactor fuel no longer releases as much energy as it used to. It is said to be "spent." At this point, the spent fuel is removed from the reactor and replaced with fresh fuel.

Once removed from the reactor, the spent fuel poses a problem. Currently it is regarded as high-level nuclear waste which must be stored on the site of the reactor, initially in a swimming pool-sized tank and later, once the radioactivity levels have subsided a bit, in "dry casks." Until 2009, there was a plan to dispose of the casks permanently in a subterranean repository being built at Yucca Mountain, Nevada, but the Obama administration has proposed ending funding for Yucca Mountain (see Dan Charles, "A Lifetime of Work Gone to Waste?" *Science,* March 20, 2009).

It is worth noting that spent fuel still contains useful components. Not all the uranium-235 has been burned up, and the plutonium created as fuel is burned, which can itself be used as fuel. When spent fuel is treated as waste, these components of the waste are discarded. Early in the Nuclear Age, it was thought that if these components could be recovered, the amount of waste to be disposed could be reduced; the fuel supply could also be extended. In fact, because plutonium is made from otherwise useless uranium-238, new fuel could be created. Reactors designed to maximize plutonium creation, known as "breeder" reactors because they "breed" fuel, were built and are still in use as power plants in Europe. In the U.S., breeder reactors have been built and operated only by the Department of Defense, because plutonium extracted from spent fuel is required for making nuclear bombs. They have not seen civilian use in part because of the fear that bomb-grade material could fall into the wrong hands.

The separation and recycling of unused fuel from spent fuel is known as reprocessing. In the United States, a reprocessing plant operated in West Valley, New York, from 1966 to 1972 (see "Plutonium Recovery from Spent Fuel Reprocessing by Nuclear Fuel Services at West Valley, New York, from 1966 to 1972" (DOE, 1996; https://www.osti.gov/opennet/forms.jsp?formurl=document/purecov/nfsrepo.html). After the Nuclear Nonproliferation Treaty went into force in 1970, it became U.S. policy not to reprocess spent nuclear fuel and thereby to limit the availability of bomb-grade material. As a consequence, spent fuel was not recycled; it was regarded as high-level waste to be disposed of, and the waste continued to accumulate.

The Yucca Mountain nuclear waste disposal site has proven to be controversial, and there remains a great deal of resistance to storing waste there. But the need to dispose of nuclear waste is not about to go away, especially if the United States expands its reliance on nuclear power (see Issue 12). This simple truth drives much of the discussion of nuclear fuel reprocessing. In the following selections, Phillip J. Finck argues that by reprocessing spent nuclear fuel, the United States can enable nuclear power to expand its contribution to the nation's energy needs while reducing carbon emissions, nuclear waste, and the need for waste repositories such as Yucca Mountain. Charles D. Ferguson and Philip D. Reed argues that even though reprocessing can help reduce nuclear waste management problems, because as currently practiced it both poses a significant risk that weapons-grade material will fall into the wrong hands and raises the price of nuclear fuel (compared to the once-through fuel cycle), it should not be pursued at present. There is time for further research. Meanwhile, we should concentrate our efforts on safe storage of nuclear wastes.

YES

Phillip J. Finck

Statement before the House Committee on Science, Energy Subcommittee, Hearing on Nuclear Fuel Reprocessing

Summary

Management of spent nuclear fuel from commercial nuclear reactors can be addressed in a comprehensive, integrated manner to enable safe, emissions-free, nuclear electricity to make a sustained and growing contribution to the nation's energy needs. Legislation limits the capacity of the Yucca Mountain repository to 70,000 metric tons from commercial spent fuel and DOE defense-related waste. It is estimated that this amount will be accumulated by approximately 2010 at current generation rates for spent nuclear fuel. To preserve nuclear energy as a significant part of our future energy generating capability, new technologies can be implemented that allow greater use of the repository space at Yucca Mountain. By processing spent nuclear fuel and recycling the hazardous radioactive materials, we can reduce the waste disposal requirements enough to delay the need for a second repository until the next century, even in a nuclear energy growth scenario. Recent studies indicate that such a closed fuel cycle may require only minimal increases in nuclear electricity costs, and are not a major factor in the economic competitiveness of nuclear power (The University of Chicago study, "The Economic Future of Nuclear Power," August 2004). However, the benefits of a closed fuel cycle can not be measured by economics alone; resource optimization and waste minimization are also important benefits. Moving forward in 2007 with an engineering-scale demonstration of an integrated system of proliferation-resistant, advanced separations and transmutation technologies would be an excellent first step in demonstrating all of the necessary technologies for a sustainable future for nuclear energy.

Nuclear Waste and Sustainability

World energy demand is increasing at a rapid pace. In order to satisfy the demand and protect the environment for future generations, energy sources must evolve from the current dominance of fossil fuels to a more balanced,

U.S. House of Representatives, June 16, 2005.

sustainable approach. This new approach must be based on abundant, clean, and economical energy sources. Furthermore, because of the growing worldwide demand and competition for energy, the United States vitally needs to establish energy sources that allow for energy independence.

Nuclear energy is a carbon-free, secure, and reliable energy source for today and for the future. In addition to electricity production, nuclear energy has the promise to become a critical resource for process heat in the production of transportation fuels, such as hydrogen and synthetic fuels, and desalinated water. New nuclear plants are imperative to meet these vital needs.

To ensure a sustainable future for nuclear energy, several requirements must be met. These include safety and efficiency, proliferation resistance, sound nuclear materials management, and minimal environmental impacts. While some of these requirements are already being satisfied, the United States needs to adopt a more comprehensive approach to nuclear waste management. The environmental benefits of resource optimization and waste minimization for nuclear power must be pursued with targeted research and development to develop a successful integrated system with minimal economic impact. Alternative nuclear fuel cycle options that employ separations, transmutation, and refined disposal (e.g., conservation of geologic repository space) must be contrasted with the current planned approach of direct disposal, taking into account the complete set of potential benefits and penalties. In many ways, this is not unlike the premium homeowners pay to recycle municipal waste.

The spent nuclear fuel situation in the United States can be put in perspective with a few numbers. Currently, the country's 103 commercial nuclear reactors produce more than 2000 metric tons of spent nuclear fuel per year (masses are measured in heavy metal content of the fuel, including uranium and heavier elements). The Yucca Mountain repository has a legislative capacity of 70,000 metric tons, including spent nuclear fuel and DOE defense-related wastes. By approximately 2010 the accumulated spent nuclear fuel generated by these reactors and the defense-related waste will meet this capacity, even before the repository starts accepting any spent nuclear fuel. The ultimate technical capacity of Yucca Mountain is expected to be around 120,000 metric tons, using the current understanding of the Yucca Mountain site geologic and hydrologic characteristics. This limit will be reached by including the spent fuel from current reactors operating over their lifetime. Assuming nuclear growth at a rate of 1.8% per year after 2010, the 120,000 metric ton capacity will be reached around 2030. At that projected nuclear growth rate, the U.S. will need up to nine Yucca Mountain-type repositories by the end of this century. Until Yucca Mountain starts accepting waste, spent nuclear fuel must be stored in temporary facilities, either storage pools or above ground storage casks.

Today, many consider repository space a scarce resource that should be managed as such. While disposal costs in a geologic repository are currently quite affordable for U.S. electric utilities, accounting for only a few percent of the total cost of electricity, the availability of U.S. repository space will likely remain limited.

Only three options are available for the disposal of accumulating spent nuclear fuel:

- Build more ultimate disposal sites like Yucca Mountain.
- Use interim storage technologies as a temporary solution.
- Develop and implement advanced fuel cycles, consisting of separation technologies that separate the constituents of spent nuclear fuel into elemental streams, and transmutation technologies that destroy selected elements and greatly reduce repository needs.

A responsible approach to using nuclear power must always consider its whole life cycle, including final disposal. We consider that temporary solutions, while useful as a stockpile management tool, can never be considered as ultimate solutions. It seems prudent that the U.S. always have at least one set of technologies available to avoid expanding geologic disposal sites.

Spent Nuclear Fuel

The composition of spent nuclear fuel poses specific problems that make its ultimate disposal challenging. Fresh nuclear fuel is composed of uranium dioxide (about 96% U238, and 4% U235). During irradiation, most of the U235 is fissioned, and a small fraction of the U238 is transmuted into heavier elements (known as "transuranics"). The spent nuclear fuel contains about 93% uranium (mostly U238), about 1% plutonium, less than 1% minor actinides (neptunium, americium, and curium), and 5% fission products. Uranium, if separated from the other elements, is relatively benign, and could be disposed of as low-level waste or stored for later use. Some of the other elements raise significant concerns:

- The fissile isotopes of plutonium, americium, and neptunium are potentially usable in weapons and, therefore, raise proliferation concerns. Because spent nuclear fuel is protected from theft for about one hundred years by its intense radioactivity, it is difficult to separate these isotopes without remote handling facilities.
- Three isotopes, which are linked through a decay process (Pu241, Am241, and Np237), are the major contributors to the estimated dose for releases from the repository, typically occurring between 100,000 and 1 million years, and also to the long-term heat generation that limits the amount of waste that can be placed in the repository.
- Certain fission products (cesium, strontium) are major contributors to the repository's shortterm heat load, but their effects can be mitigated by providing better ventilation to the repository or by providing a cooling-off period before placing them in the repository.
- Other fission products (Tc99 and I129) also contribute to the estimated dose.

The time scales required to mitigate these concerns are daunting: several of the isotopes of concern will not decay to safe levels for hundreds of thousands of years. Thus, the solutions to long-term disposal of spent nuclear fuel are limited to three options: the search for a geologic environment that will remain stable for that period; the search for waste forms that can contain these elements for that period; or the destruction of these isotopes. These three options underlie the major fuel cycle strategies that are currently being developed and deployed in the U.S. and other countries.

Options for Disposing of Spent Nuclear Fuel

Three options are being considered for disposing of spent nuclear fuel: the once-through cycle is the U.S. reference; limited recycle has been implemented in France and elsewhere and is being deployed in Japan; and full recycle (also known as the closed fuel cycle) is being researched in the U.S., France, Japan, and elsewhere.

1. Once-through Fuel Cycle

This is the U.S. reference option where spent nuclear fuel is sent to the geologic repository that must contain the constituents of the spent nuclear fuel for hundreds of thousands of years. Several countries have programs to develop these repositories, with the U.S. having the most advanced program. This approach is considered safe, provided suitable repository locations and space can be found. It should be noted that other ultimate disposal options have been researched (e.g., deep sea disposal; boreholes and disposal in the sun) and abandoned. The challenges of long-term geologic disposal of spent nuclear fuel are well recognized, and are related to the uncertainty about both the long-term behavior of spent nuclear fuel and the geologic media in which it is placed.

2. Limited Recycle

Limited recycle options are commercially available in France, Japan, and the United Kingdom. They use the PUREX process, which separates uranium and plutonium, and directs the remaining transuranics to vitrified waste, along with all the fission products. The uranium is stored for eventual reuse. The plutonium is used to fabricate mixed-oxide fuel that can be used in conventional reactors. Spent mixed-oxide fuel is currently not reprocessed, though the feasibility of mixed-oxide reprocessing has been demonstrated. It is typically stored or eventually sent to a geologic repository for disposal. Note that a reactor partially loaded with mixed-oxide fuel can destroy as much plutonium as it creates. Nevertheless, this approach always results in increased production of americium, a key contributor to the heat generation in a repository. This approach has two significant advantages:

- It can help manage the accumulation of plutonium.
- It can help significantly reduce the volume of spent nuclear fuel (the French examples indicate that volume decreases by a factor of 4).

Several disadvantages have been noted:

- It results in a small economic penalty by increasing the net cost of electricity a few percent.

- The separation of pure plutonium in the PUREX process is considered by some to be a proliferation risk; when mixed-oxide use is insufficient, this material is stored for future use as fuel.
- This process does not significantly improve the use of the repository space (the improvement is around 10%, as compared to a factor of 100 for closed fuel cycles).
- This process does not significantly improve the use of natural uranium (the improvement is around 15%, as compared to a factor of 100 for closed fuel cycles).

3. Full Recycle (the Closed Fuel Cycle)

Full recycle approaches are being researched in France, Japan, and the United States. This approach typically comprises three successive steps: an advanced separations step based on the UREX+ technology that mitigates the perceived disadvantages of PUREX, partial recycle in conventional reactors, and closure of the fuel cycle in fast reactors.

The first step, UREX+ technology, allows for the separations and subsequent management of highly pure product streams. These streams are:

- Uranium, which can be stored for future use or disposed of as low-level waste.
- A mixture of plutonium and neptunium, which is intended for partial recycle in conventional reactors followed by recycle in fast reactors.
- Separated fission products intended for short-term storage, possibly for transmutation, and for long-term storage in specialized waste forms.
- The minor actinides (americium and curium) for transmutation in fast reactors.

The UREX+ approach has several advantages:

- It produces minimal liquid waste forms, and eliminates the issue of the "waste tank farms."
- Through advanced monitoring, simulation and modeling, it provides significant opportunities to detect misuse and diversion of weapons-usable materials.
- It provides the opportunity for significant cost reduction.
- Finally and most importantly, it provides the critical first step in managing all hazardous elements present in the spent nuclear fuel.

The second step—partial recycle in conventional reactors—can expand the opportunities offered by the conventional mixed-oxide approach. In particular, it is expected that with significant R&D effort, new fuel forms can be developed that burn up to 50% of the plutonium and neptunium present in spent nuclear fuel. (Note that some studies also suggest that it might be possible to recycle fuel in these reactors many times—i.e., reprocess and recycle the irradiated advanced fuel—and further destroy plutonium and neptunium; other studies also suggest possibilities for transmuting americium in these reactors. Nevertheless, the practicality of these schemes is not yet established and requires additional scientific and engineering research.) The advantage of the second step is that it reduces the overall cost of the closed fuel cycle by burning plutonium in conventional reactors, thereby reducing the number of fast reactors needed to complete the transmutation mission of minimizing hazardous waste. This step can be entirely bypassed, and all transmutation performed in advanced fast reactors, if recycle in conventional reactors is judged to be undesirable.

The third step, closure of the fuel cycle using fast reactors to transmute the fuel constituents into much less hazardous elements, and pyroprocessing technologies to recycle the fast reactor fuel, constitutes the ultimate step in reaching sustainable nuclear energy. This process will effectively destroy the transuranic elements, resulting in waste forms that contain only a very small fraction of the transuranics (less than 1%) and all fission products. These technologies are being developed at Argonne National Laboratory and Idaho National Laboratory, with parallel development in Japan, France, and Russia.

The full recycle approach has significant benefits:

- It can effectively increase use of repository space by a factor of more than 100.
- It can effectively increase the use of natural uranium by a factor of 100.
- It eliminates the uncontrolled buildup of all isotopes that are a proliferation risk.
- The fast reactors and the processing plant can be deployed in small co-located facilities that minimize the risk of material diversion during transportation.
- The fast reactor does not require the use of very pure weapons usable materials, thus increasing their proliferation resistance.
- It finally can usher the way towards full sustainability to prepare for a time when uranium supplies will become increasingly difficult to ensure.
- These processes would have limited economic impact; the increase in the cost of electricity would be less than 10% (ref: OECD).
- Assuming that demonstrations of these processes are started by 2007, commercial operations are possible starting in 2025; this will require adequate funding for demonstrating the separations, recycle, and reactor technologies.
- The systems can be designed and implemented to ensure that the mass of accumulated spent nuclear fuel in the U.S. would always remain below 100,000 metric tons—less than the technical capacity of Yucca Mountain—thus delaying, or even avoiding, the need for a second repository in the U.S.

Conclusion

A well engineered recycling program for spent nuclear fuel will provide the United States with a long-term, affordable, carbon-free energy source with low environmental impact. This new paradigm for nuclear power will allow us to manage nuclear waste and reduce proliferation risks while creating a sustainable energy supply. It is possible that the cost of recycling will be slightly higher than direct disposal of spent nuclear fuel, but the nation will only need one geologic repository for the ultimate disposal of the residual waste.

Charles D. Ferguson  **NO**

An Assessment of the Proliferation Risks of Spent Fuel Reprocessing and Alternative Nuclear Waste Management Strategies

U.S. leadership is essential for charting a constructive and cooperative international course to prevent nuclear proliferation. An essential aspect of that leadership involves U.S. policy on reprocessing spent nuclear fuel. The United States has sought to prevent the spread of reprocessing facilities to other countries and to encourage countries with existing stockpiles of separated plutonium from reprocessing facilities to draw down those stockpiles. The previous administration launched the Global Nuclear Energy Partnership (GNEP), which proposed offering complete nuclear fuel services, including provision of fuel and waste management, from fuel service states to client states in order to discourage the latter group from enriching uranium or reprocessing spent nuclear fuel—activities that would contribute to giving these countries latent nuclear weapons programs. The current administration and the Congress seek to determine the best course for U.S. nuclear energy policy with the focus of this hearing on recycling or reprocessing of spent fuel and nuclear waste management strategies.

Here at the start, I give a brief summary of the testimony's salient points:

- Reprocessing of the type currently practiced in a handful of countries poses a significant proliferation threat because of the separation of plutonium from highly radioactive fission products. A thief, if he had access, could easily carry away separated plutonium. Fortunately, this reprocessing is confined to nuclear-armed states except for Japan. If this practice spreads to other non-nuclear-weapon states the consequences for national and international security could be dire. Presently, the vast majority of the 31 states with nuclear power programs do not have reprocessing plants.
- The types of reprocessing examined under GNEP do not appear to offer substantial proliferation-resistant benefits, according to research sponsored by the Department of Energy. However, more research is needed to determine what additional safeguards, if any, could provide greater assurances that reprocessing methods are not misused in weapons

U.S. Senate, June 17, 2009.

programs and whether it is possible to have assurances of timely detection of a diversion of a significant quantity of plutonium or other fissile material.

- Time is on the side of the United States. There is no need to rush toward development and deployment of recycling of spent nuclear fuel. Based on the foreseeable price for uranium and uranium enrichment services, this practice is presently far more expensive than the once-through uranium fuel cycle. Nonetheless, more research is needed to determine the costs and benefits of recycling techniques coupled with fast-neutron reactors or other types of reactor technologies. This cost versus benefit analysis would concentrate on the capability of these technologies to help alleviate the nuclear waste management challenge.

- In related research, there is a need to better understand the safeguards challenges in the use of fast reactors. Such reactors are dual-use in the sense that they can burn transuranic material and can breed new plutonium. In the former operation, they could provide a needed nuclear waste management benefit. In the latter operation, they can pose a serious proliferation threat.

Proliferation Risks

Reprocessing involves extraction of plutonium and/or other fissile materials from spent nuclear fuel in order to recycle these materials into new fuel for nuclear reactors. As discussed below, many reprocessing techniques are available for use. Regardless of the particular technique, fissile material is removed from all or almost all of the highly radioactive fission products, which provide a protective barrier against theft or diversion of plutonium in spent nuclear fuel. Plutonium-239 is the most prevalent fissile isotope of plutonium in spent nuclear fuel. The greater the concentration of this isotope the more weapons-usable is the plutonium mixture. Weapons-grade plutonium typically contains greater than 90 percent plutonium-239 whereas reactor-grade plutonium from commercial thermal-neutron reactors has usually less than 60 percent plutonium-239, depending on the characteristics of the reactor that produced the plutonium. The presence of non-plutonium-239 isotopes complicates production of nuclear weapons from the plutonium mixture, but the challenges are surmountable.[1] According to an unclassified U.S. Department of Energy report, reactor-grade plutonium is weapons-usable.[2]

The potential proliferation threats from reprocessing of spent nuclear fuel are twofold. First, a state operating a reprocessing plant could use that technology to divert weapons-usable fissile material into a nuclear weapons program or alternatively it could use the skills learned in operating that plant to build a clandestine reprocessing plant to extract fissile material. Second, a non-state actor such as a terrorist group could seize enough fissile material produced by a reprocessing facility in order to make an improvised nuclear device—a crude, but devastating, nuclear weapon. Such a non-state group may obtain help from insiders at the facility. While commercial reprocessing facilities have typically been well-guarded, some facilities such as those at Sellafield in the United Kingdom and Tokai-mura in Japan have not been able to account for several weapons' worth of plutonium. This lack of accountability does not mean that the fissile material was diverted into a state or non-state weapons program. The discrepancy was most likely due to plutonium caked on piping. But an insider could exploit such a discrepancy. For commercial bulk handing facilities, several tons of plutonium can be processed annually. Thus, if even one tenth of one percent of this material were accounted for, an insider could conceivably divert about one weapon's worth of plutonium every year.

Location matters when determining the proliferation risk of a reprocessing program. That is, a commercial reprocessing plant in a nuclear-armed state such as France, Russia, or the United Kingdom poses no risk of state diversion (but could pose a risk of non-state access) because this type of state, by definition, already has a weapons program. Notably, Japan is the only non-nuclear-armed state that has reprocessing facilities. Japan has applied the Additional Protocol to its International Atomic Energy Agency safeguards, but its large stockpile of reactor-grade plutonium could provide a significant breakout capability for a weapons program. (Chinese officials and analysts occasionally express concern about Japan's plutonium stockpile.) Since the Ford and Carter administrations, when the United States decided against reprocessing on proliferation and economic grounds, the United States has made stopping the spread of further reprocessing facilities especially to non-nuclear weapon states a top priority.

Another top priority of U.S. policy on reprocessing is to encourage countries with stockpiles of separated plutonium to draw down these stockpiles quickly. This drawdown can be done either through consuming the plutonium as fuel or surrounding it with highly radioactive fission products. Global stockpiles of civilian plutonium are growing and now at about 250 metric tons—equivalent to tens of thousands of nuclear bombs—are comparable to the global stockpile of military plutonium. More than 1,000 metric tons of plutonium is contained in spent nuclear fuel in about thirty countries.

While no country has used a commercial nuclear power program to make plutonium for nuclear weapons, certain countries have used research reactor programs to produce plutonium. India, notably, used a research reactor supplied by Canada to produce plutonium for its first nuclear explosive test in 1974. North Korea, similarly, has employed a research-type reactor to produce plutonium for its weapons program. Although nonproliferation efforts with Iran has focused on its uranium enrichment program, which could make fissile material for weapons, its construction of a heavy water research reactor, which when operational (perhaps early next decade) could produce at least one weapon's worth of plutonium annually, poses a latent proliferation threat. To date, Iran is not known to have constructed a reprocessing facility that would be needed to extract plutonium from this reactor's spent fuel. Further activities could take place in the Middle East and other regions. For instance, according to the U.S. government, Syria received assistance from North Korea in building a plutonium production reactor. In September 2009, Israel bombed this construction site.

The United States has been trying to balance the perceived need by many states in the Middle East for nuclear power plants versus restricting these states' access to enrichment and reprocessing technologies. Presently, as an outstanding example, the U.S.-UAE bilateral nuclear cooperation agreement is before the U.S. Congress. Proponents of this agreement tout the commitment made by the UAE to refrain from acquiring enrichment and reprocessing technologies and to rely on market mechanisms to purchase nuclear fuel. However, the last clause in the agreement appears to open the door for the UAE to engage in such activities in the future:

> **EQUAL TERMS AND CONDITIONS FOR COOPERATION**
> The Government of the United States of America confirms that the fields of cooperation, terms and conditions accorded by the United States of America to the United Arab Emirates for cooperation in the peaceful uses of nuclear energy shall be no less favorable in scope and effect than those which may be accorded, from time to time, to any other non-nuclear-weapon State in the Middle East in a peaceful nuclear cooperation agreement. If this is, at any time, not the case, at the request of the Government of the United Arab Emirates the Government of the United States of America will provide full details of the improved terms agreed with another non-nuclear-weapon State in the Middle East, to the extent consistent with its national legislation and regulations and any relevant agreements with such other non-nuclear-weapon State, and if requested by the Government of the United Arab Emirates, will consult with the Government of the United Arab Emirates regarding the possibility of amending this Agreement so that the position described above is restored.[3]

Such a request for amendment could be around the corner because Jordan is seeking to conclude a bilateral nuclear cooperation agreement with the United States, and it has expressed interest in keeping open the option to enrich uranium. Jordan has discovered large quantities of indigenous uranium and may want to "add value" to that uranium through enrichment. Jordan or any other Middle Eastern state has not yet expressed interest in reprocessing. U.S. leadership and practice in this issue will serve as an example for other states interested in acquiring new nuclear power programs.

Proliferation-Resistant Reprocessing

Can reprocessing be made more proliferation-resistant? "Proliferation resistance is that characteristic of a nuclear energy system that impedes the diversion or undeclared production of nuclear material or misuse of technology by the host state seeking to acquire nuclear weapons or other nuclear explosive devices."[4] No nuclear energy system is proliferation proof because nuclear technologies are dual-use. Enrichment and reprocessing can be used either for peaceful or military purposes. However, through a defense-in-depth approach, greater proliferation-resistance may be achieved. Both intrinsic features (for example, physical and engineering characteristics of a nuclear technology) and extrinsic features (for example, safeguards and physical barriers) complement each other to deter misuse of nuclear technologies and materials in weapons programs. The potential threats that proliferation-resistance tries to guard against are:

- "Concealed diversion of declared materials;
- Concealed misuse of declared facilities;
- Overt misuse of facilities or diversion of declared materials; and
- Clandestine declared facilities."[5]

For each of these threats, a detailed proliferation pathway analysis can be done in order to measure the proliferation risk and to determine the needed, if any, additional safeguards. The U.S. Department of Energy has sponsored such analysis for proposed reprocessing techniques considered under GNEP.[6] These techniques include UREX+, COEX, NUEX, and Pyroprocessing, and they have been compared to the PUREX technique, which is the commercially used method. PUREX separates plutonium and uranium from highly radioactive fission products. It is an aqueous separations process and thus generates sizable amounts of liquid radioactive waste. UREX+, COEX, and NUEX are also aqueous processes. UREX+ is a suite of chemical processes in which pure plutonium is not separated but different product streams can be produced depending on the reactor fuel requirements. COEX and NUEX are related processes. COEX co-extracts uranium and plutonium (and possibly neptunium) into one recycling stream; another stream contains pure uranium, which can be recycled; and a final stream contains fission products. NUEX separates into three streams: uranium, transuranics (including plutonium), and fission products. Pyroprocessing uses electrorefining techniques to extract plutonium in combination with other transuranic elements, some of the rare earth fission products, and uranium. This fuel mixture would be intended for use in fast-neutron reactors, which have yet to be proven commercially viable.

Can these reprocessing techniques meet the highest proliferation-resistance standard of the "spent fuel standard" in which plutonium in its final form should be as hard to acquire, process, and use in weapons as is plutonium embedded in spent fuel?[7] The brief answer is "no" because the act of separating most or all of the highly radioactive fission products makes the fuel product less protected than the intrinsic protection provided by spent fuel. In fact, Dr. E.D. Collins of Oak Ridge National Laboratory has shown that the radiation emission from these reprocessed products is 100 times less than the spent fuel standard.[8] In other words, a thief could carry these products and not suffer a lethal radiation dose whereas the same thief would experience a lethal dose in less than one hour of exposure to plutonium surrounded by highly radioactive fission products. But these methods may still be worth pursuing depending on a detailed systems analysis factoring in security risks on site and during transportation, the final disposition of the material once it has been recycled as fuel, as well as the costs and benefits of nuclear waste management.

According to DOE's draft nonproliferation assessment of GNEP, "for a state with preexisting PUREX or equivalent capability (or more broadly the capability to design and operate a reprocessing plant of this complexity), there is minimal proliferation resistance to be found by [using the examined reprocessing techniques] considering the potential for diversion, misuse, and breakout scenarios."[9] Moreover, the DOE assessment points out that these techniques pose additional safeguards challenges. For example, it is difficult to do an accurate accounting of the amount of plutonium in a bulk handling reprocessing facility that produces plutonium mixed with other transuranic elements.[10] This challenge raises the probability of diversion of plutonium by insiders.[11]

Another set of considerations is the choice of reactors to burn up the transuranic elements. The DOE draft assessment examined several choices including light water reactors, heavy water reactors, high temperature gas reactors, and fast-neutron reactors. Only the fast-neutron reactors offered the most benefits in terms of net consumption of transuranic material. This material would have to be recycled multiple times in fast reactors to consume almost all of it. This is called a full actinide recycle in contrast to a partial actinide recycle with the other reactor methods. The benefit from a waste management perspective is that the amount of time required for spent fuel's radiotoxicity to reduce to that of natural uranium goes from more than tens of thousands of years for partial actinide recycle to about 400 years for the full actinide recycle.

The challenge of the full actinide route, however, is that fast reactors can relatively easily be changed from a burner mode to a breeder mode. That is, these reactors can breed more plutonium by the insertion of uranium target material. The perceived need for breeder reactors has driven a few countries such as France, India, Japan, and Russia to develop reprocessing programs.

Alternative Nuclear Waste Management Programs of Other Nations

Has reprocessing programs, to date, helped certain nations solve their nuclear waste problems? The short answer is, "no." Before explicating that further, it is worth briefly examining why these countries began these programs. About fifty years ago, when the commercial nuclear industry was just starting, concerns were raised about the availability of enough natural uranium to fuel the thousands of reactors that were anticipated. Natural uranium contains 0.71 percent uranium-235, 99.28 percent uranium-238, and less than 0.1 percent uranium-234. Uranium-235 is the fissile isotope and thus is needed for sustaining a chain reaction. However, uranium-238 is a fertile isotope and can be used to breed plutonium-239, a fissile isotope that does not occur naturally. Thus, if uranium-238 can be transformed into plutonium-239, the available fissile material could be expanded by more than one hundred times, in principle. This observation motivated several countries, including the United States, to pursue reprocessing.

A related motivation was the desire for better energy security and thus less dependence on outside supplies of uranium. France and Japan, in particular, as countries with limited uranium resources, developed reprocessing plants in order to try to alleviate their dependency on external sources of uranium. They had invested in these plants before the realization that the world would not run out of uranium soon. By the late 1970s, two developments happened that alleviated the perceived pending shortfall. First, the pace of proposed nuclear power plant deployments dramatically slowed. There were plans at that time for more than 1,000 large reactors (of about 1,000 MWe power rating) by 2000, but even before the Three Mile Island accident in 1979, the number of reactor orders in the United States and other countries slackened off although France and Japan launched a reactor building boom in the 1970s that lasted through the 1980s. By 2000, there was only the equivalent of about 400 reactors of 1,000 MWe size. Second, uranium prospecting identified enough proven reserves to supply the present nuclear power demand for several decades to come.

Because there is plentiful uranium at relatively low prices and the cost of uranium enrichment has decreased, the cost of the once-through uranium cycle is significantly less than the cost of reprocessing. However, because fuel costs are a relatively small portion of the total costs of a nuclear power plant, reprocessing adds a relatively small amount to the total cost of electricity. In France, the added cost is almost six percent, and in Japan about ten percent. Nonetheless, in competitive utility markets in which consumers have choices, most countries have not chosen the reprocessing route because of the significantly greater fuel costs. France and Japan have adopted government policies in favor of reprocessing and also have sunk many billions of dollars into their reprocessing facilities. The French government owns and controls the electric utility Electricité de France (EDF) and the nuclear industry Areva. Despite this extensive government control, a 2000 French government study determined that if France stops reprocessing, it would save $4 to $5 billion over the remaining life of its reactor fleet.[12] EDF assigns a negative value to recycled plutonium.

While France's La Hague plant is operating, Japan is still struggling to start up its Rokkasho plant, which is largely based on the French design. Thus, the costs of the Japanese plant keep climbing and will likely be more than $20 billion. While the Japanese government wants to fuel up to one-third of its more than 50 reactors with plutonium-based mixed oxide fuel, local governments tend to look unfavorably on this proposal.

Only a few other nations are involved with reprocessing. Russia and the United Kingdom operate commercial-scale facilities. China and India are interested in heading down this path. But the United Kingdom is moving toward imminent shut down of its reprocessing mainly due to lack of customers. Moreover, the clean up and decommissioning costs are projected to be many billions of dollars. Russia and France also lack enough customers to keep their reprocessing plants at full capacity. In early April, I visited the French La Hague plant and was told that it is only operating at about half capacity. France only uses mixed oxide fuel in 20 of its 58 light water reactors. Presently, less than 10 percent of the world's commercial nuclear power plants burn MOX fuel. As stated earlier, the demand for MOX fuel has not kept up with the stockpiled quantities of plutonium.

With respect to nuclear waste management, an important point is that reprocessing, as currently practiced, does little or nothing to alleviate this management problem. For example, France practices a once-through recycling in which plutonium is separated once, made into MOX fuel, and the spent fuel containing this MOX is not usually recycled once (although France has done some limited recycling of MOX spent fuel). The MOX spent fuel is stored pending the further development and commercialization of fast reactors. But France admits that this full deployment of a fleet of fast reactors is projected to take place at the earliest by mid-century. France will shut down later this year its only fast reactor, the prototype Phénix. Perhaps around 2020, France may have constructed another fast reactor, but the high costs of these reactors have been prohibitive. In effect, France has shifted its nuclear waste problem from the power plants to the reprocessing plant.

France's practice of transporting plutonium hundreds of miles from the La Hague to the MOX plant at Marcoule poses a security risk. While there has never been a theft of plutonium or a major accident during the hundreds of trips to date, each shipment contains many weapons' worth of plutonium. Thus, just one theft of a shipment could be an international disaster.

No country has yet to open a permanent repository. But the country with the most promising record of accomplishment in this area is Sweden. A couple of weeks ago, Sweden announced the selection of its repository site but admits that the earliest the site will accept spent fuel is 2023. Sweden had carefully evaluated three different sites and obtained widespread community and local government involvement in the decision making process. France touts the benefits of the volume reduction of recycling in which highly radioactive fission products are formed into a glass-like compound, which is now stored at an interim storage site. By weight percentage, spent fuel typically consists of 95.6% uranium (with most of that being uranium-238), 3% stable or short-lived radioactive fission products, 0.3% cesium and strontium (the primary sources of high-level radioactive waste over a few hundred years), 0.1% long-lived iodine and technetium, 0.1% long-lived actinides (heavy radioactive elements), and 0.9% plutonium. But the critical physical factor for a repository is the heat load. For the first several hundred years of a repository the most heat emitting elements are the highly radioactive fission products. The benefit of a fast reactor recycling program could be the reduction or near elimination of the longer-lived transuranic elements that are the major heat producing elements beyond several hundred years.

Other countries may venture into reprocessing. Therefore, it is imperative for the United States to reevaluate its policies and redouble its efforts to prevent the further spread of reprocessing plants to non-nuclear-weapon states. In particular, the Republic of Korea is facing a crisis in the overcrowded conditions in the spent fuel pools at its power plants. One option is to remove older spent fuel and place it in dry storage casks, but the ROK government believes this option may cost too much because of the precedent set by the exorbitantly high price paid for a low level waste disposal facility. Another option is for the ROK to reprocess spent fuel. While this will provide significant volume reduction in the waste, it will only defer the problem to storage of MOX spent fuel, similar to the problem faced by France. This option will run counter to the agreement the ROK signed with North Korea in the early 1990s for both states to prohibit reprocessing or enrichment on the Korean Peninsula. A related option is to ship spent fuel to La Hague, but a security question is whether to ship plutonium back to the ROK. France would require shipment of the high level waste back to the ROK. Thus, the ROK will need a high level waste disposal facility. The main reason I raise this ROK issue at length is that the ROK and the United States have recently begun talks on the renewal of their peaceful nuclear cooperation agreement, which will expire in 2014. The United States has consent rights on ROK spent fuel because either it was produced with U.S.-supplied fresh fuel or U.S.-origin reactor systems. The ROK is seeking to have future spent fuel not subject to such consent rights by purchasing fresh fuel from other suppliers and by developing reactor systems that do not have critical components that are U.S.-origin or derived from U.S.-origin systems. The bottom line is that the United States is steadily losing its leverage with the ROK and other countries because of declining U.S. leadership in nuclear power plant systems and nuclear waste management.

Concerning lessons the United States can learn from other countries' nuclear waste management experience, the first lesson is that a fair political and sound scientific process is essential for selecting a permanent repository. Sweden demonstrates the effectiveness of examining multiple sites and gaining buy-in from the public and local governments. The second lesson is that reprocessing, as currently practiced, does not substantially alleviate the nuclear waste management problem. However, more research is needed to determine the costs and benefits of fast reactors for reducing transuranic waste. Any type of reprocessing will require safe and secure waste repositories.

While the United States investigates the costs and benefits of various recycling proposals through a research program, it has an opportunity now to exercise leadership in two waste management areas. First, as envisioned in GNEP, the United States should offer fuel leasing services. As part of those services, it should offer to take back spent fuel from the client countries. (Russia is offering this service to Iran's Bushehr reactor.) This spent fuel does not necessarily have to be sent to the United States. It could be sent to a third party country or location that could earn money for the spent fuel storage rental service. Spent fuel can be safely and securely stored in dry storage casks for up to 100 years. Long before this time ends, a research program will most likely determine effective means of waste management. The spent fuel leasing could be coupled to the second area where the United States can play a leadership role. That is, the United States can offer technical expertise and political support in helping to establish regional spent fuel repositories. A regional storage system would be especially helpful for countries with smaller nuclear power programs.

Recommendations

- Continue to discourage separation of plutonium from spent nuclear fuel.
- Limit the spread of reprocessing technologies to non-nuclear weapon states.
- Draw down the massive stockpile of civilian plutonium.
- Support a research program to assess the costs and benefits of various reprocessing technologies with attention focused on proliferation-resistance, safeguards, and nuclear waste management. Compare the

costs and benefits of reprocessing to enrichment, factoring in the proliferation risks of both technologies.
- Increase funding for safeguards research.
- Promote safe and secure storage of spent fuel until the time when
  reprocessing may become economically attractive.
- Evaluate multiple sites for permanent waste repositories based on
  political fairness and sound scientific assessments. Obtain buy-in from
  the public and local governments.
- Use secure interim spent fuel storage employing dry storage casks to
  relieve build up on spent fuel pools.
- Provide fuel leasing services that would include take back of spent fuel
  to either the fuel supplier state or a third party.
- Develop regional spent fuel storage facilities.
- Obtain better estimates on the remaining global reserves of uranium.
- Provide research support for developing more efficient nuclear power
  plants that would produce more electrical power per thermal power
  than today's fleet of reactors. Similarly, research more effective ways
  to make more efficient use of uranium fuel and reduce the amounts of
  plutonium-239 produced.

Notes

1. Richard L. Garwin, "Reactor-Grade Plutonium Can Be Used to Make Pow-
   erful and Reliable Nuclear Weapons," Paper for the Council on Foreign
   Relations, August 26, 1998. J. Carson Mark, "Explosive Properties of
   Reactor-Grade Plutonium," *Science and Global Security*, 4, 111–128, 1993.

2. *Nonproliferation and Arms Control Assessment of Weapons-Usable Fissile
   Material and Excess Plutonium Disposition Alternatives*, DOE/NN-0007
   (Washington, DC: U.S. Department of Energy, January 1997), pp. 38–39.

3. Agreement for Cooperation between the Government of the United
   States of America and the Government of the United Arab Emirates Concerning Peaceful Uses of Nuclear Energy, May 21, 2009.

4. Office of Nonproliferation and International Security, *A Nonproliferation
   Impact Assessment for the Global Nuclear Energy Programmatic Alternatives*,
   National Nuclear Security Administration, U.S. Department of Energy,
   Draft, December 2008, p. 26.

5. Ibid, p. 28.

6. See, for example, many of the references cited in Office of Nonproliferation and International Security, *A Nonproliferation Impact Assessment
   for the Global Nuclear Energy Programmatic Alternatives,* National Nuclear
   Security Administration, U.S. Department of Energy, Draft, December
   2008.

7. Committee on International Security and Arms Control, National Academy of Sciences, *Management and Disposition of Excess Weapons Plutonium*,
   Washington, DC: National Academy Press, 1994.

8. E.D. Collins, Oak Ridge National Laboratory, "Closing the Fuel Cycle
   Can Extend the Lifetime of the High-Level Waste Repository," American
   Nuclear Society 2005 Winter Meeting, November 17, 2005, p. 13.

9. *A Nonproliferation Impact Assessment for the Global Nuclear Energy Programmatic Alternatives*, p. 69.

10. J.E. Stewart et al., "Measurement and Accounting of the Minor Actinides Produced in Nuclear Power Reactors," Los Alamos National Laboratory, LA-13054-MS, January 1996, p. 21.

11. Ed Lyman, "U.S. Nuclear Fuel Reprocessing Initiative: DOE Research Shows Technology Does Not Reduce Risks of Nuclear Proliferation and Terrorism," Fact Sheet, Union of Concerned Scientists, February 2006.

12. *Economic Forecast Study of the Nuclear Option* (Planning Commission, Government of France, 2000), section 3.4.

POSTSCRIPT

Should the United States Reprocess Spent Nuclear Fuel?

The nuclear waste disposal problem is real and it must be dealt with. If it is not, we may face the same kinds of problems created by the former Soviet Union, which disposed of some nuclear waste simply by dumping it at sea. For a summary of the nuclear waste problem and the disposal controversy, see Michael E. Long, "Half Life: The Lethal Legacy of America's Nuclear Waste," *National Geographic* (July 2002). The need for care in nuclear waste disposal is underlined by Tom Carpenter and Clare Gilbert, "Don't Breathe the Air," *Bulletin of the Atomic Scientists* (May/June 2004); they describe the Hanford Site in Hanford, Washington, where wastes from nuclear weapons production were stored in underground tanks. Leaks from the tanks have contaminated groundwater, and an extensive cleanup program is under way. But cleanup workers are being exposed to both radioactive materials and toxic chemicals, and they are falling ill. And in June 2004, the U.S. Senate voted to ease cleanup requirements. Per F. Peterson, William E. Kastenberg, and Michael Corradini, "Nuclear Waste and the Distant Future," *Issues in Science and Technology* (Summer 2006), argue that the risks of waste disposal have been sensibly addressed and we should be focusing more attention on other risks (such as those of global warming). Behnam Taebi and Jan Kloosterman, "To Recycle or Not to Recycle? An Intergenerational Approach to Nuclear Fuel Cycles," *Science & Engineering Ethics* (June 2008), argue that the question of whether to accept reprocessing comes down to choosing between risks for the present generation and risks for future generations.

In November 2005, President Bush signed the budget for the Department of Energy, which contained $50 million to start work toward a reprocessing plant; see Eli Kintisch, "Congress Tells DOE to Take Fresh Look at Recycling Spent Reactor Fuel," *Science* (December 2, 2005). By April 2008, Senator Pete Domenici of the U.S. Senate Energy and Natural Resources Committee was working on a bill that would set up the nation's first government-backed commercialized nuclear waste reprocessing facilities. Reprocessing spent nuclear fuel will be expensive, but the costs may not be great enough to make nuclear power unacceptable; see "The Economic Future of Nuclear Power," University of Chicago (August 2004) (http://www.ne.doe.gov/np2010/reports/NuclIndustryStudy-Summary.pdf).

In February 2006, the United States Department of Energy announced the Global Nuclear Energy Partnership (GNEP), to be operated by the U.S., Russia, Great Britain, and France. It would lease nuclear fuel to other nations, reprocess spent fuel without generating material that could be diverted to making nuclear bombs, reduce the amount of waste that must be disposed of, and help meet future energy needs. See Stephanie Cooke, "Just Within Reach?" *Bulletin of the Atomic Scientists* (July/August 2006), and Jeff Johnson, "Reprocessing Key to Nuclear Plan," *Chemical & Engineering News* (June 18, 2007). Critics such as Karen Charman ("Brave Nuclear World, Parts I and II," *World Watch* (May/June and July/August 2006), insist that nuclear power is far too expensive and carries too serious risks of breakdown and exposure to wastes to rely upon, especially when cleaner, cheaper, and less dangerous alternatives exist. Early in 2009, the Department of Energy announced it was closing down GNEP.

It is an unfortunate truth that the reprocessing of nuclear spent fuel does indeed increase the risks of nuclear proliferation. Both nations and terrorists itch to possess nuclear weapons, whose destructive potential makes present members of the "nuclear club" tremble. Can the risks be controlled? John Deutch, Arnold Kanter, Ernest Moniz, and Daniel Poneman, in "Making the World Safe for Nuclear Energy," *Survival* (Winter 2004/2005), argue that present nuclear nations could supply fuel and reprocess spent fuel for other nations; nations that refuse to participate would be seen as suspect and subject to international action. Nuclear physicist and Princeton professor Frank N. von Hippel, "Rethinking Nuclear Fuel Recycling," *Scientific American* (May 2008) argues that reprocessing nuclear spent fuel is expensive and emits lethal radiation. There is also a worrisome risk that the increased availability of bomb-grade nuclear materials will increase the risk of nuclear war and terrorism. Prudence demands that spent fuel be stored until the benefits of reprocessing exceed the risks (if they ever do).

Contributors to This Volume

EDITOR

THOMAS A. EASTON is a professor of science at Thomas College in Waterville, Maine, where he has been teaching environmental science, science, technology, and society, emerging technologies, and computer science since 1983. He received a B.A. in biology from Colby College in 1966 and a Ph.D. in theoretical biology from the University of Chicago in 1971. He writes and speaks frequently on scientific and futuristic issues. His books include *Focus on Human Biology*, 2nd ed., coauthored with Carl E. Rischer (HarperCollins, 1995), *Careers in Science*, 4th ed. (VGM Career Horizons, 2004), *Taking Sides: Clashing Views in Science, Technology and Society* (McGraw-Hill, 9th ed., 2010), *Taking Sides: Clashing Views in Energy and Society* (McGraw-Hill, 2009), and *Classic Editions Sources: Environmental Studies* (McGraw-Hill, 3d ed., 2008). Dr. Easton is also a well-known writer and critic of science fiction.

AUTHORS

KATIA AVILES-VAZQUEZ is currently pursuing a Ph.D. in Geography at the University of Texas.

CATHERINE BADGLEY is an assistant professor in the Department of Ecology and Evolutionary Biology and a research scientist at the Museum of Paleontology and the Department of Geological Sciences, University of Michigan.

RONALD BAILEY is a science correspondent for *Reason* magazine. A member of the Society of Environmental Journalists, his articles have appeared in many popular publications, including the *Wall Street Journal, The Public Interest,* and *National Review.* He has produced several series and documentaries for PBS television and *ABC News,* and he was the Warren T. Brookes Fellow in Environmental Journalism at the Competitive Enterprise Institute in 1993. He is the editor of *Earth Report 2000: Revisiting the True State of the Planet* (McGraw-Hill, 1999) and the author of *Global Warming and Other Eco-Myths: How the Environmental Movement Uses False Science to Scare Us to Death* (Prima, 2002).

STEPHEN L. BAIRD is a technology education teacher for the Virginia Beach City Public School system and an adjunct assistant professor at Old Dominion University.

LESTER BROWN founded the WorldWatch Institute in 1974 and the Earth Policy Institute in 2001. He has been called one of the world's most influential thinkers. His books include *Eco-Economy: Building an Economy for the Earth* (W. W. Norton, 2001), *Outgrowing the Earth: The Food Security Challenge in an Age of Falling Water Tables and Rising Temperatures* (W. W. Norton, 2005), and *Plan B 3.0: Mobilizing to Save Civilization* (W. W. Norton, 2008).

M. JAHI CHAPPELL is currently a postdoctoral associate in Science & Technology Studies at Cornell University.

PAUL CICIO is President of the Industrial Energy Consumers of America.

JAMIE CLARK is Senior Vice President for Conservation Programs, National Wildlife Federation.

EILEEN CLAUSSEN is President of the Pew Center on Global Climate Change, an independent nonprofit, non-partisan organization dedicated to advancing practical and effective solutions and policies to address global climate change. The Pew Center is a founding member of the U. S. Climate Action Partnership, a coalition of 25 leading businesses and five environmental organizations that have come together to call on the federal government to quickly enact strong national legislation to require significant reductions of greenhouse gas emissions.

BENEDICT S. COHEN is Deputy General Counsel for Environment and Installations, Department of Defense.

GERALD D. COLEMAN is the former rector of St. Patrick's Seminary and University in Menlo Park, California. His books include several on Catholic views of sexuality.

CHARLI E. COON is Senior Policy Analyst, Thomas A. Roe Institute for Economic Policy Studies, The Heritage Foundation.

VIRGINIA H. DALE is a scientist in the Center for BioEnergy Sustainability at the Oak Ridge National Laboratory, Oak Ridge, Tennessee.

GIULIO A. De LEO is an associate professor of applied ecology and environmental impact assessment in the Dipartimento di Scienze Ambientali at the Universit degli Studi di Parma in Parma, Italy.

C. JOSH DONLAN is a research biologist at Cornell University, founder and director of Advanced Conservation Strategies, an adviser to the Galápagos National Park and to Island Conservation, and a senior fellow at the Robert and Patricia Switzer Foundation and Environmental Leadership Program.

CHARLES D. FERGUSON is the Philip D. Reed senior fellow for science and technology at the Council on Foreign Relations (CFR). His areas of expertise include arms control, climate change, energy policy, and nuclear and radiological terrorism. He has written the Council Special Report *Nuclear Energy: Balancing Benefits and Risks,* published in April 2007. He co-authored (with William Potter) the book *The Four Faces of Nuclear Terrorism* (Routledge, 2005).

PHILLIP J. FINCK is deputy associate laboratory director, applied science and technology and national security, Argonne National Laboratory.

DAVID FRIEDMAN is Research Director at the Union of Concerned Scientists.

MARINO GATTO is a professor of applied ecology in the Dipartimento di Elettronica e Informazione at Politecnico di Milano in Milan, Italy. His main research interests include ecological models and the management of renewable resources. Gato is associate editor of *Theoretical Population Biology.*

THOMAS A. GAVIN is an associate professor in the Department of Natural Resources at Cornell University.

BERNARD D. GOLDSTEIN is Professor of Environmental and Occupational Health at the University of Pittsburgh. From 2001 to 2005, he was dean of the Graduate School of Public Health. He has published a number of papers on the precautionary principle and risk assessment.

MICHAEL GOUGH, a biologist and expert on risk assessment and environmental policy, has participated in science policy issues at the Congressional Office of Technology Assessment, in Washington think tanks, and on various advisory panels. He most recently edited *Politicizing Science: The Alchemy of Policymaking* (Hoover Institution Press, 2003).

ROBERT H. HARRIS is a principal with ENVIRON International Corporation. He has over 25 years of experience in the area of environmental health and toxic chemicals, with particular emphasis on water and air pollution and hazardous waste issues. He is recognized nationally as an expert consultant on the treatment and disposal of municipal solid and hazardous waste, as well as on air, soil, and groundwater contamination.

DAVID G. HAWKINS is the director of the Climate Center of the Natural Resources Defense Council.

ALEXANDER KARSNER is the Assistant Secretary for Energy Efficiency and Renewable Energy at the U.S. Department of Energy.

KEITH KLINE is a scientist in the Center for BioEnergy Sustainability at the Oak Ridge National Laboratory, Oak Ridge, Tennessee.

RUSSELL LEE is a scientist in the Center for BioEnergy Sustainability at the Oak Ridge National Laboratory, Oak Ridge, Tennessee.

PAUL LEIBY is a scientist in the Center for BioEnergy Sustainability at the Oak Ridge National Laboratory, Oak Ridge, Tennessee.

JOHN E. LOSEY is an associate professor in the Department of Entomology at Cornell University.

SEAN McDONAGH is a Columban priest and ex-missionary to the Philippines. His latest book is *The Death of Life: The Horror of Extinction* (Columba Press, Dublin, 2004).

ANNE PLATT McGINN is a senior researcher at the Worldwatch Institute and the author of "Why Poison Ourselves? A Precautionary Approach to Synthetic Chemicals," Worldwatch Paper 153 (November 2000).

JOHN J. MILLER is a political reporter for the *National Review,* a contributing editor for *Reason* magazine, a former vice president of the Center for Equal Opportunity, and a Bradley Fellow at the Heritage Foundation. His most recent book is *A Gift of Freedom: How the John M. Olin Foundation Changed America* (Encounter, 2005).

DONALD MITCHELL is Lead Economist in the World Bank's Development Prospects Group.

JEREMY MOGHTADER is an instructor at the Department of Horticulture, Michigan State University.

IAIN MURRAY is a Director of Projects and Analysis and Senior Fellow in Energy, Science and Technology at the Competitive Enterprise Institute and the author of *The Really Inconvenient Truths: Seven Environmental Catastrophes Liberals Won't Tell You About—Because They Helped Cause Them* (Regnery, 2008).

NANCY MYERS is communications director for the Science and Environmental Health Network. She is the co-editor, with Carolyn Raffensperger, of *Precautionary Tools for Reshaping Environmental Policy* (MIT Press, 2005).

ROBERT PAARLBERG is the Betty F. Johnson '44 Professor of Political Science at Wellesley College. He is the author of *Starved for Science: How Biotechnology Is Being Kept Out of Africa* (Harvard University Press, 2008).

RANDALL PATTERSON is a journalist who writes frequently for *Mother Jones* and other magazines.

IVETTE PERFECTO is a professor at the School of Natural Resources and Environment, University of Michigan.

SUSAN PETTY is the President of AltaRock Energy, Inc., a developer of "geothermal technology to produce clean, renewable power."

EILEEN QUINTERO is a research assistant at the School of Natural Resources and Environment, University of Michigan.

JEREMY RIFKIN is the president of the Foundation on Economic Trends in Washington, D.C., and has written many books on the impact of scientific and technological changes on the economy, the workforce, society, and the environment. Among his latest books is *The European Dream: How Europe's Vision of the Future Is Quietly Eclipsing the American Dream* (Tarcher/ Penguin, 2004).

DONALD R. ROBERTS is a professor in the Division of Tropical Public Health, Department of Preventive Medicine and Biometrics, Uniformed Services University of the Health Sciences.

MARY ANNETTE ROSE is an assistant professor in the Department of Technology at Ball State University in Muncie, IN.

DANIEL I. RUBENSTEIN is professor and chair of Princeton's Department of Ecology and Evolutionary Biology and director of the Program in African Studies.

DUSTIN R. RUBENSTEIN is a behavioral and evolutionary ecologist. In 2009, he joined the Department of Ecology, Evolution and Environmental Biology at Columbia University.

ANDREA SAMULON is currently an agribusiness campaigner with the Rainforest Action Network (http://ran.org/).

CHARLES W. SCHMIDT is a freelance science writer specializing in the environment, genomics, and information technology, among other topics. In 2002, he won the National Association of Science Writers' Science-in-Society Journalism Award for his reporting on hazardous electronic waste exports to developing countries.

PAUL W. SHERMAN is a professor in the Department of Neurobiology and Behavior at Cornell University. His specialty is animal social behavior.

KRISTIN SHRADER-FRECHETTE is the O'Neill Family Professor, Department of Biological Sciences and Department of Philosophy, at the University of Notre Dame. She is the author of *Taking Action, Saving Lives: Our Duties to Protect Environmental and Public Health* (Oxford University Press, 2007).

MIKE TILCHIN is a vice president of CH2M HILL.

MICHELE L. TRANKINA is a professor of biological sciences at St. Mary's University and an adjunct associate professor of physiology at the University of Texas Health Science Center, both in San Antonio, Texas.

JAY VANDEVEN is a principal with ENVIRON International Corporation. He has 16 years of experience in the assessment and remediation of soil and groundwater contamination, contaminant fate and transport, environmental cost allocation, and environmental insurance claims.

MACE VAUGHAN is an entomologist and Conservation Director at the Xerces Society for Invertebrate Conservation (http://www.xerces.org/).

EMILY ZAKEM is at the School of Art and Design, University of Michigan.